Lecture Notes in Networks and Systems

Volume 168

The series "Lecture Notes in Networks and Systems" publishes the latest developments in Networks and Systems—quickly, informally and with high quality. Original research reported in proceedings and post-proceedings represents the core of LNNS.

Volumes published in LNNS embrace all aspects and subfields of, as well as new challenges in, Networks and Systems.

The series contains proceedings and edited volumes in systems and networks, spanning the areas of Cyber-Physical Systems, Autonomous Systems, Sensor Networks, Control Systems, Energy Systems, Automotive Systems, Biological Systems, Vehicular Networking and Connected Vehicles, Aerospace Systems, Automation, Manufacturing, Smart Grids, Nonlinear Systems, Power Systems, Robotics, Social Systems, Economic Systems and other. Of particular value to both the contributors and the readership are the short publication timeframe and the world-wide distribution and exposure which enable both a wide and rapid dissemination of research output.

The series covers the theory, applications, and perspectives on the state of the art and future developments relevant to systems and networks, decision making, control, complex processes and related areas, as embedded in the fields of interdisciplinary and applied sciences, engineering, computer science, physics, economics, social, and life sciences, as well as the paradigms and methodologies behind them.

Indexed by SCOPUS, INSPEC, WTI Frankfurt eG, zbMATH, SCImago.

All books published in the series are submitted for consideration in Web of Science.

More information about this series at http://www.springer.com/series/15179

Zakia Hammouch · Hemen Dutta ·
Said Melliani · Michael Ruzhansky
Editors

Nonlinear Analysis: Problems, Applications and Computational Methods

 Springer

Editors
Zakia Hammouch
FSTE
Moulay Ismail University
Meknes, Morocco

Said Melliani
Department of Mathematics
Sultan Moulay Slimane University
Beni Mellal, Morocco

Hemen Dutta
Department of Mathematics
Gauhati University
Guwahati, Assam, India

Michael Ruzhansky
Department of Mathematics
Ghent University
Gent, Belgium

ISSN 2367-3370   ISSN 2367-3389   (electronic)
Lecture Notes in Networks and Systems
ISBN 978-3-030-62298-5   ISBN 978-3-030-62299-2   (eBook)
https://doi.org/10.1007/978-3-030-62299-2

This Springer imprint is published by the registered company Springer Nature Switzerland AG
The registered company address is: Gewerbestrasse 11, 6330 Cham, Switzerland

Contents

Existence Results for Impulsive Partial Functional Fractional Differential Equation with State Dependent Delay

Nadjet Abada[1]([⊠]), Helima Chahdane[1], and Hadda Hammouche[2]

[1] Laboratoire MAD, Ecole Normale Superieure Assia Djebar,
Universite constantine 3, El Khroub, Algerie
n65abada@yahoo.fr, helimachahdane@yahoo.com
[2] Laboratoire LMSA, Universite Ghardaia, Bounoura, Algerie
h.hammouche@yahoo.fr

Abstract. In this paper, we study the existence of mild solutions of impulsive fractional semilinear differential equation with state dependent delay of order $0 < \alpha < 1$. We shall rely on fixed point theorem for the sum of completely continuous and contraction operators due to Burton and Kirk. An example is given to illustrate the theory.

1 Introduction

Fractional calculus is a generalization of classical differentiation and integration to an arbitrary real order. Fractional calculus is the most well known and valuable branch of mathematics which gives a good framework for biological and physical phenomena, mathematical modeling of engineering, etc. Numerous writings have showed that fractional-order differential equation could provide more methods to deal with complex problem in statistical physics and environmental issues; see the monographs of Abbas et al. [ABN12, ABN15], A. Kilbas et al. [KST06], Podlubny [P93] and Zhou [Z14] and the references therein. On the other hand, the theory of impulsive differential equations has undergone rapid development over the years and played a very important role in modern applied mathematical models of real processes rising in phenomena studied in physics, chemistry, engineering, etc.

Recently, the study of fractional differential equations with impulses has been studied by many authors (see [BHN06, HAM10, LCX12, WFZ11]).

Modivated by work [HGBA13], in this paper, we study the existence of mild solutions for fractional semilinear differential equation of the equation of the form

$$^{c}D_{t_k}^{\alpha} y(t) - Ay(t) = f(t, y_{\rho(t, y_t)}), t \in J_k := (t_k, t_{k+1}], k = 0, 1, ..m, \tag{1}$$

$$\Delta y \mid_{y=y_k} = I_k(y_{t_k}) \quad k = 1, ..., m, \tag{2}$$

$$y(t) = \phi(t), \quad t \in (-\infty, 0]. \tag{3}$$

where $^{c}D_{t_k}^{\alpha}$ is caputo fractional derivative of order $0 < \alpha < 1$, $A : D(A) \subset E \to E$ is the bounded linear operator of an α - resolvent family $S_\alpha(t) : t \geq 0$ defined on a Banach

© Springer Nature Switzerland AG 2021
Z. Hammouch et al. (Eds.): SM2A 2019, LNNS 168, pp. 1–22, 2021.
https://doi.org/10.1007/978-3-030-62299-2_1

space E, $f : J \times \mathscr{D} \to E$ is a given function, $\mathscr{D} = \{\psi : (-\infty, 0] \to E, \psi$ is continuous every where except for a finite number of points s at which $\psi(s^-), \psi(s^+)$ exist and $\psi(s^-) = \psi(s)\}$, $\phi \in D$, $(0 < r < \infty)$, $I_k : E \to E$, $(k = 0, 1, ..., m + 1)$, $0 = t_0 < t_1 < ..... < t_m < t_{m+1} = b$, $\Delta y|_{y=y_k} = y(t_k^+) - y(t_k^-)$, where $y(t_k^+) = \lim_{h \to 0^+} y(t_k + h)$ and $y(t_k^-) = \lim_{h \to 0^+} y(t_k - h)$ represent the right and left limits of $y(t)$ at $t = t_k$, respectively. We denote by y_t the element of $\mathscr{D}$ defined by $y_t(\theta) = y(t + \theta)$, $\theta \in (-\infty, 0]$. Here y_t represents the history up to the present time t of the state $y(.)$. We assume that the histories y_t belongs to some abstract phase $\mathscr{D}$, to specified later, and $\phi \in \mathscr{D}$. This paper is organized as follow, in Sect. 2 we introduce some preliminaries that will be used in the sequel, in Sect. 3 we give definition to the mild solution of problem 1–3 result inspired by works [HGBA13, HL20], also the proof of our main results is given. Finally, an example is included in Sect. 4.

2   Preliminaries

In this Section, we state some notations, definitions and properties which be used throughout this paper.

Let E be a Banach space endowed with the norm $\|.\|$, and $L(E)$ represents the Banach space of all bounded linear operators from E into E and the corresponding norm $\|.\|_{L(E)}$.

$C(J, E)$ is the Banach space of all continuous functions from J to E with the norm

$$\|u\|_{C(J,E)} = \sup\{|u(t)| : t \in J\},$$

$L^1[J, E]$ is the Banach space of measurable functions $u : J \longrightarrow E$ which are Bochner integrable normed by

$$\|u\|_{L^1} = \int_0^b |u(t)| dt.$$

Definition 1. A familly $(S_\alpha(t))_{t>0} \subset l(E)$ of bounded linear operators in E is called an α- resolvent operator function generating by A if the following conditions hold:

a) $(S_\alpha(t))_{t>0}$ is strong continuous on R_+ and $S_\alpha(0) = I$;
b) $S_\alpha(t)D(A) \subset D(A)$ and $AS_\alpha(t)x = S_\alpha(t)Ax$ for all $x \in D(A)$ and $t > 0$;
c) For all $x \in E, I_t^\alpha S_\alpha(t)x \in D(A)$ and

$$S_\alpha(t)x = x + AI_t^\alpha S_\alpha(t)x, \ t > 0;$$

d) $x \in D(A)$ and $Ax = y$ if and only if

$$S_\alpha(t)x = x + AI_t^\alpha S_\alpha(t)x, \ t > 0;$$

e) A is closed and densely defined

The generator A of $(S_\alpha(t))_{t>0}$ is defined by:

$$D(A) := \{x \in E : \lim_{t \to 0^+} \frac{S_\alpha(t)x - x}{\psi_{\alpha+1}(t)} \ exists\},$$

and

$$Ax = \lim_{t \to 0^+} \frac{S_\alpha(t)x - x}{\psi_{\alpha+1}(t)}, \quad x \in D(A),$$

where $\psi_\alpha(t) = \frac{t^{\alpha-1}}{\Gamma(\alpha)}$ for $t > 0$ and $\psi_\alpha(t) = 0$ for $t \leq 0$ and $\psi_\alpha(t) \longrightarrow \delta(t)$ as $\alpha \longrightarrow 0$, where the function delta is defined by:

$$\delta_a : D(\Omega) \longrightarrow R; \quad \phi \to \phi(a),$$

and

$$D(\Omega) = \{\phi \in C^\infty(\Omega) : \ supp\phi \subset \Omega \ is \ compact\}.$$

Definition 2. An α-ROF $(S_\alpha(t))_{t \geq 0}$ is called analytic, if the function $S_\alpha() : R+ \longrightarrow l(X)$ admits analytic extension to a sector $\Sigma(0, \theta_0)$ for some $0 < \theta_0 \leq \frac{\pi}{2}$. An analytic α-ROF (S_α) is said to be of analyticity type (ω_0, θ_0) if for each $\theta < \theta_0$ and $\omega > \omega_0$ there exists $M_1 = M_1(\omega, \theta)$ such that $\|S_\alpha(z)\| \leq M_1 e^{\omega Rez}$ for $z \in \Sigma(0, \theta)$ where Rez denotes the real part of z and $\Sigma(\omega, \theta) := \{\lambda \in C : |arg(\lambda - \omega)| < \theta, \quad \omega, \theta \in R\}$

Definition 3. An α-ROF$(S_\alpha(t))_{t \geq 0}$ is called compact for $t > 0$ if for every $t > 0$, $S_\alpha(t)$ is a compact operator.

Theorem 1. *Let A generate a compact analytic semigroup $T(t)_{t \geq 0}$ then for any α it also generates a compact analytic resolvent family $(S_\alpha(t))_{t \geq 0}$.*

Lemma 1. *Assume that α-ROF$(S_\alpha(t))_{t \geq 0}$ is compact for $t > 0$ and analytic of type (ω_0, θ_0). Then the following assertions hold:*

1. $\lim_{h \longrightarrow 0} \|S_\alpha(t+h) - S_\alpha(t)\| = 0$, *for $t > 0$.*
2. $\lim_{h \longrightarrow 0^+} \|S_\alpha(t+h) - S_\alpha(h)S_\alpha(t)\| = 0$, *for $t > 0$.*

Definition 4. An α-ROF$(S_\alpha(t))_{t \geq 0}$ is said to be exponentially bounded if there exist constants $M \geq 1$, $\omega \geq 0$ such that

$$\|S_\alpha(t)\| \leq Me^{\omega t} \ for \ t \geq 0.$$

in this case we write $A \in C_\alpha(M, \omega)$.

Definition 5. The fractional integral operator I^α of order $\alpha > 0$ of a continuous function $f(t)$ is defined by

$$I_t^\alpha f(t) := \frac{1}{\Gamma(\alpha)} \int_0^t (t-s)^{\alpha-1} f(s) ds,$$

Observe that $I_t^\alpha f(t) = f(t) * \psi_\alpha(t)$, where $\psi_\alpha(t) = \frac{t^{\alpha-1}}{\Gamma(\alpha)}$ for $t > 0$ and $\psi_\alpha(t) = 0$ for $t \leq 0$ and $\psi_\alpha(t) \longrightarrow \delta(t)$ as $\alpha \longrightarrow 0$.

Definition 6. The α-Riemann-Liouville fractional-order derivative of the function f, is defined by

$$D_a^\alpha f(t) = \frac{1}{\Gamma(n-\alpha)} \frac{d^n}{dt^n} \int_a^t (t-s)^{n-\alpha-1} f(s) ds.$$

where $n = [\alpha] + 1$ and $[\alpha]$ denotes the integer part of α.

Definition 7 [P93]. For a function f defined on the interval $[a,b]$, the Caputo fractional order derivative of order α of f, is defined by

$$({}_{a+}^c D_t^\alpha f)(t) = \frac{1}{\Gamma(n-\alpha)} \int_0^t (t-s)^{n-\alpha-1} f^{(n)}(s) ds,$$

Where $n = [\alpha] + 1$.

Therefore, for $0 < \alpha < 1$, $n = [\alpha] + 1 = 1$ and for $a = 0$, the Caputo's fractional derivative for $t \in [0,b]$ is given by

$$({}_0^c D_t^\alpha f)(t) = \frac{1}{\Gamma(1-\alpha)} \int_0^t (t-s)^{-\alpha} f'(s) ds.$$

In this paper, we will employ an axiomatic definition for the phase space D which is similar to those introduced by Hale and Kato [HK78]. Specifically, D will be a linear space of functions mapping $] -\infty, b]$ into E endowed with a semi-norm $\|.\|_D$, and satisfies the following axioms:

(A1) There exist a positive constant H and functions $K(\cdot)$, $M(\cdot) : R^+ \to R^+$ with K continuous and M locally bounded, such that for any $b > 0$, if $x : (-\infty, b] \to E$, $x \in D$, and $x(\cdot)$ is continuous on $[0,b]$, then for every $t \in [0,b]$ the following conditions hold:
  (i) x_t is in D;
  (ii) $|x(t)| \le H \|x_t\|_D$;
  (iii) $\|x_t\|_D \le K(t) \sup\{|x(s)| : 0 \le s \le t\} + M(t) \|x_0\|_D$, and H, K and M are independent of $x(\cdot)$.
  Denote

$$K_b = \sup\{K(t) : t \in J\} \text{ and } M_b = \sup\{M(t) : t \in J\}.$$

(A_2) The space D is complete.

Example 1. Let $h(.) : (-\infty, -r] \to R$ be a positive Lebesgue integrable function and $D := PC_r \times L^2(h; E), r \ge 0$, be the space formed of all classes of functions $\varphi : (-\infty, 0] \to E$ such that $\varphi|_{[-r,0]} \in PC([-r,0], E)$, $\varphi(.)$ is Lebesgue-measurable on $(-\infty, -r]$ and $h|\varphi|^p$ is Lebesgue integrable on $(-\infty, -r]$. the semi-norm in $\|.\|_D$ is defined by

$$\|\varphi\|_D = \sup_{\theta \in [-r,0]} \|\varphi(\theta)\| + \left(\int_{-\infty}^{-r} h(\theta) \|\varphi(\theta)\|^p d\theta \right)^{1/p}, \tag{4}$$

Assume that $h(.)$ satisfies conditions (g–6) and (g–7) in the terminology of [HMN91]. proceeding as in the proof of [[HMN91]. Theorem 1.3.8] it follows that $\mathscr{D}$ is a phase space which verifies the axioms (A1)–(A2) and (A3). Moreover, when $r = 0$ this space coincides with $C^0 \times L^2(h, E)$ and the parameters $H = 1; M(t) = \gamma(-t)^{1/2}$ and $K(t) = 1 + \left(\int_{-r}^0 h(\xi) d\xi \right)^{1/2}$, for $t \ge 0$ (see [HMN91]).

Definition 8. A map $f : [0,b] \times \mathscr{D} \to E$ is said to be carathéodory if

1. the function $t \mapsto f(t,y)$ is measurable for each $y \in \mathscr{D}$;
2. the function $t \mapsto f(t,y)$ is continuous for almost all $t \in J_k := (t_k, t_{k+1}], k = 0, 1, ..m$.

In order to define the mild solution 1–3, we consider the following space

$$PC(J,E) = \{y : [0,b] \longrightarrow E : y \text{ is continous at } t \neq t_k, y(t_k^-) = y(t_k),$$

$$\text{and } y(t_k^+) \text{ exists, for all } k = 1, ..., m\}$$

which is a Banach space with the norm

$$\|y\| = max\{\|y_k\|_\infty; k = 1, 2, .., m\},$$

and

$$D_b = \{y :] - \infty, b] \longrightarrow E : y|_{]-\infty,0]} \in D \text{ and } y|_J \in PC(J,E)\}.$$

Let $\|.\|_b$ be the semi norm in D_b defined by

$$\|y\|_b = \|y_0\|_{D_b} + sup\{|y(s)| : 0 \le s \le b\}, \quad y \in D_b.$$

Let us introduce the definition of Caputo's derivative in each interval $(t_k, t_{k+1}], k = 0, ..., m$,

$$(^cD_{t_k}^\alpha f)(t) = \frac{1}{\Gamma(1-\alpha)} \int_{t_k}^t (t-s)^{-\alpha} f'(s) ds.$$

3   Main Result

Before starting and proving our main result, we give the meaning of mild solution of our problem 1–3.

Definition 9. A function $y \in PC((-\infty, b], E)$ is said to be mild solution of our problem if $y(t) = \phi(t)$, for all $t \in (-\infty, 0], \Delta y|_{y=y_k} = I_k(y_{t_i}), k = 1, 2, ..., m$ and such that y satisfies the following integral equation:

$$y(t) = \begin{cases} S_\alpha(t)\phi(0) + \int_0^t S_\alpha(t-s)f(s,y_{\rho(s,y_s)})ds; & \text{if } t \in [0,t_1], \\ S_\alpha(t-t_k)\prod_{i=1}^k S_\alpha(t_i - t_{i-1})\phi(0) \\ +\sum_{i=1}^k \int_{t_{i-1}}^{t_i} S_\alpha(t-t_k)\prod_{j=i}^{k-1} S_\alpha(t_{j+1}-t_j)S_\alpha(t_i-s)f(s,y_{\rho(s,y_s)}) \\ +\int_{t_k}^t S_\alpha(t-s)f(s,y_{\rho(s,y_s)})ds \\ +\sum_{i=1}^k S_\alpha(t-t_k)\prod_{j=i}^{k-1} S_\alpha(t_{j+1}-t_j)I_i(y_{t_i}); & \text{if } t \in (t_k, t_{k+1}]. \end{cases}$$

Set

$$\mathscr{R}(\rho^-) = \{\rho(s,\varphi) : (s,\varphi) \in J \times \mathscr{D}, \rho(s,\varphi) \le 0\}.$$

We always assume that $\rho : J \times \mathscr{D} \to (-\infty, b]$ is continuous. Additionally, we introduce the following hypothesis:

- (H_φ) The function $t \to \varphi_t$ is continuous from $\mathscr{R}(\rho^-)$ into $\mathscr{D}$ and there exists a continuous and bounded function $L^\phi : \mathscr{R}(\rho^-) \to (0, \infty)$ such that

$$\|\phi_t\|_{\mathscr{D}} \le L^\phi(t) \|\phi\|_{\mathscr{D}}, \qquad \text{for every} \quad t \in \mathscr{R}(\rho^-).$$

Lemma 2 *([HPL06]). If $y :] -\infty, b] \longrightarrow E$ is a function such that $y_0 = \phi$, then*

$$\|y_t\|_D \le (M_b + L^\phi)\|\phi\|_{\mathscr{D}} + K_b \sup\{|y(s)|; s \in [0, \max\{0, t\}]\},$$

where $M_b = \sup_{t \in J} M(t)$, $K_b = \sup_{t \in J} K(t)$ and $L^\phi = \sup_{t \in \mathscr{R}(\rho^-)} L^\phi(t)$.

Our main result in this section is based upon the following fixed point theorem due to Burton and Kirk [BK98].

Theorem 2 *([BK98]). Let X be a Banach space and $A, B : X \longrightarrow X$ be two operators satisfying:*

1. *A is a contraction,*
2. *B is completely continuous,*

Then, either;

1. *the operator equation $y = Ay + By$ has a solution, or*
2. *the set $\Upsilon = \{u \in X : \lambda A(\frac{u}{\lambda}) + \lambda B(u) = u, \lambda \in (0,1)\}$ is unbounded.*

We introduce the following hypotheses:

(H_1) A generate a compact and analytic α-ROF $(S_\alpha(t))_{t \ge 0}$ which is exponentially bounded i.e there exist constants $M \ge 1, \omega \ge 0$ such that

$$\|S_\alpha(t)\| \le Me^{\omega t}; \quad t \ge 0.$$

(H_2) The functions $I_k : E \longrightarrow E$ are Lipschitz. Let M_k, for $k = 1, 2, 3, \ldots m$, be such that

$$\|I_k(y) - I_k(x)\| \le M_k \|y - x\|; \quad \text{for each } y, x \in E.$$

(H_3) the function $f : J \times D \longrightarrow E$ is Caratheodory.
(H_4) There exists a function $p \in L^1(J, R_+)$ and a continuous nondecreasing function $\psi : [0, +\infty[\longrightarrow [0, +\infty[$ such that

$$|f(t, y)| \le p(t)\psi(\|y\|)_D,$$

a.e, $t \in J$, for all $y \in D$, with

$$\int_{C_0}^\infty \frac{du}{\psi(u)} = \infty,$$

and

$$\int_{C_3}^\infty \frac{du}{\psi(u)} = \infty,$$

where
$$C_0 = C, \quad C_3 = min(C_1, C_2),$$

$$C = (M_b + L^\phi + K_b M e^{\omega b}) \|\phi\|_{\mathscr{D}_b^0},$$

$$C_1 = \frac{K_b \left(M^{k+1} e^{\omega b} |\phi(0)| + \sum_{i=1}^{k} M^{k-i+1} e^{\omega(b-t_i)} (|I_i(0)| + C) \right)}{1 - K_b \sum_{i=1}^{k} M^{k-i+1} e^{\omega(b-t_i)} M_i}$$
$$+ \frac{K_b \sum_{i=1}^{k} M^{k-i+2} e^{\omega(b-t_{k-1})} \int_{t_{i-1}}^{t_i} e^{-\omega s} p(s) \psi(\mu(s)) ds}{1 - K_b \sum_{i=1}^{k} M^{k-i+1} e^{\omega(b-t_i)} M_i} + C \quad ,$$

$$C_2 = \frac{K_b M e^{\omega b}}{\left(1 - K_b \sum_{i=1}^{k} M^{k-i+1} e^{\omega(b-t_i)} M_i\right)} \quad .$$

Theorem 3. *Assume that Hypotheses* $(H_\varphi), (A_1), (A_2), (H_1), (H_4)$ *are satisfied with*

$$K_b \sum_{i=1}^{k} M^{k-i+1} e^{\omega(b-t_i)} M_i < 1,$$

then the problem (1.1)–(1.3) has at least one mild solution on $] -\infty, b]$.

Proof. Transform the problem (1.1)–(1.3) into a fixed point problem. Consider the operator $N : \mathscr{D}_b \rightarrow \mathscr{D}_b$ defined by

$$N(y)(t) = \begin{cases} \phi(t); & t \in (-\infty, 0], \\ S_\alpha(t)\phi(0) + \int_0^t S_\alpha(t-s) f(s, y_{\rho(s,y_s)}) ds; & t \in [0, t_1], \\ S_\alpha(t-t_k) \prod_{i=1}^{k} S_\alpha(t_i - t_{i-1}) \phi(0) & \\ + \sum_{i=1}^{k} \int_{t_{i-1}}^{t_i} S_\alpha(t-t_k) \prod_{j=i}^{k-1} S_\alpha(t_{j+1} - t_j) S_\alpha(t_i - s) f(s, y_{\rho(s,y_s)}) & \\ + \int_{t_k}^t S_\alpha(t-s) f(s, y_{\rho(s,y_s)}) ds & \\ + \sum_{i=1}^{k} S_\alpha(t-t_k) \prod_{j=i}^{k-1} S_\alpha(t_{j+1} - t_j) I_i(y_{t_i}); & t \in (t_k, t_{k+1}]. \end{cases}$$

Let $x(.) :] -\infty, b] \longrightarrow E$, be the function defined by

$$x(t) = \begin{cases} \phi(t), & \text{if } t \in] -\infty, 0], \\ S_\alpha(t)\phi(0), & \text{if } t \in [0, t_1], \\ 0, & \text{if } t \in (t_k, t_{k+1}]. \end{cases}$$

Then $x_0 = \phi$. For each $z \in \mathscr{D}_b$ with $z(0) = 0$, we denote by $\bar{z}$ the function defined by

$$\bar{z}(t) = \begin{cases} 0, & \text{if } t \in] -\infty, 0], \\ \int_0^t S_\alpha(t-s) f(s, x_{\rho(s,x_s+\bar{z}_s)} + \bar{z}_{\rho(s,x_s+\bar{z}_s)}) + \bar{z}_{\rho(s,x_s+\bar{z}_s)}) ds, & \text{if } t \in [0, t_1], \\ S_\alpha(t-t_k) \prod_{i=1}^{k} S_\alpha(t_i - t_{i-1}) \phi(0) & \\ + \sum_{i=1}^{k} \int_{t_{i-1}}^{t_i} S_\alpha(t-t_k) \prod_{j=i}^{k-1} S_\alpha(t_{j+1} - t_j) S_\alpha(t_i - s) f(s, x_{\rho(s,x_s+\bar{z}_s)} + \bar{z}_{\rho(s,x_s+\bar{z}_s)}) ds & \\ + \int_{t_k}^t S_\alpha(t-s) f(s, y_{\rho(s,y_s)}) ds & \\ + \sum_{i=1}^{k} S_\alpha(t-t_k) \prod_{j=i}^{k-1} S_\alpha(t_{j+1} - t_j) I_i(x_{t_i} + \bar{z}_{t_i}), & \text{if } t \in (t_k, t_{k+1}]. \end{cases} \tag{5}$$

If $y(.)$ satisfies (3), we can decompose it as $y(t) = x(t) + z(t)$, $0 \le t \le b$, which implies $y_t = z_t + x_t$ for every $0 \le t \le b$ and the function $z(.)$ satisfies

$$z^*(t) = \begin{cases} 0, & \text{if } t \in]-\infty, 0], \\ z(t), & \text{if } t \in [0, b]. \end{cases}$$

where

$$z(t) = \begin{cases} \int_0^t S_\alpha(t-s) f(s, x_{\rho(s,x_s+\bar{z}_s)} + \bar{z}_{\rho(s,x_s+\bar{z}_s)})) ds, & \text{if } t \in [0, t_1], \\ S_\alpha(t-t_k) \prod_{i=1}^k S_\alpha(t_i - t_{i-1}) \phi(0) \\ + \sum_{i=1}^k \int_{t_{i-1}}^{t_i} S_\alpha(t-t_k) \prod_{j=i}^{k-1} S_\alpha(t_{j+1} - t_j) S_\alpha(t_i - s) f(s, x_{\rho(s,x_s+\bar{z}_s)} + \bar{z}_{\rho(s,x_s+\bar{z}_s)}) \\ + \int_{t_k}^t S_\alpha(t-s) f(s, x_{\rho(s,x_s+\bar{z}_s)} + \bar{z}_{\rho(s,x_s+\bar{z}_s)})) ds \\ + \sum_{i=1}^k S_\alpha(t-t_k) \prod_{j=i}^{k-1} S_\alpha(t_{j+1} - t_j) I_i(x_{t_i} + \bar{z}_{t_i}), & \text{if } t \in (t_k, t_{k+1}]. \end{cases}$$

Set

$$\mathscr{D}_b^0 := \{z \in \mathscr{D}_b : z_0 = 0\}.$$

and let $\|.\|_b$ be the seminorm in $\mathscr{D}_b^0$ defined by

$$\|z\|_b = \|z_0\|_+ \sup\{|z(t)| : 0 \le t \le b\}$$
$$= \sup\{|z(t)| : 0 \le t \le b\}.$$

$\mathscr{D}_b^0$ is Banach space with the norm $\|.\|_b$.

Transform the problem 1–3 into a fixed point problem. Consider the two operators

$$\mathscr{A}, \mathscr{B} : \mathscr{D}_b^0 \longrightarrow \mathscr{D}_b^0,$$

defined by

$$\mathscr{A}z(t) = \begin{cases} 0, & \text{if } t \in [0, t_1], \\ S_\alpha(t-t_k) \prod_{i=1}^k S_\alpha(t_i - t_{i-1}) \phi(0) \\ + \sum_{i=1}^k S_\alpha(t-t_k) \prod_{j=i}^{k-1} S_\alpha(t_{j+1} - t_j) I_i(x_{t_i} + \bar{z}_{t_i}), & \text{if } t \in (t_k, t_{k+1}]. \end{cases}$$

and

$$\mathscr{B}z(t) = \begin{cases} \int_0^t S_\alpha(t-s) f(s, x_{\rho(s,x_s+\bar{z}_s)} + \bar{z}_{\rho(s,x_s+\bar{z}_s)})) ds, & \text{if } t \in [0, t_1], \\ \sum_{i=1}^k \int_{t_{i-1}}^{t_i} S_\alpha(t-t_k) \prod_{j=i}^{k-1} S_\alpha(t_{j+1} - t_j) S_\alpha(t_i - s) f(s, x_{\rho(s,x_s+\bar{z}_s^n)} + \bar{z}_{\rho(s,x_s+\bar{z}_s^n)}) \\ + \int_{t_k}^t S_\alpha(t-s) f(s, x_{\rho(s,x_s+\bar{z}_s)} + \bar{z}_{\rho(s,x_s+\bar{z}_s)})) ds, & \text{if } t \in (t_k, t_{k+1}]. \end{cases}$$

Then the problem of finding the solution of the problem 1–3 is reduced to finding the solution of operator equation $\mathscr{A}z(t) + \mathscr{B}z(t) = z(t)$, $t \in (-\infty, b]$, we shall that the operators $\mathscr{A}$ and $\mathscr{B}$ satisfy all the conditions of theorem 3.

We give the proof into a sequence of steps.

Step 1: $\mathscr{B}$ is continuous.

Let $(z^n)_{n \ge 0}$ be a sequence such that $z^n \longrightarrow z$ in $\mathscr{D}_b^0$. Since f satisfies (H3), we get

$$f(s, x_s + \bar{z}_s^n) \to f(s, x_s + \bar{z}_s) \qquad \text{as} \quad n \to \infty.$$

Then

1. For $t \in [0, t_1]$, we have

$$|\mathscr{B}(z^n)(t) - \mathscr{B}(z)(t)|$$
$$= |\int_0^t S_\alpha(t-s)[f(s, x_{\rho(s,x_s+\bar{z}_s^n)} + \bar{z}_{\rho(s,x_s+\bar{z}_s^n)}^n) - f(s, x_{\rho(s,x_s+\bar{z}_s)} + \bar{z}_{\rho(s,x_s+\bar{z}_s)})]ds$$
$$\leq \int_0^t \|S_\alpha(t-s)\| \|f(s, x_{\rho(s,x_s+\bar{z}_s^n)} + \bar{z}_{\rho(s,x_s+\bar{z}_s^n)}^n)) - f(s, x_{\rho(s,x_s+\bar{z}_s)} + \bar{z}_{\rho(s,x_s+\bar{z}_s)}))|ds$$
$$\leq M e^{\omega t} \int_0^t e^{-\omega s} |f(s, x_{\rho(s,x_s+\bar{z}_s^n)} + \bar{z}_{\rho(s,x_s+\bar{z}_s^n)}^n) - f(s, x_{\rho(s,x_s+\bar{z}_s)} + \bar{z}_{\rho(s,x_s+\bar{z}_s)}))|ds \longrightarrow 0.$$

2. For $t \in (t_k, t_{k+1}]$,

$$|\mathscr{B}(z^n)(t) - \mathscr{B}(z)(t)|$$

$$= |\sum_{i=1}^k \int_{t_{i-1}}^{t_i} S_\alpha(t-t_k) \prod_{j=i}^{k-1} S_\alpha(t_{j+1}-t_j) S_\alpha(t_i-s)[f(s, x_{\rho(s,x_s+\bar{z}_s^n)} + \bar{z}_{\rho(s,x_s+\bar{z}_s^n)}^n)$$
$$-f(s, x_{\rho(s,x_s+\bar{z}_s)} + \bar{z}_{\rho(s,x_s+\bar{z}_s)})]ds + \int_{t_k}^t S_\alpha(t-s)[f(s, x_{\rho(s,x_s+\bar{z}_s^n)} + \bar{z}_{\rho(s,x_s+\bar{z}_s^n)}^n) - f(s, x_{\rho(s,x_s+\bar{z}_s)} + \bar{z}_{\rho(s,x_s+\bar{z}_s)})]|$$
$$\leq \sum_{i=1}^k \int_{t_{i-1}}^{t_i} \|S_\alpha(t-t_k)\| \prod_{j=i}^{k-1} \|S_\alpha(t_{j+1}-t_j)\| \|S_\alpha(t_i-s)\| \times |[f(s, x_{\rho(s,x_s+\bar{z}_s^n)} + \bar{z}_{\rho(s,x_s+\bar{z}_s^n)}^n)$$
$$-f(s, x_{\rho(s,x_s+\bar{z}_s)} + \bar{z}_{\rho(s,x_s+\bar{z}_s)})]ds + \int_{t_k}^t \|S_\alpha(t-s)\| \|[f(s, x_{\rho(s,x_s+\bar{z}_s^n)} + \bar{z}_{\rho(s,x_s+\bar{z}_s^n)}^n)$$
$$-f(s, x_{\rho(s,x_s+\bar{z}_s)} + \bar{z}_{\rho(s,x_s+\bar{z}_s)})]|ds$$
$$\leq \sum_{i=1}^k \int_{t_{i-1}}^{t_i} M e^{\omega(t-t_k)} \prod_{j=i}^{k-1} M e^{\omega(t_{j+1}-t_j)} M e^{\omega(t_i-s)} |[f(s, x_{\rho(s,x_s+\bar{z}_s^n)} + \bar{z}_{\rho(s,x_s+\bar{z}_s^n)}^n)$$
$$-f(s, x_{\rho(s,x_s+\bar{z}_s)} + \bar{z}_{\rho(s,x_s+\bar{z}_s)})]|ds + \int_{t_k}^t M e^{\omega(t-s)} |[f(s, x_{\rho(s,x_s+\bar{z}_s^n)} + \bar{z}_{\rho(s,x_s+\bar{z}_s^n)}^n) -$$
$$f(s, x_{\rho(s,x_s+\bar{z}_s)} + \bar{z}_{\rho(s,x_s+\bar{z}_s)})]|ds$$
$$\leq \sum_{i=1}^k \int_{t_{i-1}}^{t_i} M e^{\omega(t-t_k)} [M e^{\omega(t_{i+1}-t_i)} \times M e^{\omega(t_{i+2}-t_{i+1})} \times M e^{\omega(t_{i+3}-t_{i+2})}$$
$$\times ... \times M e^{\omega(t_k-t_{k-1})}] M e^{\omega(t_i-s)} |[f(s, x_{\rho(s,x_s+\bar{z}_s^n)} + \bar{z}_{\rho(s,x_s+\bar{z}_s^n)}^n)$$
$$-f(s, x_{\rho(s,x_s+\bar{z}_s)} + \bar{z}_{\rho(s,x_s+\bar{z}_s)})]|ds$$
$$+ \int_{t_k}^t M e^{\omega(t-s)} |[f(s, x_{\rho(s,x_s+\bar{z}_s^n)} + \bar{z}_{\rho(s,x_s+\bar{z}_s^n)}^n)$$
$$-f(s, x_{\rho(s,x_s+\bar{z}_s)} + \bar{z}_{\rho(s,x_s+\bar{z}_s)})]|ds$$
$$\leq \sum_{i=1}^k \int_{t_{i-1}}^{t_i} M e^{\omega t} [M^{k-1-i+1}] M e^{-\omega s} |[f(s, x_{\rho(s,x_s+\bar{z}_s^n)} + \bar{z}_{\rho(s,x_s+\bar{z}_s^n)}^n) - f(s, x_{\rho(s,x_s+\bar{z}_s)} + \bar{z}_{\rho(s,x_s+\bar{z}_s)})]|ds$$
$$+ M e^{\omega t} \int_{t_k}^t e^{-\omega s} |[f(s, x_{\rho(s,x_s+\bar{z}_s^n)} + \bar{z}_{\rho(s,x_s+\bar{z}_s^n)}^n)$$
$$-f(s, x_{\rho(s,x_s+\bar{z}_s)} + \bar{z}_{\rho(s,x_s+\bar{z}_s)})]|ds$$
$$\leq \sum_{i=1}^k M^{k-i+2} e^{\omega t} \int_{t_{i-1}}^{t_i} e^{-\omega s} |[f(s, x_{\rho(s,x_s+\bar{z}_s^n)} + \bar{z}_{\rho(s,x_s+\bar{z}_s^n)}^n) - f(s, x_{\rho(s,x_s+\bar{z}_s)} + \bar{z}_{\rho(s,x_s+\bar{z}_s)})]|ds$$
$$+ M e^{\omega t} \int_{t_k}^t e^{-\omega s} |[f(s, x_{\rho(s,x_s+\bar{z}_s^n)} + \bar{z}_{\rho(s,x_s+\bar{z}_s^n)}^n) - f(s, x_{\rho(s,x_s+\bar{z}_s)} + \bar{z}_{\rho(s,x_s+\bar{z}_s)})]|ds \longrightarrow 0.$$

We get

$$\|\mathscr{B}(z^n)(t) - \mathscr{B}(z)(t)\|_{\mathscr{D}_b^0} \longrightarrow 0.$$

as $n \longrightarrow +\infty$.

This means that $\mathscr{B}$ is continuous.

Step 2: $\mathscr{B}$ maps bounded sets into bounded sets in $\mathscr{D}_b^0$.

A linear operator $\mathscr{B} : \mathscr{D}_b^0 \longrightarrow \mathscr{D}_b^0$ is bounded if only it maps bounded sets into bounded sets; i.e it is enough to show that for any $q > 0$, there exists a positive constant $l_k; k = 1, 2, ..., m$ such that for each $z \in B_q = \{z \in \mathscr{D}_b^0 : \|z\| \leq q\}$, we have $\|\mathscr{B}(z)\| \leq l_k$.

Let $z \in B_q$. Then,

$$|\mathscr{B}z(t)| \leq \begin{cases} \int_0^t \|S_\alpha(t-s)\| |f(s, x_{\rho(s,x_s+\bar{z}_s)} + \bar{z}_{\rho(s,x_s+\bar{z}_s)})| ds, & \text{if } t \in [0, t_1], \\ \sum_{i=1}^k \int_{t_{i-1}}^{t_i} \|S_\alpha(t-t_k)\| \\ \times \prod_{j=i}^{k-1} \|S_\alpha(t_{j+1}-t_j)\| \|S_\alpha(t_i-s)\| |f(s, x_{\rho(s,x_s+\bar{z}_s)} + \bar{z}_{\rho(s,x_s+\bar{z}_s)})| ds \\ + \int_{t_k}^t \|S_\alpha(t-s)\| |f(s, x_{\rho(s,x_s+\bar{z}_s)} + \bar{z}_{\rho(s,x_s+\bar{z}_s)})| ds, & \text{if } t \in (t_k, t_{k+1}]. \end{cases}$$

$$|\mathscr{B}z(t)| \leq \begin{cases} \int_0^t \|S_\alpha(t-s)\| p(s) \psi(\|x_{\rho(s,x_s+\bar{z}_s)} + \bar{z}_{\rho(s,x_s+\bar{z}_s)}\|) ds, & \text{if } t \in [0, t_1], \\ \sum_{i=1}^k \int_{t_{i-1}}^{t_i} \|S_\alpha(t-t_k)\| \\ \times \prod_{j=i}^{k-1} \|S_\alpha(t_{j+1}-t_j)\| \|S_\alpha(t_i-s)\| p(s) \psi(\|x_{\rho(s,x_s+\bar{z}_s)} + \bar{z}_{\rho(s,x_s+\bar{z}_s)}\|) ds \\ + \int_{t_k}^t \|S_\alpha(t-s)\| p(s) \psi(\|x_{\rho(s,x_s+\bar{z}_s)} + \bar{z}_{\rho(s,x_s+\bar{z}_s)}\|) ds, & \text{if } t \in (t_k, t_{k+1}]. \end{cases}$$

Using Lemma 3.1, we get

$$\|x_{\rho(s,x_s+\bar{z}_s)} + \bar{z}_{\rho(s,x_s+\bar{z}_s)}\|_{\mathscr{D}_b^0} \leq K_b M e^{\omega t_1} |\phi(0)| + (M_b + L^\phi) \|\phi\|_{\mathscr{D}_b^0} + K_b |z(s)|.$$

Then

$$\|x_{\rho(s,x_s+\bar{z}_s)} + \bar{z}_{\rho(s,x_s+\bar{z}_s)}\|_{\mathscr{D}_b^0} \leq (M_b + L^\phi + K_b M e^{\omega b}) \|\phi\|_{\mathscr{D}_b^0} + K_b q = q^*.$$

Set $C = (M_b + L^\phi + K_b M e^{\omega b}) \|\phi\|_{\mathscr{D}_b^0}$. Then we obtain

$$\|x_{\rho(s,x_s+\bar{z}_s)} + \bar{z}_{\rho(s,x_s+\bar{z}_s)}\|_{\mathscr{D}_b^0} \leq K_b |z(s)| + C.$$

$$|\mathscr{B}z(t)| \leq \begin{cases} M e^{\omega t_1} \psi(q_1^*) \int_0^t e^{-\omega s} p(s) ds, & \text{if } t \in [0, t_1], \\ \sum_{i=1}^k M e^{\omega(t-t_k)} [M e^{\omega(t_{i-1}-t_i)} M e^{\omega(t_{i+2}-t_{i+1})} \dots M e^{\omega(t_{k-1}-t_{k-2})} \\ \times M e^{\omega(t_k-t_{k-1})}] M e^{\omega t_i} \times \psi(q^*) \int_{t_{i-1}}^{t_i} p(s) e^{-\omega s} ds \\ + M e^{\omega t} \psi(q^*) \int_{t_k}^t p(s) e^{-\omega s} ds, & \text{if } t \in (t_k, t_{k+1}]. \end{cases}$$

Then

$$|\mathscr{B}z(t)| \leq \begin{cases} M e^{\omega t_1} \psi(q^*)) \int_0^t e^{-\omega s} p(s) ds, & \text{if } t \in [0, t_1], \\ \sum_{i=1}^k M^{k-i+2} e^{\omega(t-t_k)+t_{i-1}-t_i+t_{i+2}-t_{i+1}\dots+t_{k-1}-t_{k-2}+t_k-t_{k-1}+t_i)} \\ \times \psi(q^*) \int_{t_{i-1}}^{t_i} p(s) e^{-\omega s} ds + M e^{\omega t} \psi(q^*) \int_{t_k}^t p(s) e^{-\omega s} ds, & \text{if } t \in (t_k, t_{k+1}]. \end{cases}$$

Using characteristic of the exponential function, we get

$$|\mathscr{B}z(t)| \leq \begin{cases} M e^{\omega t_1} \psi(q_1^*) \int_0^t e^{-\omega s} p(s) ds, & \text{if } t \in [0, t_1], \\ \sum_{i=1}^k M^{k-i+2} e^{\omega(t-t_{k-1}} \times \psi(q_2^*) \int_{t_{i-1}}^{t_i} p(s) e^{-\omega s} ds \\ + M e^{\omega t} \psi(q_2^*) \int_{t_k}^t p(s) e^{-\omega s} ds, & \text{if } t \in (t_k, t_{k+1}]. \end{cases}$$

Finally, we obtain

$$|\mathscr{B}z(t)| \le \begin{cases} Me^{\omega t_1}\psi(q_1^*)\int_0^t e^{-\omega s}p(s)ds = l_1, & \text{if } t \in [0,t_1], \\ \sum_{i=1}^k M^{k-i+2}e^{\omega(t_{k+1}-t_{k-1})} \times \psi(q_2^*)\int_{t_{i-1}}^{t_i} p(s)e^{-\omega s}ds \\ +Me^{\omega t_{k+1}}\psi(q_2^*)\int_{t_k}^t p(s)e^{-\omega s}ds = l_k, & k=2,3,...,m, \quad \text{if } t \in (t_k,t_{k+1}]. \end{cases}$$

$$|\mathscr{B}z(t)| \le \begin{cases} Me^{\omega t_1}\psi(q^*)\int_0^t e^{-\omega s}p(s)ds = l_1, & \text{if } t \in [0,t_1], \\ \sum_{i=1}^k M^{k-i+2}e^{\omega(t_{k+1}-t_{k-1})} \times \psi(q^*)\int_{t_{i-1}}^{t_i} p(s)e^{-\omega s}ds \\ +Me^{\omega t_{k+1}}\psi(q^*)\int_{t_k}^t p(s)e^{-\omega s}ds = l_k, & k=2,3,...,m, \quad \text{if } t \in (t_k,t_{k+1}]. \end{cases}$$

Step 3: $\mathscr{B}$ maps bounded sets into equicontinuous sets of $\mathscr{D}_b^0$.

Let $\tau_1, \tau_2 \in J\backslash\{t_1,t_2,...,t_m\}$ with $\tau_1 < \tau_2$, let B_q be a bounded set in $\mathscr{D}_b^0$, and let $z \in B_q$.

• If $\tau_1, \tau_2 \in [0,t_1]$, we have

$$|\mathscr{B}z(\tau_2) - \mathscr{B}z(\tau_1)|$$

$$= |\int_0^{\tau_2} S_\alpha(\tau_2 - s)f(s,x_{\rho(s,x_s+\bar{z}_s)} + \bar{z}_{\rho(s,x_s+\bar{z}_s)})ds - \int_0^{\tau_1} S_\alpha(\tau_1 - s)f(s,x_{\rho(s,x_s+\bar{z}_s)} + \bar{z}_{\rho(s,x_s+\bar{z}_s)})ds|.$$

Using the linearity of integral operator and hypotheses H_4, we get

$$|\mathscr{B}z(\tau_2) - \mathscr{B}z(\tau_1)|$$

$$= |\int_0^{\tau_1} S_\alpha(\tau_2 - s)f(s,x_{\rho(s,x_s+\bar{z}_s)} + \bar{z}_{\rho(s,x_s+\bar{z}_s)})ds + \int_{\tau_1}^{\tau_2} S_\alpha(\tau_2 - s)f(s,x_{\rho(s,x_s+\bar{z}_s)} + \bar{z}_{\rho(s,x_s+\bar{z}_s)})ds$$

$$- \int_0^{\tau_1} S_\alpha(\tau_1 - s)f(s,x_{\rho(s,x_s+\bar{z}_s)} + \bar{z}_{\rho(s,x_s+\bar{z}_s)})ds|$$

$$= |\int_0^{\tau_1} (S_\alpha(\tau_2 - s) - S_\alpha(\tau_1 - s))f(s,\bar{z}_t(s) + x_t(s))ds + \int_{\tau_1}^{\tau_2} S_\alpha(\tau_2 - s)f(s,x_{\rho(s,x_s+\bar{z}_s)} + \bar{z}_{\rho(s,x_s+\bar{z}_s)})ds|$$

$$\le \int_0^{\tau_1} \|S_\alpha(\tau_2 - s) - S_\alpha(\tau_1 - s)\|\|f(s,x_{\rho(s,x_s+\bar{z}_s)} + \bar{z}_{\rho(s,x_s+\bar{z}_s)})\|ds$$

$$+ \int_{\tau_1}^{\tau_2} \|S_\alpha(\tau_2 - s)\|\|f(s,x_{\rho(s,x_s+\bar{z}_s)} + \bar{z}_{\rho(s,x_s+\bar{z}_s)})\|ds$$

$$\le \psi(q_1^*)\int_0^{\tau_1} \|S_\alpha(\tau_2 - s) - S_\alpha(\tau_1 - s)\|p(s)ds + Me^{\omega\tau_2}\psi(q^*)\int_{\tau_1}^{\tau_2} e^{-\omega s}p(s)ds.$$

If $\tau_1 = 0$, the right-hand side of previous inequality tends to zero as $\tau_2 \longrightarrow 0$ uniformly for $z \in \mathscr{D}_b^0$.

If $0 < \tau_1 < \tau_2$, for $\varepsilon > 0$ whit $\varepsilon < \tau_1 < \tau_2$, we have

$$|\mathscr{B}(z(\tau_2)) - \mathscr{B}(z(\tau_1))| \le \int_0^{\tau_1-\varepsilon} \|S_\alpha(\tau_2 - s) - S_\alpha(\tau_1 - s)\|\|f(s,x_{\rho(s,x_s+\bar{z}_s)} + \bar{z}_{\rho(s,x_s+\bar{z}_s)})\|ds$$

$$+ \int_{\tau_1 - \varepsilon}^{\tau_1} \|S_\alpha(\tau_2 - s) - S_\alpha(\tau_1 - s)\| |f(s, x_{\rho(s,x_s+\bar{z}_s)} + \bar{z}_{\rho(s,x_s+\bar{z}_s)})| ds$$

$$+ \int_{\tau_1}^{\tau_2} \|S_\alpha(\tau_2 - s)\| |f(s, x_{\rho(s,x_s+\bar{z}_s)} + \bar{z}_{\rho(s,x_s+\bar{z}_s)})| ds$$

$$\leq \psi(q_1^*) \int_0^{\tau_1 - \varepsilon} \|S_\alpha(\tau_2 - s) - S_\alpha(\tau_1 - s)\| p(s) ds$$

$$+ \psi(q_1^*) \int_{\tau_1 - \varepsilon}^{\tau_1} \|S_\alpha(\tau_2 - s) - S_\alpha(\tau_1 - s)\| p(s) ds$$

$$+ M e^{\omega \tau_2} \psi(q^*) \int_{\tau_1}^{\tau_2} e^{-\omega s} p(s) ds.$$

From lemma 1, the operator $S_\alpha(t)$ is a uniformly continuous operator for $t \in [\varepsilon, t_1]$. Combining this and the arbitrariness of ε with the above estimation on $|\mathscr{B}(z(\tau_2)) - \mathscr{B}(z(\tau_1))|$, we can conclude that

$$lim_{[\tau_1,\tau_2] \to 0} |\mathscr{B}(z(\tau_2)) - B(z(\tau_1))| = 0.$$

Thus the operator $\mathscr{B}$ is equicontinous on $[0, t_1]$.

- If $\tau_1, \tau_2 \in (t_k, t_{k+1}]$,

$$|\mathscr{B}(z(\tau_2)) - \mathscr{B}(z(\tau_1))|$$

$$= \| \sum_{i=1}^{k} \int_{t_{i-1}]}^{t_i} S_\alpha(\tau_2 - t_k) \prod_{j=i}^{k-1} S_\alpha(t_{j+1} - t_j) S_\alpha(t_i - s) f(s, x_{\rho(s,x_s+\bar{z}_s)} + \bar{z}_{\rho(s,x_s+\bar{z}_s)}) ds$$

$$- \sum_{i=1}^{k} \int_{t_{i-1}]}^{t_i} S_\alpha(\tau_1 - t_k) \prod_{j=i}^{k-1} S_\alpha(t_{j+1} - t_j) S_\alpha(t_i - s) f(s, x_{\rho(s,x_s+\bar{z}_s)} + \bar{z}_{\rho(s,x_s+\bar{z}_s)}) ds$$

$$+ \int_{t_k}^{\tau_2} S_\alpha(\tau_2 - s) f(s, x_{\rho(s,x_s+\bar{z}_s)} + \bar{z}_{\rho(s,x_s+\bar{z}_s)}) ds - \int_{t_k}^{\tau_1} S_\alpha(\tau_1 - s) f(s, x_{\rho(s,x_s+\bar{z}_s)} + \bar{z}_{\rho(s,x_s+\bar{z}_s)}) ds\|.$$

Then

$$|\mathscr{B}z(\tau_2) - \mathscr{B}z(\tau_1)|$$

$$\leq \sum_{i=1}^{k} \int_{t_{i-1}}^{t_i} \|S_\alpha(\tau_2 - t_k) - S_\alpha(\tau_1 - t_k)\| \prod_{j=i}^{k-1} \|S_\alpha(t_{j+1} - t_j)\|$$

$$\times \|S_\alpha(t_i - s)\| |f(s, x_{\rho(s,x_s+\bar{z}_s)} + \bar{z}_{\rho(s,x_s+\bar{z}_s)})| ds + \| \int_{t_k}^{\tau_1} S_\alpha(\tau_2 - s) f(s, x_{\rho(s,x_s+\bar{z}_s)} + \bar{z}_{\rho(s,x_s+\bar{z}_s)}) ds$$

$$+ \int_{\tau_1}^{\tau_2} S_\alpha(\tau_2 - s) f(s, x_{\rho(s,x_s+\bar{z}_s)} + \bar{z}_{\rho(s,x_s+\bar{z}_s)}) ds - \int_{t_k}^{\tau_1} S_\alpha(\tau_1 - s) f(s, x_{\rho(s,x_s+\bar{z}_s)} + \bar{z}_{\rho(s,x_s+\bar{z}_s)}) ds\|.$$

Which gives

$$|\mathscr{B}z(\tau_2) - \mathscr{B}z(\tau_1)| \leq \sum_{i=1}^{k} \int_{t_{i-1}}^{t_i} \|S_\alpha(\tau_2 - t_k) - S_\alpha(\tau_1 - t_k)\| \prod_{j=i}^{k-1} \|S_\alpha(t_{j+1} - t_j)\|$$

$$\times \|S_\alpha(t_i-s)\| |f(s, x_{\rho(s,x_s+\bar{z}_s)}+\bar{z}_{\rho(s,x_s+\bar{z}_s)})|ds$$

$$+\int_{t_k}^{\tau_1} \|S_\alpha(\tau_2-s)-S_\alpha(\tau_1-s)\| |f(s, x_{\rho(s,x_s+\bar{z}_s)}+\bar{z}_{\rho(s,x_s+\bar{z}_s)})|ds$$

$$+\int_{\tau_1}^{\tau_2} \|S_\alpha(\tau_2-s)\| |f(s, x_{\rho(s,x_s+\bar{z}_s)}+\bar{z}_{\rho(s,x_s+\bar{z}_s)})|ds.$$

Under the hypothesis H_4, and lemma, we obtain

$$|\mathscr{B}z(\tau_2)-\mathscr{B}z(\tau_1)| \leq \sum_{i=1}^{k} \psi(q_2^*) \int_{t_{i-1}}^{t_i} \|S_\alpha(\tau_2-t_k)-S_\alpha(\tau_1-t_k)\|$$

$$\prod_{j=i}^{k-1} \|S_\alpha(t_{j+1}-t_j)\| \times \|S_\alpha(t_i-s)\| p(s)ds$$

$$+\psi(q^*) \int_{t_k}^{\tau_1-\varepsilon} \|S_\alpha(\tau_2-s)-S_\alpha(\tau_1-s)\| p(s)ds$$

$$+\psi(q^*) \int_{\tau_1-\varepsilon}^{\tau_1} \|S_\alpha(\tau_2-s)-S_\alpha(\tau_1-s)\| p(s)ds$$

$$+M\psi(q^*)e^{\omega\tau_2} \int_{\tau_1}^{\tau_2} e^{-\omega s} p(s)ds.$$

As $\tau_1 \longrightarrow \tau_2$ and ε becomes sufficiently small, the right-hand side of the above inequality tends to zero, since S_α is analytic operator and the compactness of $S_\alpha(t)$ for $t > 0$ implies the continuity in the uniform operator topology. This proves the equicontinuity for the case where $t \neq t_i, i = 1, ..., m+1$.

Now, it remains to examine equicontinuity at $t = t_l$. We have for $z \in B_q$, for each $t \in J$.
First, we prove equicontinuity at $t = t_l^-$.
Fix $\delta_1 > 0$ such that $\{t_k, k \neq l\} \cap [t_l - \delta_1, t_l - \delta_1] = \emptyset$.
For $0 < h < \delta_1$, we have
• if $l = 1$ i.e $t_1 - h, t_1 \in [0, t_1]$,

$$|\mathscr{B}(z(t_1-h))-\mathscr{B}(z(t_1))| \leq \psi(q_1^*) \int_0^{t_1-h} \|S_\alpha(t_1-s)-S_\alpha(t_1-h-s)\| p(s)ds$$

$$+Me^{\omega t_1} \psi(q^*) \int_{t_1-h}^{t_1} e^{-\omega s} p(s).ds$$

Which tends to zero as $h \longrightarrow 0$ since $S_\alpha(t)$ is uinformly continuous operator for $t \in [0, t_1]$ thus the operator B is equicontinuous at $t = t_1^-$.

• if $t_l - h, t_l \in [t_k, t_{k+1}]$.
Then:

$$|\mathscr{B}(z)(t_l-h)-\mathscr{B}(z)(t_l)| \leq \sum_{i=1}^{k} \psi(q^*) \int_{t_{i-1}}^{t_i} \|S_\alpha(t_l-t_k)-S_\alpha(t_l-h-t_k)\|$$

$$\times \prod_{j=i}^{k-1} \|S_\alpha(t_{j+1}-t_j)\| \|S_\alpha(t_i-s)\| p(s)ds$$

$$+\psi(q^*) \int_{t_k}^{t_l-h} \|S_\alpha(t_l-s)-S_\alpha(t_l-h-s)\| p(s)ds$$

$$+M\psi(q^*)e^{\omega t_l} \int_{t_l-h}^{t_l} e^{-\omega s} p(s)ds.$$

The right-hand side of the previous inequality tends to zero as $h \longrightarrow 0$.
So the operator $\mathscr{B}$ is equicontinuous at t_l^-.
Now, define

$$\widehat{\mathscr{B}_0}(z)(t) = \mathscr{B}(z)(t), if \ t \in [0, t_1],$$

and

$$\widehat{\mathscr{B}_i}(z)(t) = \begin{cases} \mathscr{B}(z)(t), & \text{if } t \in (t_i, t_{i+1}), \\ \mathscr{B}(z)(t_i^+), & \text{if } t = t_i. \end{cases}$$

Next, we prove equicontinuity at $t = t_i^+$.

Fix $\delta_2 > 0$ such that $\{t_k, k \neq i\} \cap [t_i - \delta_2, t_i + \delta_2] = \emptyset$.

First, we study the equicontinuity at $t = 0^+$.

If $t \in [0, t_1]$, we have

$$\widehat{\mathscr{B}_0}(z)(t) = \begin{cases} \mathscr{B}z(t), & \text{if } t \in [0, t_1], \\ 0, & \text{if } t = 0. \end{cases}$$

For $0 < h < \delta_2$, we have

$$
\begin{aligned}
|\widehat{\mathscr{B}_0}(z)(h) - \widehat{\mathscr{B}_0}(z)(0)| \quad &= |\mathscr{B}(z)(h)| \\
&= \| \int_0^h S_\alpha(h-s) f(s, x_{\rho(s, x_s + \bar{z}_s)} + \bar{z}_{\rho(s, x_s + \bar{z}_s)}) ds \| \\
&\leq \int_0^h \|S_\alpha(h-s)\| \|f(s, x_{\rho(s, x_s + \bar{z}_s)} + \bar{z}_{\rho(s, x_s + \bar{z}_s)})\| ds \\
&\leq \psi(q^*) e^{\omega h} \int_0^h e^{-\omega s} p(s) ds.
\end{aligned}
$$

The right-hand side tends to zero as $h \longrightarrow 0$.

Now, we study the equicontinuity at $t_1^+, t_2^+, ..., t_m^+ (t_l^+, 1 \leq l \leq m)$.

For $0 < h < \delta_2$, we have

$$
\begin{aligned}
|\mathscr{B}(z)(t_l + h) - \mathscr{B}(z)(t_l)| \quad &\leq \sum_{i=1}^k \psi(q^*) \int_{t_{i-1}}^{t_i} \|S_\alpha(h) - S_\alpha(0)\| \\
&\times \prod_{j=i}^{k-1} \|S_\alpha(t_{j+1} - t_j)\| \|S_\alpha(t_i - s)\| p(s) ds \\
&+ M \psi(q^*) e^{\omega(t_l + h)} \int_{t_l}^{t_l + h} e^{-\omega s} p(s) ds.
\end{aligned}
$$

It is clear that the right-hand side tends to zero as $h \longrightarrow 0$.

Then $\mathscr{B}$ is equicontinuous at $t_l^+, (1 \leq l \leq m)$. The equicontinuity for the cases $\tau_1 < \tau_2 \leq 0$ and $\tau_1 \leq 0 \leq \tau_2$ follows from the uniform continuity of ϕ on the interval $]-\infty, 0]$. As a consequence of steps 1 and 3 toghether with Arzel-Ascoli Theorem it suffices to show that $\mathscr{B}z$ maps B_q into a precompact set in E i.e.: we show that the set $\{\mathscr{B}z(t), z \in B_q\}$ is precompact in E for every $t \in [0, b]$.

Now, let $x \in B_q$ and let ε be a positive real number satisfying $0 < \varepsilon < t \leq b$.

For $z \in B_q$ and $t \in [0, t_1]$.

we have if $t = 0$ the set $\{\mathscr{B}z(0); z \in B_q\} = \{0\}$ which is precompact as a finite set.

For $0 < \varepsilon < t \leq t_1$, we have

$$
\begin{aligned}
\mathscr{B}(z)(t) \quad &= \int_0^t S_\alpha(t-s) f(s, x_{\rho(s, x_s + \bar{z}_s)} + \bar{z}_{\rho(s, x_s + \bar{z}_s)}) ds \\
&= \int_0^{t-\varepsilon} S_\alpha(t-s) f(s, x_{\rho(s, x_s + \bar{z}_s)} + \bar{z}_{\rho(s, x_s + \bar{z}_s)}) ds + \int_{t-\varepsilon}^t S_\alpha(t-s) f(s, x_{\rho(s, x_s + \bar{z}_s)} + \bar{z}_{\rho(s, x_s + \bar{z}_s)}) ds.
\end{aligned}
$$

Set $F_0 := \{ S_\alpha(t - \theta) f(\theta, x_{\rho(\theta, x_\theta + \bar{z}_\theta)} + \bar{z}_{\rho(\theta, x_\theta + \bar{z}_\theta)})); \theta \in [0, t-\varepsilon], z \in B_q \}$, from the mean value Theorem for the Bochner integral, we have

$$\int_0^{t-\varepsilon} S_\alpha(t-s) f(s, \bar{z}_t(s) + x_t(s)) ds \in (t-\varepsilon) Conv(\overline{F_0}). \tag{6}$$

On the other hand, using hypotheses (H_1) and (H_4), we obtain

$$\int_{t-\varepsilon}^{t} |S_\alpha(t-s)f(s,\bar{z}_t(s)+x_t(s))|ds \leq Me^{\omega t}\psi(q^*)\int_{t-\varepsilon}^{t} e^{-\omega s}p(s)ds.$$

Let C_ε^0 the circle who's diameter d_ε^0 is such that

$$d_\varepsilon^0 \leq Me^{\omega t}\psi(q_1^*)\int_{t-\varepsilon}^{t} e^{-\omega s}p(s)ds. \tag{7}$$

As a consequence of (6) and (7), we conclude that

$$Bz(t) \in (t-\varepsilon)Conv(\overline{F_0})+C_\varepsilon^0, \quad \forall 0 < \varepsilon < t \leq t_1. \tag{8}$$

For $t_k < \varepsilon < t < t_{k+1}$ and $z \in B$, we have

$$Bz(t) = \sum_{i=1}^{k}\int_{t_{i-1}}^{t_i} S_\alpha(t-t_k)\prod_{j=i}^{k-1} S_\alpha(t_{j+1}-t_j)S_\alpha(t_i-s)f(s,\bar{z}_t(s)+x_t(s))ds. \tag{9}$$

$$+ \int_{t_k}^{t-\varepsilon} S_\alpha(t-s)f(s,\bar{z}_t(s)+x_t(s))ds + \int_{t-\varepsilon}^{t} S_\alpha(t-s)f(s,\bar{z}_t(s)+x_t(s))ds.$$

Set $F_k := \{S_\alpha(t-\theta)f(\theta,\bar{z}_t(\theta)+x_t(\theta)); \theta \in (t_k,t_{k+1}), z \in B_q\}$, from the mean value theorem for the Bochner integral, we have

$$\int_{t_k}^{t-\varepsilon} S_\alpha(t-s)f(s,x_{\rho(s,x_s+\bar{z}_s)}+\bar{z}_{\rho(s,x_s+\bar{z}_s)})ds \in (t-t_k-\varepsilon)Conv(\overline{F_k}). \tag{10}$$

From $(H_1),(H_4)$ we obtain

$$\sum_{i=1}^{k}\int_{t_{i-1}}^{t_i} S_\alpha(t-t_k)\prod_{j=i}^{k-1} S_\alpha(t_{j+1}-t_j)S_\alpha(t_i-s)f(s,x_{\rho(s,x_s+\bar{z}_s)}+\bar{z}_{\rho(s,x_s+\bar{z}_s)})ds$$
$$+ \int_{t_k}^{t-\varepsilon} S_\alpha(t-s)f(s,\bar{z}_t(s)+x_t(s))ds$$
$$\leq \psi(q^*)\sum_{i=1}^{k} M^{k-i+2}e^{\omega(t-t_{k-1})}\int_{t_{i-1}}^{t_i} e^{-\omega s}p(s)ds + M\psi(q^*)e^{\omega t}\int_{t-\varepsilon}^{t} e^{-\omega s}p(s)ds.$$

Let C_ε^k the circle who's diameter d_ε^k is such that

$$d_\varepsilon^k \leq \psi(q^*)\sum_{i=1}^{k} M^{k-i+2}e^{\omega(t-t_{k-1})}\int_{t_{i-1}}^{t_i} e^{-\omega s}p(s)ds + M\psi(q^*)e^{\omega t}\int_{t-\varepsilon}^{t} e^{-\omega s}p(s)ds. \tag{11}$$

From (9) and (11), it follows that

$$\mathscr{B}z(t) \in (t-t_k-\varepsilon)Conv(\overline{F_k})+C_\varepsilon^k, \quad \forall t_k < \varepsilon < t < t_{k+1}. \tag{12}$$

From 8 and 12, we conclude that $\mathscr{B}z(t)$ is precompact in E. From Step1– Step 3, we deduce that $\mathscr{B}$ is completey continuous.

Step 4: $\mathscr{A}$ is a contraction.

For $t \in]-\infty, t_1]$, we have

$$|\mathscr{A}z_1(t) - \mathscr{A}z_2(t)| = 0.$$

which implies that $\mathscr{A}$ is contraction for all $t \in]-\infty, t_1]$. It remains to prove that A is a contraction operator for $t \in [t_k, t_{k+1}], k \geq 1$

$$
\begin{aligned}
|\mathscr{A}z_1(t) - \mathscr{A}z_2(t)| \quad &= |\textstyle\sum_{i=1}^{k} S_\alpha(t - t_k) \prod_{j=i}^{k-1} S_\alpha(t_{j+1} - t_j) \\
&\quad \times [(I_i(x_{t_i} + \bar{z}_{t_i}^1) - I_i(x_{t_i} + \bar{z}_{t_i}^2)| \\
&\leq \textstyle\sum_{i=1}^{k} \|S_\alpha(t - t_k)\| \prod_{j=i}^{k-1} \|S_\alpha(t_{j+1} - t_j)\| \|[I_i(\bar{z}_{t_i}^1) - I_i(\bar{z}_{t_i}^2)| \\
&\leq \textstyle\sum_{i=1}^{k} M e^{\omega(t - t_k)} \prod_{j=i}^{k-1} M e^{\omega(t_{j+1} - t_j)} \|[I_i(\bar{z}_{t_i}^1) - I_i(\bar{z}_{t_i}^2)| \\
&\leq \textstyle\sum_{i=1}^{k} M e^{\omega(t - t_k)} [M e^{\omega(t_{i+1} - t_i)} M e^{\omega(t_{i+2} - t_{i+1})} \ldots M e^{\omega(t_{k-1} - t_{k-2})} \\
&\quad \times M e^{\omega(t_k - t_{k-1})}] \times \|[I_i(\bar{z}_{t_i}^1) - I_i(\bar{z}_{t_i}^2)| \\
&\leq \textstyle\sum_{i=1}^{k} M^{k-i+1} e^{\omega(t - t_i)} \|[I_i(\bar{z}_{t_i}^1) - I_i(\bar{z}_{t_i}^2)|.
\end{aligned}
$$

Since $t \in J := [0, b]$ and the functions $I_k; k = 1, 2, \ldots, m$. Lipschitz; Then

$$|\mathscr{A}z_1(t) - \mathscr{A}z_2(t)| \leq \sum_{i=1}^{k} M^{k-i+1} e^{\omega(b - t_i)} M_i \|z_{t_i}^1 - z_{t_i}^2\|_{\mathscr{D}}.$$

It fallows that

$$\|\mathscr{A}z_1 - \mathscr{A}z_2\| \leq K_b \sum_{i=1}^{k} M^{k-i+1} e^{\omega(b - t_i)} M_i \|z^1 - z^2\|_{\mathscr{D}}.$$

Thus the operator $\mathscr{A}$ is a contraction, since

$$K_b \sum_{i=1}^{k} M^{k-i+1} e^{\omega(b - t_i)} M_i < 1.$$

Step 5: A *priori* bounds.

Now it remains to show that the set

$$\Upsilon = \{z \in PC(]-\infty, b], E) : z = \lambda \mathscr{B}(z) + \lambda \mathscr{A}(\tfrac{z}{\lambda}), \quad \text{for some} \quad 0 < \lambda < 1\}.$$

is bounded.

Let $z \in \Upsilon$ be any element, then $z = \lambda \mathscr{B}(z) + \lambda \mathscr{A}(\frac{z}{\lambda})$,
for some $0 < \lambda < 1$.
First, for each $t \in [0, t_1]$,

$$
\begin{aligned}
|z(t)| \quad &= |\lambda \textstyle\int_0^t S_\alpha(t - s) f(s, x_{\rho(s, x_s + z_s)} + \bar{z}_{\rho(s, x_s + z_s)}) ds| \\
&\leq M e^{\omega t} \textstyle\int_0^t e^{-\omega s} |f(s, x_{\rho(s, x_s + z_s)} + \bar{z}_{\rho(s, x_s + z_s)})| ds \\
&\leq M e^{\omega t_1} \textstyle\int_0^{t_1} e^{-\omega s} p(s) \psi(\|x_{\rho(s, x_s + z_s)} + \bar{z}_{\rho(s, x_s + z_s)}\|) ds \\
&\leq M e^{\omega t_1} \textstyle\int_0^{t_1} e^{-\omega s} p(s) \psi(K_b |z(s)| + (M_b + M K_b e^{\omega t_1} + L^\phi) \|\phi\|) ds.
\end{aligned}
$$

On the other hand, for each $t \in (t_k, t_{k+1}]$, we have

$$|z(t)| = \|\lambda \left(\sum_{i=1}^{k} \int_{t_{i-1}}^{t_i} S_\alpha(t-t_k) \prod_{j=i}^{k-1} S_\alpha(t_{j+1}-t_j) S_\alpha(t_i-s) f(s, x_{\rho(s,x_s+z_s)} + \overline{z}_{\rho(s,x_s+z_s)}) ds$$
$$+ \int_{t_k}^{t} S_\alpha(t-s) f(s, x_{\rho(s,x_s+z_s)} + \overline{z}_{\rho(s,x_s+z_s)}) \right) + \lambda \left(S_\alpha(t-t_k) \prod_{i=1}^{k} S_\alpha(t_i-t_{i-1}) \phi(0) \right.$$
$$\left. + \sum_{i=1}^{k} S_\alpha(t-t_k) \prod_{j=i}^{k-1} S_\alpha(t_{j+1}-t_j) I_i(\tfrac{z_{t_i}}{\lambda} + x_{t_i}) \right)\|.$$

From (H_1), (H_2) and since $\lambda < 1$, we obtain

$$|z(t)| \leq M^{k+1} e^{\omega t} |\phi(0)| + \sum_{i=1}^{k} M^{k-i+1} e^{\omega(t-t_i)} |I_i(0)|$$
$$+ \sum_{i=1}^{k} M^{k-i+2} e^{\omega(t-t_{k-1})} \int_{t_{i-1}}^{t_i} e^{-\omega s} p(s) \psi(K_b|z(s)|+C) ds$$
$$+ M e^{\omega t} \int_{t_k}^{t} e^{-\omega s} p(s) \psi(K_b|z(s)|+C) ds$$
$$+ \sum_{i=1}^{k} M^{k-i+1} e^{\omega(t-t_i)} |I_i(\tfrac{z_{t_i}}{\lambda} + x_{t_i}) - I_i(0)|.$$

Since I_i are Lipschitz, then

$$|z(t)| \leq M^{k+1} e^{\omega t} |\phi(0)| + \sum_{i=1}^{k} M^{k-i+1} e^{\omega(t-t_i)} |I_i(0)|$$
$$+ \sum_{i=1}^{k} M^{k-i+2} e^{\omega(t-t_{k-1})} \int_{t_{i-1}}^{t_i} e^{-\omega s} p(s) \psi(K_b|z(s)|+C) ds$$
$$+ M e^{\omega t} \int_{t_k}^{t} e^{-\omega s} p(s) \psi(K_b|z(s)|+C) ds$$
$$+ \sum_{i=1}^{k} M^{k-i+1} e^{\omega(t-t_i)} M_i \|z(t_i^-)\|$$
$$\leq M^{k+1} e^{\omega t} |\phi(0)| + \sum_{i=1}^{k} M^{k-i+1} e^{\omega(t-t_i)} |I_i(0)|$$
$$+ \sum_{i=1}^{k} M^{k-i+2} e^{\omega(t-t_{k-1})} \int_{t_{i-1}}^{t_i} e^{-\omega s} p(s) \psi(K_b|z(s)|+C) ds$$
$$+ M e^{\omega t} \int_{t_k}^{t} e^{-\omega s} p(s) \psi(K_b|z(s)|+C) ds$$
$$+ \sum_{i=1}^{k} M^{k-i+1} e^{\omega(t-t_i)} M_i (K_b|z(t)|+C).$$

Therefore

$$[1 - K_b \sum_{i=1}^{k} M^{k-i+1} e^{\omega(b-t_i)} M_i] |z(t)| \leq M^{k+1} e^{\omega t} \|\phi(0)\| + \sum_{i=1}^{k} M^{k-i+1} e^{\omega(b-t_i)} |I_i(0)|$$
$$+ \sum_{i=1}^{k} M^{k-i+2} e^{\omega(t-t_{k-1})} \int_{t_{i-1}}^{t_i} e^{-\omega s} p(s) \psi(K_b|z(s)|+C) ds$$
$$+ M e^{\omega b} \int_{t_k}^{t} e^{-\omega s} p(s) \psi(K_b|z(s)|+C) ds + C \sum_{i=1}^{k} M^{k-i+1} e^{\omega(b-t_i)} M_i.$$

Thus we have

$$|z(t)| \leq \left(M^{k+1} e^{\omega b} |\phi(0)| + \sum_{i=1}^{k} M^{k-i+1} e^{\omega(b-t_i)} (|I_i(0)|+C) \right.$$
$$\left. + \sum_{i=1}^{k} M^{k-i+2} e^{\omega(b-t_{k-1})} \int_{t_{i-1}}^{t_i} e^{-\omega s} p(s) \psi(K_b|z(s)|+C) ds \right) \Big/ \left(1 - K_b \sum_{i=1}^{k} M^{k-i+1} e^{\omega(b-t_i)} M_i \right)$$
$$+ \frac{M e^{\omega b}}{(1 - K_b \sum_{i=1}^{k} M^{k-i+1} e^{\omega(b-t_i)} M_i)} \int_{t_k}^{t} e^{-\omega s} p(s) \psi(K_b|z(s)|+C) ds.$$

We consider the function $\mu(t)$ defined by

$$\mu(t) = \sup\{K_b|z(s)| + C; 0 \leq s \leq t\}, \quad 0 \leq t \leq b.$$

Let $t^* \in [0,t]$ be such that $\mu(t) = K_b|z(t^*)|+C$. If $t^* \in J$, by the previous inequality, we have for $t \in J$.

- if $t \in [0,t_1]$,

$$\mu(t) \leq K_b M e^{\omega b} \int_0^t e^{-\omega s} p(s)\psi(\mu(s))ds + C.$$

- if $t \in (t_k,t_{k+1}]$,

$$|\mu(t)| \leq K_b\left(M^{k+1}e^{\omega b}|\phi(0)| + \sum_{i=1}^k M^{k-i+1}e^{\omega(b-t_i)}(|I_i(0)|+C)\right)$$
$$+ \sum_{i=1}^k M^{k-i+2}e^{\omega(b-t_{k-1})}\int_{t_{i-1}}^{t_i}e^{-\omega s}p(s)\psi(|\mu(s)|)ds\Big/\left(1 - K_b\sum_{i=1}^k M^{k-i+1}e^{\omega(b-t_i)}M_i\right)$$
$$+ \frac{K_b M e^{\omega b}}{(1-K_b\sum_{i=1}^k M^{k-i+1}e^{\omega(b-t_i)}M_i)}\int_{t_k}^t e^{-\omega s}p(s)\psi(|\mu(s)|)ds + C.$$

Then

$$\mu(t) \leq C_1 + C_2\int_{t_k}^t e^{-\omega s}p(s)\psi(\mu(s)ds.$$

Where

$$C_1 = \frac{K_b\left(M^{k+1}e^{\omega b}|\phi(0)| + \sum_{i=1}^k M^{k-i+1}e^{\omega(b-t_i)}(|I_i(0)|+C)\right)}{1 - K_b\sum_{i=1}^k M^{k-i+1}e^{\omega(b-t_i)}M_i}$$
$$+ \frac{K_b\sum_{i=1}^k M^{k-i+2}e^{\omega(b-t_{k-1})}\int_{t_{i-1}}^{t_i}e^{-\omega s}p(s)\psi(\mu(s))ds}{1 - K_b\sum_{i=1}^k M^{k-i+1}e^{\omega(b-t_i)}M_i} + C.$$

$$C_2 = \frac{K_b M e^{\omega b}}{(1 - K_b\sum_{i=1}^k M^{k-i+1}e^{\omega(b-t_i)}M_i)}.$$

It follows that

$$\mu(t) \leq \begin{cases} C + K_b M e^{\omega b}\int_0^t e^{-\omega s}p(s)\psi(\mu(s))ds, & if\ t \in [0,t_1] \\ C_1 + C_2\int_{t_k}^t e^{-\omega s}p(s)\psi(\mu(s))ds, & if\ t \in (t_k,t_{k+1}). \end{cases}$$

Let us take the right-hand side of the above inequality as $\vartheta(t)$,

$$\mu(t) \leq \vartheta(t).$$

and

$$\vartheta(t) := \begin{cases} C + K_b M e^{\omega b}\int_0^t e^{-\omega s}p(s)\psi(\mu(s))ds, & if\ t \in [0,t_1] \\ C_1 + C_2\int_{t_k}^t e^{-\omega s}p(s)\psi(\mu(s))ds, & if\ t \in (t_k,t_{k+1}). \end{cases}$$

$$\begin{cases} \vartheta(0) = C, \\ \vartheta(t_k) = C_1, & k = 1,2,...,m. \end{cases}$$

And differentiating both sides of the above equality, we obtain

$$\vartheta'(t) := \begin{cases} K_b M e^{\omega(b-t)}p(t)\psi(\mu(t)), & if\ t \in [0,t_1], \\ C_2 e^{-\omega t}p(t)\psi(\mu(t)), & if\ t \in (t_k,t_{k+1}]. \end{cases}$$

Using the non decreasing character of function ψ, i.e

$$\mu(t) \le \vartheta(t) \Rightarrow \psi(\mu(t)) \le \psi(\vartheta(t))$$

We have

$$\vartheta'(t) \le \begin{cases} K_b M e^{\omega(b-t)} p(t) \psi(\vartheta(t)), & if \ t \in [0,t_1], \\ C_2 e^{-\omega t} p(t) \psi(\vartheta(t)), & if \ t \in (t_k,t_{k+1}]. \end{cases}$$

It gives

$$\frac{\vartheta'(t)}{\psi(\vartheta(t))} \le \begin{cases} K_b M e^{\omega(b-t)} p(t), & if \ t \in [0,t_1], \\ C_2 e^{-\omega t} p(t), & if \ t \in (t_k,t_{k+1}). \end{cases}$$

• Integrating from 0 to t, if $t \in [0,t_1]$, we get

$$\int_0^t \frac{\vartheta'(s)}{\psi(\vartheta(s))} ds \le K_b M e^{\omega b} \int_0^t e^{-\omega s} p(s) ds.$$

By change of variable $(\vartheta(s) = u)(s : 0 \longrightarrow t, u : C \longrightarrow \vartheta(t))$:

$$\int_C^{v(t)} \frac{du}{\psi(u)} \le M e^{\omega b} \int_C^t e^{-\omega s} p(s) ds \le \int_0^\infty \frac{du}{\psi(u)}.$$

Hence, there exists a constant η_1 such that

$$\mu(t) \le \vartheta(t) \le \eta_1.$$

• Now, integrating from t_k to t if $t \in (t_k,t_{k+1}]$, we get

$$\int_{t_k}^t \frac{\vartheta'(s)}{\psi(\vartheta(s))} ds \le C_2 \int_{t_k}^t e^{-\omega s} p(s) ds.$$

By change of variable $(\vartheta(s) = u)(s : t_k \longrightarrow t, u : C_1 \longrightarrow \vartheta(t))$:

$$\int_0^{v(t)} \frac{du}{\psi(u)} \le C_2 \int_{t_k}^t e^{-\omega s} p(s) ds \le \int_{C_3}^{v(t)} \frac{du}{\psi(u)}.$$

Where $C_3 = min(C,C_2)$. Henc, there existe a constant η_2 such that

$$\mu(t) \le \vartheta(t) \le \eta_2, \quad t \in (t_k,t_{k+1})$$

In conclusion, there exists $\eta = min(\eta_1,\eta_2)$ such that

$$\mu(t) \le \vartheta(t) \le \eta, \ for \ all \quad t \in J.$$

Now from the definition of μ it follows that, there exist $\eta^* > 0$ such that

$$\|z\|_{\mathscr{D}_b^0} \le \eta^*, \quad \forall z \in \Upsilon.$$

This shows that the set Υ is bounded.

As a consequence of theorem, we deduce that $\mathscr{A} + \mathscr{B}$ has a fixed point z^*. Then $y^*(t) = z^*(t) + x(t), t \in (-\infty, b]$ is a fixed point of the operator N and hence the problem have a mild solution on interval $(-\infty, b]$. This completes the proof.

4 Application

We consider the following impulsive fractional differential equation of the form:

$$\frac{\partial_t^q}{\partial t^q} v(t,x) = \frac{\partial^2}{\partial x^2} v(t,x) + \int_{-\infty}^t a_1(s-t)v(s-\rho_1(t)\rho_2(|v(t)|,\xi)ds, \ x \in [0,\pi], \ t \in [0,b]\setminus\{t_1,...,t_m\},$$

(13)

$$\Delta v(t_i)(x) = \int_{-\infty}^t d_i(t_i-s)v(s,x)ds, \ x \in [0,\pi], i = 1,...,m,$$

(14)

$$v(t,0) = v(t,\pi) = 0, t \in [0,b],$$

(15)

$$v(t,x) = v_0(\theta,x), \ \theta \in]-\infty,0], x \in [0,\pi].$$

(16)

where $0 < q < 1$, $d_i : R \to R, i = 1,2,...,m$, and $a_1 : (-\infty,0] \to R$, $\rho_i : [0,+\infty) \to [0,+\infty), i = 1,2$ are continuous functions.

Set $E - L^?([0,\pi])$ and let $D(A) \subset E \to E$ be the operator $Au = u''$ with the domaine

$$D(A) = \{u \in H_0^1(0,\pi) \cap H^2(0,\pi)\}.$$

The operator A is the infinitesimal generator of analytic semi-group $S(t)$.
Set $\gamma > 0$. For the phase space, we choose $\mathscr{D}$ to defined by:

$$\mathscr{D} = PC^\gamma = \{\Phi \in PC((-\infty,0],E) : \lim_{\theta \in (-\infty,0]} \exp(\gamma^\theta)\Phi(\theta) \quad \text{exists in} \quad E\}.$$

with norm

$$\|\phi\|_\gamma = \sup_{\theta \in (-\infty,0]} \exp(\gamma^\theta)|\phi(\theta)|, \qquad \phi \in PC^\gamma.$$

For this space, axioms $(A1),(A2)$ are satisfied. the problem (4.1)–(4.4) takes the abstract form (1.1)–(1.3) by making the following change of variables.

$$y(t)(x) = v(t,x), \quad x \in [0,\pi], t \in J = [0,1].$$

$$\phi(\theta)(x) = v_0(\theta,x), \quad x \in [0,\pi], \theta \leq 0.$$

$$f(t,\varphi)(x) = \int_{-\infty}^t a_1\varphi(s,x)ds.$$

$$\rho(t,\varphi) = s - \rho_1(t)\rho_2(|\varphi(0)|).$$

$$I_i(\varphi)(x) = \int_{-\infty}^0 d_i(-\theta)\varphi(\theta)(x)d\theta$$

Theorem 4. *Let $\varphi \in \mathscr{B}$ such that H_φ holds, the problem (4.1)–(4.4) has at least one mild solution.*

5   Conclusion

In this work, we provided the existence of mild solutions and with sufficient conditions for some differential fractional equations. The main tool of this paper is the fixed point theory combined with resolvent famillies. To our knowlege, there are few works using this technique. The obtained results have a contribution to the related literature and extend the results in [HL20, HGBA13].

References

[AABH08] Abada, N., Agarwal, R.P., Benchora, M., Hammouche, H.: Existence results for nondensely defined impulsive semilinear functional differential equations with state-dependent delay. (Asian Eur. J. Math. **1**(4), 449–468 (2008)

[ABN10] Agarwal, R.P., Benchohra, M., Hamani, S.: A survey on existence results for boundary value problems of nonlinear fractional differential equations and inclusions. Acta Applicandae Mathematicae **109**(3), 973–1033 (2010)

[ABN12] Abbas, S., Benchohra, M.: N'guérékata. Topics in fractional differential equations. Springer-Verlag, New York, G.M. (2012)

[ABN15] Abbas, S., Benchohra, M.: N'guérékata. Advanced fractional differential and integral equations. Nova Science Publishers, New York, G.M. (2015)

[ABN] Abbas, S., Benchohra, M., N'guérékata, G.M.: Topics in Fractional Differential Equations. Springer Science & Business Media, New York (2012)

[BHN06] Benchohra, M., Henderson, J., Ntouyas, S.: Impulsive Differential Equations and Inclusions Contemporary Mathematics and Its Applications. Hindawi Publishing Corporation, New York (2006)

[BK98] Burton, T.A., Kirk, C.: A fixed point theorem of Krasnoselskii-Schaefer type. Math. Nachr **189**, 23–31 (1998)

[F14] Fan, Z.: Characterization of compactness for resolvents and applications. Appl. Math. Comput. **231**, 60–67 (2014)

[FM13] Fan, Z., Mophou, G.: Nonlocal problems for fractional differential equations via resolvent operators. Int. J. Difference Equ. (2013)

[HAM10] Hernandez, E., Anguraj, A., Mallika Arjunan, M.: Existence results for an impulsive second order differential equation with state-dependent delay. Dyn. Contin. Discrete Impuls. Syst. Ser. A Math. Anal. **17**, 287–301 (2010)

[HGBA13] Hammouche, H., Guerbati, K., Benchohra, M., Abada, N.: Existence Results for Impulsive semilinear Fractional Differential inclusions with delay in Banach spaces. Differential Incl. Control opt. **33**, 149–170 (2013)

[HK78] Hale, J., Kato, J.: Phase space for retarded equations with infinite delay. Funkcial. Ekvac **21**, 11–41 (1978)

[HL20] Hammouche, H., Lemkeddem, M., Guerbati, K., Ezzinbi, K.: Existence results for some impulsive partial functional fractional differential equations. Asian Eur. J. Math. **13**(1), 2050074 (2020)

[HMN91] Hino, Y., Murakami, S., Naito, T.: Funtional Differential Equations with Unbounded Delay. Springer-Verlag, Berlin (1991)

[HPL06] Hernandez, E., Prokopczyk, A., Ladeira, L.: A note on partial functional Differential equations with stade-dependent delay in Banach spaces. Nolinear Anal. RWA **7**, 510–519 (2006)

[KST06] Kilbas, A.A., Srivastava, H.M., Trujillo, J.J.: Theory and Applications of Fractional Differential Equations. North-Holland Mathematics Studies. Elsevier Science, Amsterdam (2006)

[LCX12] Li, X., Chen, F.: Generalized anti-periodic boundary value problems of impulsive fractional differ- ential equations. Commun. Nonlinear Sci. Numer. Simul. **18**, 28–41 (2013)

[P83] Pazy, A.: Semigroups of linear Operators and Applications to Partial Differential Equations. Springer Verlag, New York (1983)

[P93] Podlubny, I.: Fractional Differential Equations. Acadmic press, New York, USA (1993)

[WFZ11] Wang, J., Feckan, M., Zhou, Y.: On the new concept of solutions and existence results for impulsive fractional evolution equations. Dyn. Partial Differ. Equ. **8**, 345–361 (2011)

[Z14] Zhou, Y.: Theory of fractional differential equations. World Scientific, Singapore (1983)

A Novel Method for Solving Nonlinear Jerk Equations

Ali Akgül$^{(\boxtimes)}$ and Esra Karatas Akgül

Art and Science Faculty, Department of Mathematics,
Siirt University, 56100 Siirt, Turkey
aliakgul00727@gmail.com

Abstract. In this article, reproducing kernel method for solving Jerk equations is given. Convergence of the solution is shown. This method is applied to the equation for chosen values of the parameters that seem in the model and some numerical experiments prove that the reproducing kernel method is very effective method.

Keywords: Jerk equations · Reproducing kernel functions · Bounded linear operator

2000 Mathematics Subject Classification: 47B32 · 46E22 · 30E25

1 Introduction

Although most of the efforts on dynamical systems are related to second-order differential equations, the behavior of some dynamical systems is governed by nonlinear jerk (third-order) differential equations. Jerk is the rate of acceleration change in physics; that is, the time derivative of acceleration, and as such the second velocity derivative, or the third time position derivative. The jerk is significant in some mechanics and acoustics implementations. Many geometric features of the Jerk vector are founded for plane motion utilizing the aberrancy features of curves [1]. Nonlinear third-order differential equations, known as nonlinear Jerk equations, including the third temporal displacement derivative, are of great interest in investigating the structures which exhibit rotating and translating movements, such as robots or machine tools, where excessive Jerk leads to accelerated wear of transmissions and bearing elements, noisy operations and large contouring errors in discontinuities (such as corners) in the machining path [2]. The jerk equations are the minimal setting for solutions showing chaotic behaviour. The numerical solutions of the Jerk equation have been worked by many investigators [3].

Hu et al. [4] have investigated the iteration calculations of periodic solutions to nonlinear Jerk equations. Liu et al. [5] have obtained the periods and periodic solutions of nonlinear Jerk equations by an iterative algorithm based on a shape function method. Rahman et al. [6] have worked on modified harmonic balance method for the solution of nonlinear Jerk equations.

© Springer Nature Switzerland AG 2021
Z. Hammouch et al. (Eds.): SM2A 2019, LNNS 168, pp. 23–33, 2021.
https://doi.org/10.1007/978-3-030-62299-2_2

We investigate the Jerk equation by reproducing kernel method in this paper. Reproducing kernel method (RKM) is very accurate and reliable method. There are many papers related to the reproducing kernel method in the literature. We apply this method to a new problem in this work. Akgül [7–9] has worked on reproducing kernel Hilbert space method based on reproducing kernel functions for investigating boundary layer flow of a Powell-Eyring non-Newtonian fluid, new reproducing kernel functions and the solutions of variable-order fractional differential equations by reproducing kernel method. Akgül et al. [10] have investigated the numerical solutions of fractional differential equations of Lane-Emden type by an accurate technique. Aronszajn [11] has studied the theory of reproducing kernels. Arqub [12–14] has investigated the approximate solutions of DASs with nonclassical boundary conditions using novel reproducing kernel algorithm, the reproducing kernel algorithm for handling differential algebraic systems of ordinary differential equations and the fitted reproducing kernel Hilbert space method for the solutions of some certain classes of time-fractional partial differential equations subject to initial and Neumann boundary conditions. Azarnavid et al. [15] have worked on an iterative reproducing kernel method in Hilbert space for the multi-point boundary value problems.

In this work, we studied a Jerk equation as:

$$(1.1) \qquad\qquad y''' = J(y, y', y''),$$

with initial conditions

$$(1.2) \qquad\qquad y(0) = 0, y(0) = B, y(0) = 0.$$

We define the most general function of Jerk as;

$$(1.3) \qquad y''' + \alpha y y' y'' + \beta y' y''^2 + \delta y^2 y' + \varepsilon y^3 + \gamma y' = 0,$$

where the parameters α, β, δ, ε and γ are constants.

This paper is organized as follows. In Sect. 2, the RKM is discussed with some preliminary concept and definition. The solution procedure and approximate solutions of Eqs. (1.2)–(1.3) are presented in this section. Numerical experiments are demonstrated in Sect. 3. Conclusion is given in the last section.

2 Reproducing Kernel Method

First of all we will construct the reproducing kernel Hilbert spaces that we need to solve our problem.

Definition 2.1. The first reproducing kernel Hilbert space that we will use is $M_2^3[0, 1]$

$$M_2^3[0, 1] = \{a \in AC[0, 1] : a', a'' \in AC[0, 1], \ a^{(3)} \in L^2[0, 1]\}.$$

We have the inner product and the norm for this space as:

$$\langle a, b\rangle_{M_2^3} = a(0)b(0) + a'(0)b'(0) + a''(0)b''(0) + \int_0^1 a'''(x)b'''(x)\mathrm{d}x, \quad a, b \in M_2^3[0, 1]$$

and

$$\|a\|_{M_2^3} = \sqrt{\langle a, a\rangle_{M_2^3}}, \quad a \in M_2^3[0, 1].$$

Lemma 2.2. $M_2^3[0, 1]$ *is a reproducing kernel Hilbert space. We get the reproducing kernel function* K_y *by [16]:*

$$K_y(x) = \begin{cases} 1 + xy + \frac{x^2y^2}{4} + \frac{x^3y^2}{12} - \frac{x^4y}{24} + \frac{x^5}{120}, & x \le y, \\ \\ 1 + yx + \frac{y^2x^2}{4} + \frac{y^3x^2}{12} - \frac{y^4x}{24} + \frac{y^5}{120}, & x > y. \end{cases}$$

Definition 2.3. We construct the reproducing kernel Hilbert space $M_2^4[0, 1]$ as:

$$M_2^4[0, 1] = \{a \in AC[0, 1] : a', a'', a''' \in AC[0, 1],\ a^{(4)} \in L^2[0, 1], a(0) = a'(0) = a''(0) = 0\}.$$

We have the inner product and the norm for this special Hilbert space by:

$$\langle a, b\rangle_{M_2^4} = \sum_{i=0}^{3} a^{(i)}(0)b^{(i)}(0) + \int_0^1 a^{(4)}(x)b^{(4)}(x)\mathrm{d}x, \quad a, b \in M_2^4[0, 1]$$

and

$$\|a\|_{M_2^4} = \sqrt{\langle a, a\rangle_{M_2^4}}, \quad a \in M_2^4[0, 1].$$

Theorem 2.4. *We find the reproducing kernel function for the reproducing kernel Hilbert space* $M_2^4[0, 1]$ *as:*

$$L_y(x) = \begin{cases} h_y(x), & x \le y, \\ \\ g_y(x), & x > y. \end{cases}$$

where,

$$h_y(x) = -\frac{x^7}{5040} + \frac{x^6y}{720} - \frac{x^5y^2}{240} + \frac{x^4y^3}{144} + \frac{x^3y^3}{36},$$

$$g_y(x) = -\frac{y^7}{5040} + \frac{y^6x}{720} - \frac{y^5x^2}{240} + \frac{y^4x^3}{144} + \frac{y^3x^3}{36}.$$

Proof. We have

$$\langle b, L_y \rangle_{M_2^4} = \sum_{i=0}^{3} L_y^{(i)}(0) b^{(i)}(0) + \int_0^1 L_y^{(4)}(x) b^{(4)}(x) \mathrm{d}x,$$

by Definition 2.3. We obtain

$$\begin{aligned}
\langle b, L_y \rangle_{M_2^4} = {} & L_y(0)b(0) + L_y'(0)b'(0) + L_y''(0)b''(0) + L_y'''(0)b'''(0) \\
& + L_y^{(4)}(1)b'''(1) - L_y^{(4)}(0)b'''(0) - L_y^{(5)}(1)b''(1) \\
& + L_y^{(5)}(0)b''(0) + L_y^{(6)}(1)b'(1) - L_y^{(6)}(0)b'(0) \\
& - \int_0^1 L_y^{(7)}(x)b'(x)\mathrm{d}x,
\end{aligned}$$

by integration by parts. Since $b(0) = b'(0) = b''(0) = 0$ we get

$$\langle b, L_y \rangle_{M_2^4} = L_y'''(0)b'''(0) + L_y^{(4)}(1)b'''(1) - L_y^{(4)}(0)b'''(0) - L_y^{(5)}(1)b''(1)$$

$$L_y^{(6)}(1)b'(1) - \int_0^1 L_y^{(7)}(x)b'(x)\mathrm{d}x.$$

We have

$$L_y'''(0) = \frac{y^3}{6},$$

$$L_y^{(4)}(0) = \frac{y^3}{6},$$

$$L_y^{(4)}(1) = 0,$$

$$L_y^{(5)}(1) = 0,$$

$$L_y^{(6)}(1) = 0,$$

Therefore, we obtain

$$\langle b, L_y \rangle_{M_2^4} = -\int_0^y L_y^{(7)}(x)b'(x)\mathrm{d}x - \int_y^1 L_y^{(7)}(x)b'(x)\mathrm{d}x.$$

We know

$$L_y^{(7)}(x) = \begin{cases} -1, & x < y, \\ \\ 0, & x > y. \end{cases}$$

Then, we reach

$$\langle b, L_y \rangle_{M_2^4} = \int_0^y b'(x)\,\mathrm{d}x.$$

Thus, we get

$$\langle b, L_y \rangle_{M_2^4} = b(y).$$

This completes the proof. □

We consider the solutions of the problem (1.3) in the reproducing kernel Hilbert space $M_2^4[0,1]$. We denote the bounded linear operator $T : M_2^4[0,1] \to M_2^3[0,1]$ as:

$$(2.1) \quad Tu = u'''(x) + \alpha x B^2 u''(x) + 2\delta x B^2 u(x) + \delta x^2 B^2 u'(x) + 3\varepsilon x^2 B^2 u(x) + T\gamma u'(x),$$

we have the following problem.

$$(2.2) \qquad\qquad Tu = D(r,u),$$

with the initial conditions

$$(2.3) \qquad\qquad u(0) = u'(0) = u''(0) = 0,$$

where

$$\begin{aligned}
D(r,u) = & -\alpha u(x)u'(x)u''(x) - \alpha Bu(x)u''(x) - \alpha Bxu'(x)u''(x) - \alpha B^2 xu''(x) \\
& -\beta u'(x)u''(x)^2 - \beta Bu''(x)^2 - \delta u'(x)u(x)^2 - B\delta u(x)^2 \\
& -2\delta Bxu'(x)u(x) - \delta B^3 x^2 - \varepsilon u(x)^3 - 3\varepsilon Bxu(x)^2 - \varepsilon B^3 x^3 - \gamma B.
\end{aligned}$$

Lemma 2.5. *T is a bounded linear operator.*

Proof. We need to prove

$$\|Tu\|^2_{M_2^3[0,1]} \le A \|u\|^2_{M_2^4[0,1]},$$

where A is a positive constant. We have
(2.4)
$$\|Tu\|^2_{M_2^3[0,1]} = \langle Tu, Tu \rangle_{M_2^3[0,1]} = [Tu(0)]^2 + [Tu(0)]'^2 + [Tu(0)]''^2 + \int_0^1 [Tu(y)]'''^2 \, dy.$$

By reproducing property, we have

$$u(y) = \langle u(\cdot), L_y(\cdot) \rangle_{M_2^4[0,1]}$$

and

$$Tu(y) = \langle u(\cdot), TL_y(\cdot) \rangle_{M_2^4[0,1]},$$

$$(Tu(y))' = \langle u(\cdot), (TL_y(\cdot))' \rangle_{M_2^4[0,1]},$$

$$(Tu(y))'' = \langle u(\cdot), (TL_y(\cdot))'' \rangle_{M_2^4[0,1]},$$

so

$$|Tu| \le \|u\|_{M_2^4[0,1]} \|TL_y\|_{M_2^4[0,1]} = A_1 \|u\|_{M_2^4[0,1]},$$

thus

$$[Tu]^2 \le A_1^2 \|u\|_{M_2^4[0,1]}^2.$$

Since

$$(Tu)'(y) = \langle u(\cdot), (TL_y)'(\cdot) \rangle_{M_2^4[0,1]},$$

we get

$$|(Tu)'| \le \|u\|_{M_2^4[0,1]} \|(TF_y)'\|_{M_2^4[0,1]} = A_2 \|u\|_{M_2^4[0,1]},$$

so, we have

$$[Tu]'^2 \le A_2^2 \|u\|_{M_2^4[0,1]}^2,$$

$$[Tu]''^2 \le A_3^2 \|u\|_{M_2^4[0,1]}^2,$$

$$[Tu]'''^2 \le A_4^2 \|u\|_{M_2^4[0,1]}^2,$$

that is,

$$\|Tu\|_{M_2^3[0,1]}^2 = [Tu(0)]^2 + [Tu(0)]'^2 + [Tu(0)]''^2 + \int_0^1 [(Tu)'''(y)]^2 \, dy$$

$$\le (A_1^2 + A_2^2 + A_3^2 + A_4^2) \|u\|_{M_2^4[0,1]}^2,$$

where $A = A_1^2 + A_2^2 + A_3^2 + A_4^2$ is a positive constant.   □

We construct $\varsigma_i(x) = K_{x_i}(x)$ and $\psi_i(x) = T^* \varsigma_i(x)$, where T^* is conjugate operator of T. The orthonormal system $\left\{ \widehat{\psi}_i(x) \right\}_{i=1}^{\infty}$ of $M_2^4[0,1]$ can be acquired by Gram-Schmidt orthogonalization operation of $\{\psi_i(x)\}_{i=1}^{\infty}$,

$$(2.5) \qquad \widehat{\psi}_i(x) = \sum_{k=1}^{i} \beta_{ik} \psi_k(x), \quad (\beta_{ii} > 0, \quad i = 1, 2, \ldots).$$

Theorem 2.6. *Let* $\{x_i\}_{i=1}^{\infty}$ *be dense in* $[0, 1]$ *and* $\psi_i(x) = T_y L_x(y)|_{y=u_i}$. *Then the sequence* $\{\psi_i(x)\}_{i=1}^{\infty}$ *is a complete system in* $M_2^4[0,1]$.

Proof. We get

$$\psi_i(x) = (T^*\varsigma_i)(x) = \langle (T^*\varsigma_i)(y), L_x(y) \rangle = \langle (\varsigma_i)(y), T_y L_x(y) \rangle = T_y L_x(y)|_{y=x_i} \, .$$

Let $\langle u(x), \psi_i(x) \rangle = 0$, $(i = 1, 2, \ldots)$, which means that,

$$\langle u(x), (T^*\varsigma_i)(x) \rangle = \langle Tu(\cdot), \varsigma_i(\cdot) \rangle = (Tu)(x_i) = 0.$$

$\{x_i\}_{i=1}^\infty$ is dense in $[0,1]$. Therefore, $(Tu)(x) = 0$. $u \equiv 0$ by T^{-1}.   □

Theorem 2.7. *If $u(x)$ is the exact solution of* (2.2)*, then we acquire*

$$(2.6) \qquad u(x) = \sum_{i=1}^\infty \sum_{k=1}^i \beta_{ik} D(x_k, u_k) \widehat{\psi}_i(x).$$

where $\{x_i\}_{i=1}^\infty$ is dense in $[0,1]$.

Proof. We get

$$u(x) = \sum_{i=1}^\infty \left\langle u(x), \widehat{\psi}_i(x) \right\rangle_{M_2^4[0,1]} \widehat{\psi}_i(x)$$

$$= \sum_{i=1}^\infty \sum_{k=1}^i \beta_{ik} \langle u(x), \psi_k(x) \rangle_{M_2^4[0,1]} \widehat{\psi}_i(x)$$

$$= \sum_{i=1}^\infty \sum_{k=1}^i \beta_{ik} \langle u(x), T^*\varsigma_k(x) \rangle_{M_2^4[0,1]} \widehat{\psi}_i(x)$$

$$= \sum_{i=1}^\infty \sum_{k=1}^i \beta_{ik} \langle Tu(x), \varsigma_k(x) \rangle_{M_2^3[0,1]} \widehat{\psi}_i(x)$$

$$= \sum_{i=1}^\infty \sum_{k=1}^i \beta_{ik} \langle D(r,u) K_{x_k} \rangle_{M_2^3[0,1]} \widehat{\psi}_i(x)$$

$$= \sum_{i=1}^\infty \sum_{k=1}^i \beta_{ik} D(x_k, u_k) \widehat{\psi}_i(x).$$

by uniqueness of solution of (2.2). This completes the proof.   □

The approximate solution $u_n(x)$ can be obtained as:

$$(2.7) \qquad u_n(x) = \sum_{i=1}^n \sum_{k=1}^i \beta_{ik} D(x_k, u_k) \widehat{\psi}_i(x).$$

3   Numerical Experiments

We apply the RKM for the approximate analytical solution of the Jerk Equation (1.1)–(1.2).

Example 3.1. We take into consideration

$$(3.1) \qquad\qquad y''' = -y' + yy'y'',$$

with initial conditions $y(0) = 0, y'(0) = B, y''(0) = 0$. We have the exact solution as:

$$(3.2) \quad y(x) = \frac{B}{\Omega}\sin(\Omega x) + \frac{B}{96\Omega^3}\left((-9B^2 - 48 + 48\Omega^2)\sin(\Omega x) - B^2\sin(3\Omega x)\right)$$

where $\Omega = \frac{1}{2}\sqrt{B^2 + 4}$.

Table 1 shows the absolute errors for $B = 0.2, 0.3, 0.4$ respectively. Table 2 shows the relative errors for $B = 0.2, 0.3, 0.4$ respectively.

Example 3.2. We investigate

$$(3.3) \qquad\qquad y''' = -y' - y'(y'')^2,$$

with initial conditions $y(0) = 0, y'(0) = B, y''(0) = 0$. We get exact solution as:

$$y(x) = \frac{B}{\Omega}\sin(\Omega x) + \frac{B}{96\Omega^3}\left((-9B^2\Omega^2 - 48 + 48\Omega^2)\sin(\Omega x)\right.$$
$$\left. + ((12B^2(\Omega)^3 - 48\Omega + 48(\Omega)^3)x\cos(\Omega x) - B^2\Omega^2\sin(3\Omega x))\right)$$

where $\Omega = 2\sqrt{\frac{1}{4-B^2}}$.

Table 3 shows the absolute errors for $B = 0.2, 0.3, 0.4$ respectively. Table 4 shows the relative errors for $B = 0.2, 0.3, 0.4$ respectively.

Table 1. Absolute Errors for the first example.

x	$B = 0.2$	$B = 0.3$	$B = 0.4$
0.125	2.8088×10^{-7}	4.2077×10^{-7}	5.6041×10^{-7}
0.250	0.0000011935	0.0000017943	0.0000024143
0.375	0.0000027482	0.0000041708	0.0000057546
0.500	0.0000049504	0.0000076345	0.0000109688
0.625	0.0000078006	0.0000122943	0.0000185799
0.750	0.0000112879	0.0000182374	0.0000290702
0.875	0.0000153830	0.0000254769	0.0000426444
1.00	0.0000200317	0.0000339035	0.0000590332
1.125	0.0000233859	0.0000405546	0.0000736859
1.250	0.0001115761	0.0000089211	0.0000352990
1.375	0.0001115761	0.0001651283	0.0002024070

Table 2. Relative Errors for the first example.

x	$B = 0.2$	$B = 0.3$	$B = 0.4$
0.125	0.00001126451361	0.00001124981105	0.00001123744895
0.250	0.00002412134243	0.00002417518187	0.00002439700359
0.375	0.00003751627443	0.00003795830294	0.00003928026318
0.500	0.00005163110050	0.00005308586566	0.00005720732970
0.625	0.00006666731728	0.00007005716478	0.00007942052378
0.750	0.00008281647176	0.00008922519004	0.00010670764110
0.875	0.00010024618200	0.00011073531080	0.00013910935150
1.00	0.00011910109620	0.00013449058700	0.00017583060030
1.125	0.00012972006220	0.00015015336190	0.00020497726770
1.250	0.00000985305713	0.00003143790375	0.00009353950721
1.375	0.00056991392360	0.00056376716640	0.00052021705710

Table 3. Absolute Errors for the second example.

x	$B = 0.2$	$B = 0.3$	$B = 0.4$
0.125	2.8088×10^{-7}	4.2077×10^{-7}	5.6041×10^{-7}
0.250	0.0000011935	0.0000017943	0.0000024143
0.375	0.0000027482	0.0000041708	0.0000057546
0.500	0.0000049504	0.0000076345	0.0000109688
0.625	0.0000078006	0.0000122943	0.0000185799
0.750	0.0000112879	0.0000182374	0.0000290702
0.875	0.0000153830	0.0000254769	0.0000426444
1.00	0.0000200317	0.0000339035	0.0000590332
1.125	0.0000233859	0.0000405546	0.0000736859
1.250	0.0001115761	0.0000089211	0.0000352990
1.375	0.0001115761	0.0001651283	0.0002024070

Table 4. Relative Errors for the second example.

x	$B = 0.2$	$B = 0.3$	$B = 0.4$
0.125	0.00001126451361	0.00001124981105	0.00001123744895
0.250	0.00002412134243	0.00002417518187	0.00002439700359
0.375	0.00003751627443	0.00003795830294	0.00003928026318
0.500	0.00005163110050	0.00005308586566	0.00005720732970
0.625	0.00006666731728	0.00007005716478	0.00007942052378
0.750	0.00008281647176	0.00008922519004	0.00010670764110
0.875	0.00010024618200	0.00011073531080	0.00013910935150
1.00	0.00011910109620	0.00013449058700	0.00017583060030
1.125	0.00012972006220	0.00015015336190	0.00020497726770
1.250	0.00000985305713	0.00003143790375	0.00009353950721
1.375	0.00056991392360	0.00056376716640	0.00052021705710

4 Conclusions

In this paper, we investigated the nonlinear Jerk equations by the reproducing kernel method. We constructed very useful reproducing kernel Hilbert spaces and we found some important reproducing kernel functions in these spaces. We found a bounded linear operator to get the results for the problems. We obtained absolute errors and relative errors for some numerical experiments. We proved the efficiency of the proposed method in the work.

References

1. Schot, S.H.: Jerk: the time rate of change of acceleration. Am. J. Phys. **46**, 1090–1094 (1978)
2. Anu, N., Marinca, V.: Approximate analytical solutions to jerk equations. In: Dynamical Systems : Theoretical and Experimental Analysis. vol. 182. pp.169–176, Springer, Cham (2016)
3. Kashkari, B.S.H., Alqarni, S.: Two-step hybrid block method for solving nonlinear jerk equations. J. Appl. Math. Phys. **7**, 1893–1910 (2019)
4. Hu, H., Zheng, M.Y., Guo, Y.J.: Iteration calculations of periodic solutions to nonlinear jerk equations. Acta Mech. **209**(3–4), 269–274 (2010)
5. Liu, C.S., Chang, J.R.: The periods and periodic solutions of nonlinear jerk equations solved by an iterative algorithm based on a shape function method. Appl. Math. Lett. **102**, 106151 (2020)
6. Rahman, M.S., Hasan, A.S.M.Z.: Modified harmonic balance method for the solution of nonlinear jerk equations. Results Phys. **8**, 893–897 (2018)
7. Akgül, A.: Reproducing kernel Hilbert space method based on reproducing kernel functions for investigating boundary layer flow of a Powell-Eyring non-Newtonian fluid. J. Taibah Univ. Sci. **13**(1), 858–863 (2019)
8. Akgül, A.: New reproducing kernel functions. Math. Probl. Eng. **2015**, 158134 (2015)

9. Akgül, A.: On solutions of variable-order fractional differential equations. Int. J. Optim. Control Theor. Appl. (IJOCTA) **7**(1), 112–116 (2017)

10. Akgül, A., Karatas, E., Baleanu, D.: Numerical solutions of fractional differential equations of Lane-Emden type by an accurate technique. Adv. Differ. Equ. pp. 1–12 (2015)

11. Aronszajn, N.: Theory of reproducing kernels. Trans. Am. Math. Soc. **68**, 337–404 (1950)

12. Arqub, O.A.: Approximate solutions of DASs with nonclassical boundary conditions using novel reproducing kernel algorithm. Fundam. Inform. **146**(3), 231–254 (2016)

13. Arqub, O.A.: The reproducing kernel algorithm for handling differential algebraic systems of ordinary differential equations. Math. Methods Appl. Sci. **39**(15), 4549–4562 (2016)

14. Arqub, O.A.: Fitted reproducing kernel Hilbert space method for the solutions of some certain classes of time-fractional partial differential equations subject to initial and Neumann boundary conditions. Comput. Math. Appl. **73**(6), 1243–1261 (2017)

15. Azarnavid, B., Parand, K.: An iterative reproducing kernel method in Hilbert space for the multi-point boundary value problems. J. Comput. Appl. Math. **328**, 151–163 (2018)

16. Cui, M., Yingzhen, L.: Nonlinear Numerical Analysis in the Reproducing Kernel Space. Nova Science Publishers Inc., New York (2009)

Solving a New Type of Fractional Differential Equation by Reproducing Kernel Method

Ali Akgül[1,2(✉)] and Esra Karatas Akgül[1,2]

[1] Art and Science Faculty, Department of Mathematics, Siirt University,
56100 Siirt, Turkey
aliakgul00727@gmail.com
[2] Faculty of Education, Department of Mathematics, Siirt University,
56100 Siirt, Turkey

Abstract. The aim of this work is to get the solutions of the fractional counterpart of a boundary value problem by implementing the reproducing kernel Hilbert space method. Convergence of the solution problem discussed has been shown. The efficiency of the proposed technique is demonstrated by some tables.

1 Introduction

We consider the following problem [2]:

$$(1.1) \qquad \frac{d^2\omega}{dr^2} + \frac{1}{r}\frac{d\omega}{dr} + Ha^2\left(1 - \frac{\omega}{1-\alpha\omega}\right) = 0, \quad 0 \langle r \langle 1,$$

where $\omega(r)$ is the velocity of the fluid, r is the radial distance from the cylindrical conduit centre, Ha is the Hartmann electric number and α is the magnitude of the power of non-linearity. We have the boundary conditions as:

$$(1.2) \qquad \omega'(0) = 0, \quad \omega(1) = 0.$$

The existence and uniqueness of a solution to the problem have been investigated in [3]. Mastroberardino [4] has investigated the problem by the homotopy analysis method. Moghtadaei et al. [5] have applied a spectral method to investigate the problem. Chebyshev spectral collocation method has been used to solve the problem in [6]. Alomari et al. [7] have investigated fractional version of a singular boundary value problem.

In this paper, we consider the following problem.

$$(1.3) \qquad \frac{d^\gamma\omega}{dr^\gamma} + \frac{1}{r}\frac{d^\beta\omega}{dr^\beta} + Ha^2\left(1 - \frac{\omega}{1-\alpha\omega}\right) = 0,$$

where d^γ/dr^γ and d^β/dr^β are the fractional derivative operators in the Caputo sense. $\gamma \in (1,2]$ and $\beta \in (0,1]$ are parameters defining the order of the fractional

© Springer Nature Switzerland AG 2021
Z. Hammouch et al. (Eds.): SM2A 2019, LNNS 168, pp. 34–43, 2021.
https://doi.org/10.1007/978-3-030-62299-2_3

derivative with the property $\gamma - \beta \geq 1$ and subject to the boundary conditions (1.2). Abbas et al. [8] have presented some cases that show the fractional models present better approximate results. Iyiolaa et al. [9] have worked the cancer tumor model of fractional order which demonstrates better approximate results.

We apply reproducing kernel method (RKM) to get the approximate solutions of Eq. (1.3). Reproducing kernel space is a special Hilbert space. Many investigators have applied the RKM to many problems [10]. Arqub et al. [11,12] have investigated some interesting problems by RKM.

This paper is organized as follows. In Sect. 2, RKM is discussed with some preliminary concepts and definitions. The solution procedure and approximate solutions of Eqs. (1.2)–(1.3) are presented in this section. Numerical experiments are demonstrated in Sect. 3. Conclusion is given in the last section.

2   Reproducing Kernel Method

First of all we will construct the reproducing kernel Hilbert spaces that we need to solve our problem.

Definition 2.1. The first reproducing kernel Hilbert space that we will use is $E_2^1 [\, 0,1\,]$

$$E_2^1 [\, 0,1\,] = \{ s \in AC\, [\, 0,1\,] : \ s' \in L^2\, [\, 0,1\,] \}.$$

We have the inner product and the norm for this space as:

$$\langle s,p \rangle_{E_2^1} = s(0)p(0) + \int_0^1 s'(\tau)p'(\tau)\mathrm{d}\tau, \quad s,p \in E_2^1 [\, 0,1\,]$$

and

$$\| s \|_{E_2^1} = \sqrt{\langle s,s \rangle_{E_2^1}}, \quad s \in E_2^1 [\, 0,1\,].$$

Lemma 2.2. $E_2^1 [\, 0,1\,]$ *is a reproducing kernel Hilbert space. We get the reproducing kernel function* G_z *by* [10]:

$$G_z(\tau) = \begin{cases} 1 + \tau, & \tau \leq z, \\ 1 + z, & \tau \big) z. \end{cases}$$

Definition 2.3. We construct the reproducing kernel Hilbert space $E_2^3 [\, 0,1\,]$ as:

$$E_2^3 [\, 0,1\,] = \{ s \in AC\, [\, 0,1\,] : \ s', s'' \in AC\, [\, 0,1\,], \ s^{(3)} \in L^2\, [\, 0,1\,], s'(0) = 0 = s(1) \}.$$

We have the inner product and the norm for this special Hilbert space by:

$$\langle s,p \rangle_{E_2^3} = \sum_{i=0}^{2} s^{(i)}(0)p^{(i)}(0) + \int_0^1 s^{(3)}(\tau)p^{(3)}(\tau)\mathrm{d}\tau, \quad s,p \in E_2^3 [\, 0,1\,]$$

and

$$\| s \|_{E_2^3} = \sqrt{\langle s,s \rangle_{E_2^3}}, \quad s \in E_2^3 [\, 0,1\,].$$

Theorem 2.4. *We find the reproducing kernel function for the reproducing kernel Hilbert space* $E_2^3\,[\,0,1\,]$ *as:*

$$F_z(\tau) = \begin{cases} h_z(\tau), & \tau \le z, \\ \\ g_z(\tau), & \tau \,\rangle\, z. \end{cases}$$

Where,

$$h_z(\tau) = -\frac{\tau^5 z^2}{624} + \frac{\tau^5 z^4}{3744} - \frac{\tau^5 z^3}{1872} + \frac{5\tau^4 z^3}{1872} + \frac{5\tau^4 z^2}{624} - \frac{\tau^5 z^5}{18720} + \frac{\tau^4 z^5}{3744}$$

$$-\frac{5\tau^4 z^4}{3744} + \frac{5\tau^3 z^4}{1872} - \frac{5\tau^3 z^3}{936} - \frac{\tau^4 z}{24} - \frac{5\tau^2 z^3}{312} + \frac{7\tau^3 z^2}{104} - \frac{\tau^3 z^5}{1872}$$

$$-\frac{\tau^2 z^5}{624} + \frac{5\tau^2 z^4}{624} + \frac{21\tau^2 z^2}{104} - \frac{5z^2}{26} + \frac{\tau^5}{520} + \frac{5z^5}{156} - \frac{5z^3}{78} + \frac{5\tau^4}{156}$$

$$-\frac{z^5}{156} - \frac{5\tau^3}{78} - \frac{5\tau^2}{26} + \frac{3}{13},$$

$$g_z(\tau) = -\frac{z^5\tau^2}{624} + \frac{z^5\tau^4}{3744} - \frac{z^5\tau^3}{1872} + \frac{5z^4\tau^3}{1872} + \frac{5z^4\tau^2}{624} - \frac{z^5\tau^5}{18720} + \frac{z^4\tau^5}{3744}$$

$$-\frac{5z^4\tau^4}{3744} + \frac{5z^3\tau^4}{1872} - \frac{5z^3\tau^3}{936} - \frac{z^4\tau}{24} - \frac{5z^2\tau^3}{312} + \frac{7z^3\tau^2}{104} - \frac{z^3\tau^5}{1872}$$

$$-\frac{z^2\tau^5}{624} + \frac{5z^2\tau^4}{624} + \frac{21z^2\tau^2}{104} - \frac{5\tau^2}{26} + \frac{z^5}{520} + \frac{5\tau^5}{156} - \frac{5\tau^3}{78} + \frac{5z^4}{156}$$

$$-\frac{\tau^5}{156} - \frac{5z^3}{78} - \frac{5z^2}{26} + \frac{3}{13}.$$

Proof. We have

$$\langle p, F_z\rangle_{E_2^3} = \sum_{i=0}^{2} F_z^{(i)}(0)p^{(i)}(0) + \int_0^1 F_z^{(3)}(\tau)p^{(3)}(\tau)\mathrm{d}\tau,$$

by Definition 2.3. We obtain

$$\langle p, F_z\rangle_{E_2^3} = F_z(0)p(0) + F_z'(0)p'(0) + F_z''(0)p''(0)$$
$$+F_z'''(1)p''(1) - F_z'''(0)p''(0) - F_z^{(4)}(1)p'(1)$$
$$+F_z^{(4)}(0)p'(0) + \int_0^1 F_z^{(5)}(\tau)p'(\tau)\mathrm{d}\tau,$$

by integration by parts. Since $p'(0) = 0 = p(1)$, we get

$$\langle p, F_z \rangle_{E_2^3} = F_z(0)p(0) + F_z''(0)p''(0) + F_z'''(1)p''(1)$$
$$-F_z'''(0)p''(0) - F_z^{(4)}(1)p'(1)$$
$$+ \int_0^1 F_z^{(5)}(\tau)p'(\tau)d\tau.$$

We have

$$F_z(0) = -\frac{z^5}{156} + \frac{5z^4}{156} - \frac{5z^3}{178} - \frac{5z^2}{26} + \frac{3}{13},$$

$$F_z''(0) = \frac{21z^2}{52} - \frac{z^5}{312} + \frac{5z^4}{312} - \frac{5z^3}{156} - \frac{5}{13},$$

$$F_z'''(0) = \frac{21z^2}{52} - \frac{z^5}{312} + \frac{5z^4}{312} - \frac{5z^3}{156} - \frac{5}{13},$$

$$F_z'''(1) = F_z^{(4)}(1) = 0,$$

Therefore, we obtain

$$\langle p, F_z \rangle_{E_2^3} = (-\frac{z^5}{156} + \frac{5z^4}{156} - \frac{5z^3}{178} - \frac{5z^2}{26} + \frac{3}{13})p(0)$$
$$+ \int_0^z F_z^{(5)}(\tau)p'(\tau)d\tau + \int_z^1 F_z^{(5)}(\tau)p'(\tau)d\tau.$$

We know

$$F_z^{(5)}(\tau) = \begin{cases} -\frac{z^5}{156} + \frac{5z^4}{156} - \frac{5z^3}{178} - \frac{5z^2}{26} + \frac{3}{13}, & \tau \langle z, \\[2mm] -\frac{z^5}{156} + \frac{5z^4}{156} - \frac{5z^3}{178} - \frac{5z^2}{26} - \frac{10}{13}, & \tau \rangle z. \end{cases}$$

Then, we reach

$$\langle p, F_z \rangle_{E_2^3} = (-\frac{z^5}{156} + \frac{5z^4}{156} - \frac{5z^3}{178} - \frac{5z^2}{26} + \frac{3}{13})p(0)$$
$$+ \int_0^z (-\frac{z^5}{156} + \frac{5z^4}{156} - \frac{5z^3}{178} - \frac{5z^2}{26} + \frac{3}{13})p'(\tau)d\tau$$
$$+ \int_z^1 (-\frac{z^5}{156} + \frac{5z^4}{156} - \frac{5z^3}{178} - \frac{5z^2}{26} - \frac{10}{13})p'(\tau)d\tau.$$

Thus, we obtain

$$
\begin{aligned}
\langle p, F_z \rangle_{E_2^3} = {} & (-\frac{z^5}{156} + \frac{5z^4}{156} - \frac{5z^3}{178} - \frac{5z^2}{26} + \frac{3}{13})p(0) \\
& + (-\frac{z^5}{156} + \frac{5z^4}{156} - \frac{5z^3}{178} - \frac{5z^2}{26} + \frac{3}{13})p(z) \\
& - (-\frac{z^5}{156} + \frac{5z^4}{156} - \frac{5z^3}{178} - \frac{5z^2}{26} + \frac{3}{13})p(0) \\
& + (-\frac{z^5}{156} + \frac{5z^4}{156} - \frac{5z^3}{178} - \frac{5z^2}{26} - \frac{10}{13})p(1) \\
& - (-\frac{z^5}{156} + \frac{5z^4}{156} - \frac{5z^3}{178} - \frac{5z^2}{26} - \frac{10}{13})p(z).
\end{aligned}
$$

Therefore, we obtain

$$
\langle p, F_z \rangle_{E_2^3} = p(z).
$$

This completes the proof.

We consider the solutions of the problem (1.3) in the reproducing kernel Hilbert space $E_2^3[0,1]$. We denote the bounded linear operator $X : E_2^3[0,1] \to E_2^1[0,1]$ as:

$$
(2.1) \qquad X\omega = \frac{d^\gamma \omega}{dr^\gamma} + \frac{1}{r}\frac{d^\beta \omega}{dr^\beta}.
$$

Then, we have the following problem.

$$
(2.2) \qquad X\omega = D(r, \omega),
$$

with the boundary conditions

$$
(2.3) \qquad \omega'(0) = 0 = \omega(1),
$$

where

$$
(2.4) \qquad D(r, \omega) = -Ha^2 \left(1 - \frac{\omega}{1 - \alpha\omega}\right)
$$

Lemma 2.5. X *is a bounded linear operator.*

Proof. We need to prove

$$
\|X\omega\|_{E_2^1[0,1]}^2 \le K \|\omega\|_{E_2^3[0,1]}^2,
$$

where K is a positive constant. We have

$$
(2.5) \qquad \|X\omega\|_{E_2^1[0,1]}^2 = \langle X\omega, X\omega \rangle_{E_2^1[0,1]} = [X\omega(0)]^2 + \int_0^1 [X\omega'(z)]^2\, dz.
$$

By reproducing property, we have

$$\omega(z) = \langle \omega(\cdot), F_z(\cdot) \rangle_{E_2^3[0,1]}$$

and

$$X\omega(z) = \langle \omega(\cdot), XF_z(\cdot) \rangle_{E_2^3[0,1]},$$

so

$$|X\omega| \le \|\omega\|_{E_2^3[0,1]} \|XF_z\|_{E_2^3[0,1]} = K_1 \|\omega\|_{E_2^3[0,1]},$$

thus

$$[X\omega(0)]^2 \le K_1^2 \|\omega\|_{E_2^3[0,1]}^2.$$

Since

$$(X\omega)'(z) = \langle \omega(\cdot), (XF_z)'(\cdot) \rangle_{E_2^3[0,1]},$$

we get

$$|(X\omega)'| \le \|\omega\|_{E_2^3[0,1]} \|(XF_z)'\|_{E_2^3[0,1]} = K_2 \|\omega\|_{E_2^3[0,1]},$$

so, we have

$$[X\omega]^2 \le K_2^2 \|\omega\|_{E_2^3[0,1]}^2,$$

that is,

$$\|X\omega\|_{E_2^1[0,1]}^2 = [X\omega(0)]^2 + \int_0^1 [(X\omega)'(z)]^2\, dz \le (K_1^2 + K_2^2) \|\omega\|_{E_2^3[0,1]}^2,$$

where $K = K_1^2 + K_2^2$ is a positive constant.

We construct $\varsigma_i(\tau) = G_{\tau_i}(\tau)$ and $\psi_i(\tau) = X^*\varsigma_i(\tau)$, where X^* is conjugate operator of X. The orthonormal system $\{\widehat{\psi}_i(\tau)\}_{i=1}^\infty$ of $E_2^3[0,1]$ can be acquired by Gram-Schmidt orthogonalization operation of $\{\psi_i(\tau)\}_{i=1}^\infty$,

$$(2.6) \qquad \widehat{\psi}_i(\tau) = \sum_{k=1}^i \beta_{ik}\psi_k(\tau), \quad (\beta_{ii} \rangle 0, \quad i = 1, 2, \ldots).$$

Theorem 2.6. *Let $\{\tau_i\}_{i=1}^\infty$ be dense in $[0,1]$ and $\psi_i(\tau) = X_z F_\tau(z)|_{z=\tau_i}$. Then the sequence $\{\psi_i(\tau)\}_{i=1}^\infty$ is a complete system in $E_2^3[0,1]$.*

Proof. We get

$$\psi_i(\tau) = (X^*\varsigma_i)(\tau) = \langle (X^*\varsigma_i)(z), F_\tau(z) \rangle = \langle (\varsigma_i)(z), X_z F_\tau(z) \rangle = X_z F_\tau(z)|_{z=\tau_i}.$$

Let $\langle \omega(\tau), \psi_i(\tau) \rangle = 0$, $(i = 1, 2, \ldots)$, which means that,

$$\langle \omega(\tau), (X^*\varsigma_i)(\tau) \rangle = \langle X\omega(\cdot), \varsigma_i(\cdot) \rangle = (X\omega)(\tau_i) = 0.$$

$\{\tau_i\}_{i=1}^\infty$ is dense in $[0,1]$. Therefore, $(X\omega)(\tau) = 0$. $\omega \equiv 0$ by X^{-1}.

Theorem 2.7. *If $\omega(r)$ is the exact solution of (2.2), then we acquire*

$$(2.7) \qquad \omega(r) = \sum_{i=1}^{\infty} \sum_{k=1}^{i} \beta_{ik} D(r_k, \omega_k) \widehat{\psi}_i(r).$$

where $\{r_i\}_{i=1}^{\infty}$ is dense in $[0,1]$.

Proof. We get

$$\omega(r) = \sum_{i=1}^{\infty} \left\langle \omega(r), \widehat{\psi}_i(r) \right\rangle_{E_2^3[0,1]} \widehat{\psi}_i(r)$$

$$= \sum_{i=1}^{\infty} \sum_{k=1}^{i} \beta_{ik} \left\langle \omega(r), \psi_k(r) \right\rangle_{E_2^3[0,1]} \widehat{\psi}_i(r)$$

$$= \sum_{i=1}^{\infty} \sum_{k=1}^{i} \beta_{ik} \left\langle \omega(r), X^* \varsigma_k(r) \right\rangle_{E_2^3[0,1]} \widehat{\psi}_i(r)$$

$$= \sum_{i=1}^{\infty} \sum_{k=1}^{i} \beta_{ik} \left\langle X\omega(r), \varsigma_k(r) \right\rangle_{E_2^1[0,1]} \widehat{\psi}_i(r)$$

$$= \sum_{i=1}^{\infty} \sum_{k=1}^{i} \beta_{ik} \left\langle D(r, \omega), G_{r_k} \right\rangle_{E_2^1[0,1]} \widehat{\psi}_i(r)$$

$$= \sum_{i=1}^{\infty} \sum_{k=1}^{i} \beta_{ik} D(r_k, \omega_k) \widehat{\psi}_i(r).$$

by uniqueness of solution of (2.2). This completes the proof.

The approximate solution $\omega_n(r)$ can be obtained as:

$$(2.8) \qquad \omega_n(r) = \sum_{i=1}^{n} \sum_{k=1}^{i} \beta_{ik} D(r_k, \omega_k) \widehat{\psi}_i(r).$$

3   Numerical Experiments

In Table 1, we show the solution when $\gamma = 1.9$, $\beta = 0.9$, $\alpha = 0.5$ and vary the Hartmann electric number. In Table 2, we fixed the fractional derivatives as $\gamma = 1.9$, $\beta = 0.9$, $Ha^2 = 1.0$ and vary the α. In Table 3, we give the solution with the fractional derivatives $\gamma = 1.3$, $\beta = 0.3$, $Ha^2 = 1.0$ and vary the α.

Table 1. Approximate solutions by reproducing kernel method for $\gamma = \frac{19}{10}$, $\beta = \frac{9}{10}$, $\alpha = \frac{1}{2}$ and different values of Ha^2.

r	$Ha^2 = 1$	$Ha^2 = 2$
0.0	0.2299695869	0.3825201644
0.1	0.2266810832	0.3754454339
0.2	0.2180411628	0.3600500656
0.3	0.2062671368	0.3434211029
0.4	0.1900081364	0.3158370163
0.5	0.1697602524	0.2874089253
0.6	0.1451770620	0.2475168041
0.7	0.1161364572	0.2007161568
0.8	0.0824312374	0.1443651864
0.9	0.0438131825	0.0777826865
1.0	-8.95×10^{-10}	-3.54×10^{-11}

Table 2. Approximate solutions by reproducing kernel method for $\gamma = \frac{19}{10}$, $\beta = \frac{9}{10}$, $Ha^2 = 1$ and different values of α.

r	$\alpha = 1$	$\alpha = 2$
0.0	0.2259726183	0.2093112589
0.1	0.2226475274	0.2058336031
0.2	0.2141458034	0.1979096087
0.3	0.2029104831	0.1888516297
0.4	0.1870087144	0.1744727611
0.5	0.1673526319	0.1579851722
0.6	0.1433018034	0.1361693674
0.7	0.1147939073	0.1098299577
0.8	0.0815892945	0.0785261212
0.9	0.0434202449	0.0419967019
1.0	-9.79×10^{-10}	-3.97×10^{-10}

Table 3. Approximate solutions by reproducing kernel method for $\gamma = \frac{13}{10}$, $\beta = \frac{3}{10}$, $Ha^2 = 1$ and different values of α.

r	$\alpha = 1$	$\alpha = 2$
0.0	0.4454618088	0.3386406399
0.1	0.4284599310	0.3204729467
0.2	0.3272191714	0.2609343009
0.3	0.2498402621	0.2849602586
0.4	0.2020451400	0.3044655878
0.5	0.1890369644	0.2413493934
0.6	0.2063732634	0.1921955815
0.7	0.1160557551	0.1449551875
0.8	0.0896312152	0.1026976873
0.9	0.0478590068	0.0537369610
1.0	2.77×10^{-6}	-8.159×10^{-7}

4 Conclusions

In this work, we acquired the solutions of fractional version of a singular boundary value problem occurring in the electrohydrodynamic flow in a circular cylindrical conduit based on the reproducing kernel method. We demonstrated our effective results by some tables. We investigated the effect of the Hartmann electric number and the fractional order of the problem. We concluded that the reproducing kernel method can be applied to much more complicated fractional differential equations.

References

1. McKee, S., Watson, R., Cuminato, J.A., Caldwell, J., Chen, M.S.: Calculation of electrohydrodynamic flow in a circular cylindrical conduit. Z. Angew. Math. Mech. **77**, 457–465 (1997). https://doi.org/10.1002/zamm.19970770612. 1455891
2. Ghasemi, S.E., Hatami, M., Mehdizadeh, G.R., Ganji, D.D.: Electrohydrodynamic flow analysis in a circular cylindrical conduit using least square method. J. Electrostat. **72**, 47–52 (2014). https://doi.org/10.1016/j.elstat.2013.11.005
3. Paullet, J.E.: On the solutions of electrohydrodynamic flow in a circular cylindrical conduit. Angew. Math. Mech. **79**, 357–360 (1999). https://doi.org/10.1002/(SICI)1521-4001(199905)79:5⟨357::AID-ZAMM357⟩3.0.CO;2-B. 1695270
4. Mastroberardino, A.: Homotopy analysis method applied to electrohydrodynamic flow. Commun. Nonlinear Sci. Numer. Simul. **16**, 2730–2736 (2011). https://doi.org/10.1016/j.cnsns.2010.10.004. 2772289
5. Moghtadaei, M., Nik, H.S., Abbasbandy, S.: A spectral method for the electrohydrodynamic flow in a circular cylindrical conduit. Chin. Ann. Math. Ser. B **36**, 307–322 (2015). https://doi.org/10.1007/s11401-015-0882-z. 3305711

6. Beg, O.A., Hameed, M., Beg, T.A.: Chebyshev spectral collocation simulation of nonlinear boundary value problems in electrohydrodynamics. Int. J. Comput. Methods Eng. Sci. Mech. **14**, 104–105 (2013). https://doi.org/10.1080/15502287. 2012.698707. 3040428

7. Alomari, A.K., Suat Erturk, V., Momani, S., Alsaedi, A.: An approximate solution method for the fractional version of a singular BVP occurring in the electrohydro-dynamic flow in a circular cylindrical conduit. Eur. Phys. J. Plus **134**, 158 (2019). https://doi.org/10.1140/epjp/i2019-12498-0

8. Abbas, S., Erturk, V.S., Momani, S.: Dynamical analysis of the Irving-Mullineux oscillator equation of fractional order. Signal Process. **102**, 171–176 (2014). https://doi.org/10.1016/j.sigpro.2014.03.019

9. Iyiolaa, O.S., Zaman, F.D.: A fractional diffusion equation model for cancer tumor. AIP Adv. **4**, 107121 (2014). https://doi.org/10.1063/1.4898331

10. Akgül, A.: New reproducing kernel functions. Math. Probl. Eng. **2015**, 3312767 (2015). https://doi.org/10.1155/2015/158134. Art. ID 158134, 10

11. Arqub, O.A.: Approximate solutions of DASs with nonclassical boundary conditions using novel reproducing kernel algorithm. Fundam. Inform. **146**, 231–254 (2016). https://doi.org/10.3233/FI-2016-1384. 3581119

12. Arqub, O.A.: The reproducing kernel algorithm for handling differential algebraic systems of ordinary differential equations. Math. Methods Appl. Sci. **39**, 4549–4562 (2016). https://doi.org/10.1002/mma.3884. 3549413

An Efficient Approach for the Model of Thrombin Receptor Activation Mechanism with Mittag-Leffler Function

P. Veeresha[1]([✉]), D. G. Prakasha[2], and Zakia Hammouch[3]

[1] Department of Mathematics, Karnatak University, 580003 Dharwad, India
viru0913@gmail.com
[2] Department of Mathematics Faculty of Science Shivagangothri,
Davangere University, Davangere 577007, India
prakashadg@gmail.com
[3] Department of Mathematics, Faculty of Sciences and Techniques Errachidia,
BP 509, 52000 Errachidia, Morocco
z.hammouch@fste.umi.ac.ma

Abstract. In the present work, we haired an efficient technique called, *q-homotopy analysis transform method* (*q*-HATM) in order to find the solution for the model of thrombin receptor activation mechanism (TRAM) and examine the nature of *q*-HATM solution with distinct fractional order. The considered model elucidates the TRA mechanism in calcium signalling, and this mechanism plays a vital role in the human body. We defined fractional derivative defined with Atangana-Baleanu (AB) operator and the projected scheme is an amalgamation of Laplace transform with *q*-homotopy analysis scheme. For the achieved results, to present the existence and uniqueness we hired the fixed point hypothesis. To validate and illustrate the effectiveness of the considered scheme, we examined the projected model with arbitrary order. The behaviour of the achieved results is captured in terms of plots and also showed the importance of the parameters offered by the considered solution procedure. The attained results illuminate, the projected scheme is easy to employ and more effective in order to analyse the behaviour of fractional order differential systems exemplifying real word problems associated with science and technology.

Keywords: *q*-Homotopy analysis method · Laplace transform · Thrombin receptor activation mechanism · Fixed point theorem · Atangana-Baleanu derivative

1 Introduction

The human body is mainly the composition of six elements of about 99%, and those are namely, phosphorus, calcium, nitrogen, hydrogen, carbon and oxygen. In our body, the most generous mineral is calcium (Ca) and it is about 1.5%. Ca is the most play a vital rule in muscle contraction and protein regulation and also it's very essential in the processes of contractions bones and their protection. Most of the phenomena including cell death and fertilization are achieved with the help of calcium oscillations. With the

© Springer Nature Switzerland AG 2021
Z. Hammouch et al. (Eds.): SM2A 2019, LNNS 168, pp. 44–60, 2021.
https://doi.org/10.1007/978-3-030-62299-2_4

exploit of the inositol phospholipid cascade by raising the cytosolic calcium levels produced, most of the pathways of signal transduction are arbitrated [1, 2]. Ca acts as emissary in information processing. For the analysis of enzyme phospholipase C (PLC), G protein is playing an important role.

In the present scenario, the study of most of the phenomena related to the human body like diseases and their behaviour, the essential components of our body and their functions; magnetize the attention of mathematicians and researchers associated to mathematics in order to model and analyse as well as predict its essential behaviours. In connection with this, authors in [3] nurtured the mathematical model in order to illustrate the mediated activation of human platelets, researchers in [4] analyse the cytosolic calcium dynamics by the aid of mathematical model and later by the help of fractional calculus (FC), authors in [5] present their viewpoint in order to understand the importance of FC while analysing the mathematical model stimulating the above-cited phenomenon.

The seed of fractional calculus (FC) is planted before 324 years, however lately become an essential tool for the distinct discipline of science and engineering, and hence fascinated the attention of authors. It was shortly discovered that fractional calculus is more appropriate for modelling the phenomena describing nature in a systematic manner as associated with integer order calculus. The calculus of arbitrary order turned out one of the most essential tools to describe biological phenomena. The human diseases which are modelled through derivative having fractional-order help us to incorporate the information about its present and past states. Diverse pioneering notions and fundamentals are prescribed by many senior researchers [6–11]. Recently, due to diverse applications and favourable properties, the concept of FC is widely hired to investigate real world problems [12–20]. Particularly, authors in [21] analysed the fractional order system exemplifying the fish farm model within the frame of new fractional operator and also the captured some simulating consequences associated to the model using efficient scheme, authors in [22, 23] investigated the numerical solution for the fractional order coupled special cases of KdV equations and presented some interesting results with respect to different fractional order. The epidemic model of childhood disease is analysed by the authors in [24] within the frame of fractional calculus and they presented the nature of the corresponding results for distinct arbitrary order. Authors in [19] analyse the evolution of 2019-nCoV and its dynamic structures with help of nonlocal operator and presented some numerical surface using efficient scheme.

The activated form of phospholipase $C(PLC)$ hydrolyzes the diacylglycerol (DAG), 5-trisphosphate $[Ins(I, 4, 5)P_3]$, 5-bisphosphate $[PtdIns(4, 5)P_2]$ to inositol and phosphatidylinositol 4. From the endoplasmic reticulum, the 5-trisphosphate is helps to stimulate the let out of endogenous calcium. The number of activated cell surface receptor proportional to the rate of generates of the 5-trisphosphate. The thrombin is a multiprocessing serine protease aids from the endothelial cell to take calcium transient and it acts as a ligand for the present model. Here, we consider the system of the equation which described the TRS mechanism. In endothelial cells, this model provides incite of calcium arbitrated signal transduction. The release of calcium is determined by the 5-trisphosphate cytosolic level in the calcium homeostasis and the number of active surface receptors (S) aid to generate the 5-trisphosphate. The receptor-ligand complex (C) formed due to ligand binding with surface receptors and on cleavage outcomes in activated receptors (A). The above-cited phenomenon is illustrated with the aid of the system of three differential equations and concentration of thrombin (ε) as follows [4, 5]

$$
\begin{aligned}
\frac{dS(t)}{dt} &= -\delta\varepsilon S(t) + \beta C(t) \\
\frac{dC(t)}{dt} &= \delta\varepsilon S(t) - (\beta + \lambda)C(t)\,, \\
\frac{dA(t)}{dt} &= \lambda C(t)
\end{aligned}
\tag{1}
$$

where δ and β respectively symbolise the on and off rate constant of thrombin binding.

Many nonlinear and important models are effectively and methodically examined with the assist of FC. Many senior pioneers proposed distinct definitions including, Riemann, Liouville, Caputo and Fabrizio. Soon after the invention of each notion, many researchers identify some limitations while examining specific problems. Including physical meaning of the initial conditions, kernel associated to singularity, non-locality and others associated with complex phenomena. With the assist of Mittag–Leffler function, Atangana and Baleanu [25] proposed a new fractional-order operator and overcome all the above-cited consequences which play a vital role while investigating properties of the models.

Authors in [5] presented the simulation for the fractional system with Caputo-Fabrizio derivative using perturbation iterative scheme, which poses interesting consequences. In the present framework, we consider with AB derivative and which as follows

$$
\begin{aligned}
{}_{a}^{ABC}D_t^\alpha S(t) &= -\delta\varepsilon S(t) + \beta C(t) \\
{}_{a}^{ABC}D_t^\alpha C(t) &= \delta\varepsilon S(t) - (\beta + \lambda)C(t), \qquad 0 < \alpha \leq 1, \\
{}_{a}^{ABC}D_t^\alpha A(t) &= \lambda C(t),
\end{aligned}
\tag{2}
$$

where α is fractional order of the system.

As much as impartment of modelling real-world problems, finding the solution for the corresponding system is also vital and difficult. Most of the complex and nonlinear problems don't have an analytical solution. In this connection, researchers preferred for semi-analytical or numerical schemes. One of the efficient and most widely hired methods to solve nonlinear problems is the homotopy analysis method (HAM) and which natured by *Liao Shijun* [26, 27]. This solution procedure overcomes most of the limitation arise while solving nonlinear problems with dissertation and perturbation. However, a few limitations have been pointed out by researchers in order to reduce computational work and time. The presented method is the mixture of LT with q-HAM and nurtured by Singh et al. [28]. Clearly, q-HATM is an enhanced algorithm of HAM; it does not require linearization, perturbation or discretization. Recently, many researchers hired the considered method due to its efficacy and reliability to understand physical behaviour numerous classes of nonlinear problems [29–34]. The considered scheme gives more freedom to choose problems associated with distinct initial conditions and it proposed with axillary and homotopy corresponding phenomena [35, 36].

2 Preliminaries

Here, the basic notions and definitions of FC and LT are presented [25, 37–41].

Definition 1. In Caputo and Riemann-Liouville sense, for a function $f \in H^1(a, b)$ the fractional Atangana-Baleanu-derivative are presented respectively as follows [25]:

$$
{}^{ABC}_{a}D^{\alpha}_{t}(f(t)) = \frac{\mathcal{B}[\alpha]}{1 - \alpha} \int_{a}^{t} f'(\vartheta) E_{\alpha}\left[\alpha \frac{(t - \vartheta)^{\alpha}}{\alpha - 1}\right] d\vartheta. \tag{3}
$$

$$
{}^{ABR}_{a}D^{\alpha}_{t}(f(t)) = \frac{\mathcal{B}[\alpha]}{1 - \alpha} \frac{d}{dt} \int_{a}^{t} f(\vartheta) E_{\alpha}\left[\alpha \frac{(t - \vartheta)^{\alpha}}{\alpha - 1}\right] d\vartheta, \tag{4}
$$

where $\mathcal{B}[\alpha]$ is a normalization function such that $\mathcal{B}(0) = \mathcal{B}(1) = 1$.

Definition 2. The AB integral with fractional order is presented [25] as

$$
{}^{AB}_{a}I^{\alpha}_{t}(f(t)) = \frac{1 - \alpha}{\mathcal{B}[\alpha]} f(t) + \frac{\alpha}{\mathcal{B}[\alpha]\Gamma(\alpha)} \int_{a}^{t} f(\vartheta)(t - \vartheta)^{\alpha - 1} d\vartheta. \tag{5}
$$

Definition 3. The Laplace transform (LT) Associated to AB operator is defined as

$$
L\left[{}^{ABR}_{0}D^{\alpha}_{t}(f(t))\right] = \frac{\mathcal{B}[\alpha]}{1 - \alpha} \frac{s^{\alpha}L[f(t)] - s^{\alpha-1}f(0)}{s^{\alpha} + (\alpha/(1 - \alpha))}. \tag{6}
$$

Theorem 1. The following Lipschitz conditions satisfy respectively for the Riemann-Liouville and AB derivatives [25]

$$
\left\|{}^{ABC}_{a}D^{\alpha}_{t}f_1(t) - {}^{ABC}_{a}D^{\alpha}_{t}f_2(t)\right\| < K_1\|f_1(x) - f_2(x)\|, \tag{7}
$$

and

$$
\left\|{}^{ABR}_{a}D^{\alpha}_{t}f_1(t) - {}^{ABR}_{a}D^{\alpha}_{t}f_2(t)\right\| < K_2\|f_1(x) - f_2(x)\|. \tag{8}
$$

Theorem 2. The fractional differential equation ${}^{ABC}_{a}D^{\mu}_{t}f(t) = s(t)$ has a unique solution is given by [25]

$$
f(t) = \frac{1 - \alpha}{\mathcal{B}[\alpha]} s(t) + \frac{\alpha}{\mathcal{B}[\alpha]\Gamma(\alpha)} \int_{0}^{t} s(\varsigma)(t - \varsigma)^{\alpha - 1} d\varsigma. \tag{9}
$$

3 Basic idea of q-HATM

Here, we consider the differential equation of fractional order with respectively linear $\mathcal{R}$ and nonlinear $\mathcal{N}$ differential operator form

$$_a^{ABC}D_t^\alpha v(x,t) + \mathcal{R}v(x,t) + \mathcal{N}v(x,t) = f(x,t), n-1 < \alpha \leq n, \tag{10}$$

with the initial condition

$$v(x,0) = g(x), \tag{11}$$

where $_a^{ABC}D_t^\alpha v(x,t)$ symbolise the AB derivative of $v(x,t)$, $f(x,t)$ denotes the source term. Using LT, Eq. (10) gives

$$\mathcal{L}[v(x,t)] - \frac{g(x)}{s} + \frac{1}{B[\alpha]}\left(1 - \alpha + \frac{\alpha}{s^\alpha}\right)\{\mathcal{L}[\mathcal{R}v(x,t)] + \mathcal{L}[\mathcal{N}v(x,t)] - \mathcal{L}[f(x,t)]\} = 0. \tag{12}$$

By the assist of HAM, $\mathcal{N}$ is projected as

$$\mathcal{N}[\varphi(x,t;q)] = \mathcal{L}[\varphi(x,t;q)] - \frac{g(x)}{s} \tag{13}$$
$$+ \frac{1}{B[\alpha]}\left(1 - \alpha + \frac{\alpha}{s^\alpha}\right)\{L[\mathcal{R}\varphi(x,t;q)] + L[\mathcal{N}\varphi(x,t;q)] - L[f(x,t)]\}.$$

Here, $\varphi(x,t;q)$ is the real-valued function. Now, we have

$$(1 - nq)\mathcal{L}[\varphi(x,t;q) - v_0(x,t)] = \hbar q \mathcal{N}[\varphi(x,t;q)], \tag{14}$$

where L is signifying LT, $q \in \left[0, \frac{1}{n}\right] (n \geq 1)$ is the embedding parameter and $\hbar \neq 0$ is an auxiliary parameter. For $q = 0$ and $q = \frac{1}{n}$, we have

$$\varphi(x,t;0) = v_0(x,t), \varphi\left(x,t;\frac{1}{n}\right) = v(x,t). \tag{15}$$

Thus, by intensifying q from 0 to $\frac{1}{n}$, then $\varphi(x,t;q)$ changes from $v_0(x,t)$ to $v(x,t)$. Using Taylor theorem near to q, we defining $\varphi(x,t;q)$ in series form and then we get

$$\varphi(x,t;q) = v_0(x,t) + \sum_{m=1}^\infty v_m(x,t)q^m, \tag{16}$$

where

$$v_m(x,t) = \frac{1}{m!}\frac{\partial^m \varphi(x,t;q)}{\partial q^m}\Big|_{q=0}. \tag{17}$$

For the proper chaise of $v_0(x,t), n$ and $\hbar$, the series (14) converges at $q = \frac{1}{n}$. By simplifying Eq. (14), we achieved

$$\mathcal{L}[v_m(x,t) - k_m v_{m-1}(x,t)] = \hbar \Re_m(\vec{v}_{m-1}),\tag{18}$$

where the vectors are defined as

$$\vec{v}_m = \{v_0(x,t), v_1(x,t), \ldots, v_m(x,t)\}.\tag{19}$$

On employing inverse LT on Eq. (18), we get

$$v_m(x,t) = k_m v_{m-1}(x,t) + \hbar \mathcal{L}^{-1}[\Re_m(\vec{v}_{m-1})],\tag{20}$$

where

$$\Re_m(\vec{v}_{m-1}) = L[v_{m-1}(x,t)] - \left(1 - \frac{k_m}{n}\right)\left(\frac{g(x)}{s} + \frac{1}{B[\alpha]}\left(1 - \alpha + \frac{\alpha}{s^\alpha}\right)L[f(x,t)]\right)$$
$$+ \frac{1}{B[\alpha]}\left(1 - \alpha + \frac{\alpha}{s^\alpha}\right)L[Rv_{m-1} + \mathcal{H}_{m-1}],$$

$$\tag{21}$$

and

$$k_m = \begin{cases} 0, m \le 1, \\ n, m > 1. \end{cases}\tag{22}$$

In Eq. (21), $\mathcal{H}_m$ signifies homotopy polynomial and which is defined as

$$\mathcal{H}_m = \frac{1}{m!}\left[\frac{\partial^m \varphi(x,t;q)}{\partial q^m}\right]_{q=0} \text{ and } \varphi(x,t;q) = \varphi_0 + q\varphi_1 + q^2\varphi_2 + \ldots .\tag{23}$$

By the aid of Eqs. (20) and (21), one can get

$$v_m(x,t) = (k_m + \hbar)v_{m-1}(x,t) - \left(1 - \frac{k_m}{n}\right)\mathcal{L}^{-1}\left(\frac{g(x)}{s} + \frac{1}{B[\alpha]}\left(1 - \alpha + \frac{\alpha}{s^\alpha}\right)L[f(x,t)]\right)$$
$$+ \hbar \mathcal{L}^{-1}\left\{\frac{1}{B[\alpha]}\left(1 - \alpha + \frac{\alpha}{s^\alpha}\right)L[Rv_{m-1} + \mathcal{H}_{m-1}]\right\}.$$

$$\tag{24}$$

The q-HATM solution is presented as

$$v(x,t) = v_0(x,t) + \sum_{m=1}^{\infty} v_m(x,t)\left(\frac{1}{n}\right)^m.\tag{25}$$

4 Solution for Proposed Model

To demonstrate the efficiency and solution procedure of the projected method, in here we consider system describing considered a model with arbitrary order. By the assist of Eq. (2), one can get

$$
\begin{aligned}
{}^{ABC}_{a}D^{\alpha}_t S(t) + \delta\varepsilon S(t) - \beta C(t) &= 0, \\
{}^{ABC}_{a}D^{\alpha}_t C(t) - \delta\varepsilon S(t) + (\beta + \lambda)C(t) &= 0, \qquad\qquad 0 < \alpha \le 1, \\
{}^{ABC}_{a}D^{\alpha}_t A(t) - \lambda C(t) &= 0
\end{aligned}
\tag{26}
$$

with initial conditions

$$
S(0) = S_0(t), C(0) = C_0(t), A(0) = A_0(t). \tag{27}
$$

Taking LT on Eq. (26) and then using the Eq. (27), we get

$$
\begin{aligned}
L[S(t)] &= \frac{1}{s}(S_0(t)) + \frac{1}{B[\alpha]}\left(1 - \alpha + \frac{\alpha}{s^\alpha}\right)L\{\delta\varepsilon S(t) - \beta C(t)\}, \\
L[C(t)] &= \frac{1}{s}(C_0(t)) - \frac{1}{B[\alpha]}\left(1 - \alpha + \frac{\alpha}{s^\alpha}\right)L\{\delta\varepsilon S(t) - (\beta + \lambda)C(t)\}, \\
L[A(t)] &= \frac{1}{s}(A_0(t)) - \frac{1}{B[\alpha]}\left(1 - \alpha + \frac{\alpha}{s^\alpha}\right)L\{\lambda C(t)\}.
\end{aligned}
\tag{28}
$$

Now, we define N as below

$$
\begin{aligned}
N^1[\varphi_1(t;q), \varphi_2(t;q), \varphi_3(t;q)] &= L[\varphi_1(t;q)] - \frac{1}{s}(S_0(t)) \\
&\quad + \frac{1}{B[\alpha]}\left(1 - \alpha + \frac{\alpha}{s^\alpha}\right)L\{\delta\varepsilon\varphi_1(t;q) + \beta\varphi_2(t;q)\}, \\
N^2[\varphi_1(t;q), \varphi_2(t;q), \varphi_3(t;q)] &= L[\varphi_2(t;q)] - \frac{1}{s}(C_0(t)) \\
&\quad - \frac{1}{B[\alpha]}\left(1 - \alpha + \frac{\alpha}{s^\alpha}\right)L\{\delta\varepsilon\varphi_1(t;q) + (\beta + \lambda)\varphi_2(t;q)\}, \\
N^3[\varphi_1(t;q), \varphi_2(t;q), \varphi_3(t;q)] &= L[\varphi_3(t;q)] - \frac{1}{s}(A_0(t)) \\
&\quad - \frac{1}{B[\alpha]}\left(1 - \alpha + \frac{\alpha}{s^\alpha}\right)L\{\lambda\varphi_2(t;q)\}.
\end{aligned}
\tag{29}
$$

The deformation equation of *m-th* order at $\mathcal{H}(x,t) = 1$ is defined as

$$L[S_m(t) - k_m S_{m-1}(t)] = \hbar\mathfrak{R}_{1,m}\left[\vec{S}_{m-1}, \vec{C}_{m-1}, \vec{A}_{m-1}\right],$$

$$L[C_m(t) - k_m C_{m-1}(t)] = \hbar\mathfrak{R}_{2,m}\left[\vec{S}_{m-1}, \vec{C}_{m-1}, \vec{A}_{m-1}\right], \qquad (30)$$

$$L[A_m(t) - k_m A_{m-1}(t)] = \hbar\mathfrak{R}_{3,m}\left[\vec{S}_{m-1}, \vec{C}_{m-1}, \vec{A}_{m-1}\right],$$

where

$$\mathfrak{R}_{1,m}\left[\vec{S}_{m-1}, \vec{C}_{m-1}, \vec{A}_{m-1}\right] = L[S_{m-1}(t)] - \left(1 - \frac{k_m}{n}\right)\left\{\frac{1}{s}(S_0(t))\right\}$$

$$+ \frac{1}{\mathcal{B}[\alpha]}\left(1 - \alpha + \frac{\alpha}{s^\alpha}\right)L\{\delta\varepsilon S_{m-1}(t) - \beta C_{m-1}(t)\},$$

$$\mathfrak{R}_{2,m}\left[\vec{S}_{m-1}, \vec{C}_{m-1}, \vec{A}_{m-1}\right] = L[C_{m-1}(t)] + \left(1 - \frac{k_m}{n}\right)\left\{\frac{1}{s}(C_0(t))\right\}$$

$$+ \frac{1}{\mathcal{B}[\alpha]}\left(1 - \alpha + \frac{\alpha}{s^\alpha}\right)L\{\delta\varepsilon S_{m-1}(t) - (\beta + \lambda)C_{m-1}(t)\},$$

$$\mathfrak{R}_{3,m}\left[\vec{S}_{m-1}, \vec{C}_{m-1}, \vec{A}_{m-1}\right] = L[A_{m-1}(t)] + \left(1 - \frac{k_m}{n}\right)\left\{\frac{1}{s}(A_0(t))\right\}$$

$$- \frac{1}{\mathcal{B}[\alpha]}\left(1 - \alpha + \frac{\alpha}{s^\alpha}\right)L\{\lambda C_{m-1}(t)\}.$$

$$(31)$$

Eq. (31) reduces after employing inverse *LT*, as follows

$$S_m(t) = k_m S_{m-1}(t) + \hbar L^{-1}\left\{\mathfrak{R}_{1,m}\left[\vec{S}_{m-1}, \vec{C}_{m-1}, \vec{A}_{m-1}\right]\right\},$$

$$C_m(t) = k_m C_{m-1}(t) + \hbar L^{-1}\left\{\mathfrak{R}_{2,m}\left[\vec{S}_{m-1}, \vec{C}_{m-1}, \vec{A}_{m-1}\right]\right\}, \qquad (32)$$

$$A_m(t) = k_m A_{m-1}(t) + \hbar L^{-1}\left\{\mathfrak{R}_{3,m}\left[\vec{S}_{m-1}, \vec{C}_{m-1}, \vec{A}_{m-1}\right]\right\}.$$

Using $S_0(t) = R_T, C_0 = 0$ and $A_0(t) = 0$ we can obtain the terms of the series solution with the help of the above system

$$S(t) = S_0(t) + \sum_{m=1}^{\infty} S_m(t)\left(\frac{1}{n}\right)^m,$$

$$C(t) = C_0(t) + \sum_{m=1}^{\infty} C_m(t)\left(\frac{1}{n}\right)^m, \qquad (33)$$

$$A(t) = A_0(t) + \sum_{m=1}^{\infty} A_m(t)\left(\frac{1}{n}\right)^m,$$

5 Existence of Solutions for the Proposed Problem

Here, to present the existence of the solution, we considered the fixed-point theorem. Now, the system (27) is considered as

$$
\begin{cases}
{}^{ABC}_{0}D_t^\alpha[S(t)] = \mathcal{G}_1(t, S), \\
{}^{ABC}_{0}D_t^\alpha[C(t)] = \mathcal{G}_2(t, C), \\
{}^{ABC}_{0}D_t^\alpha[A(t)] = \mathcal{G}_3(t, A).
\end{cases}
\tag{34}
$$

Using the Theorem 2, Eq. (35) is transformed to the Volterra integral equation and defined as

$$
\begin{cases}
S(t) - S(0) = \frac{(1-\alpha)}{B(\alpha)}\mathcal{G}_1(t, S) + \frac{\alpha}{B(\alpha)\Gamma(\alpha)}\int_0^t \mathcal{G}_1(\zeta, S)(t - \zeta)^{\alpha-1}d\zeta, \\
C(t) - C(0) = \frac{(1-\alpha)}{B(\alpha)}\mathcal{G}_2(t, C) + \frac{\alpha}{B(\alpha)\Gamma(\alpha)}\int_0^t \mathcal{G}_2(\zeta, C)(t - \zeta)^{\alpha-1}d\zeta, \\
A(t) - A(0) = \frac{(1-\alpha)}{B(\alpha)}\mathcal{G}_3(t, A) + \frac{\alpha}{B(\alpha)\Gamma(\alpha)}\int_0^t \mathcal{G}_3(\zeta, A)(t - \zeta)^{\alpha-1}d\zeta.
\end{cases}
\tag{35}
$$

Theorem 3. The kernel $\mathcal{G}_1$ satisfies the Lipschitz condition and contraction if $0 \leq (\delta\varepsilon - \beta\lambda_2) < 1$ holds.

Proof. We consider the two functions u and u_1 to prove the required result, as follows

$$
\begin{aligned}
\|\mathcal{G}_1(t, S) - \mathcal{G}_1(t, S_1)\| &= \|(\delta c[S(t) - S(t_1)] - \beta C(t))\| \\
&\leq \|\delta\varepsilon - \beta C(t)\| \|S(t) - S(t_1)\| \\
&\leq (\delta\varepsilon - \beta\lambda_2)\|S(t) - S(t_1)\|,
\end{aligned}
\tag{36}
$$

where $\| C(t) \| \leq \lambda_2$ be the bounded function. Putting $\eta_1 = \delta\varepsilon - \beta\lambda_2$ in Eq. (37), we have

$$
\|\mathcal{G}_1(t, S) - \mathcal{G}_1(t, S_1)\| \leq \eta_1 \|S(t) - S(t_1)\|.
\tag{37}
$$

Equation (38) signifies Lipschitz condition for $\mathcal{G}_1$. If $0 \leq (\delta\varepsilon - \beta\lambda_2) < 1$, then it gives the contraction. Similarly, we have

$$
\begin{cases}
\|\mathcal{G}_2(t, C) - \mathcal{G}_2(t, C_1)\| \leq \eta_2 \|C(t) - C(t_1)\|, \\
\|\mathcal{G}_3(t, A) - \mathcal{G}_3(t, A_1)\| \leq \eta_3 \|A(t) - A(t_1)\|.
\end{cases}
\tag{38}
$$

Now, we define the recursive form of Eq. (36) as with initial conditions

$$
\begin{cases}
S_n(t) = \frac{(1-\alpha)}{B(\alpha)}\mathcal{G}_1(t, S_{n-1}) + \frac{\alpha}{B(\alpha)\Gamma(\alpha)}\int_0^t \mathcal{G}_1(\zeta, S_{n-1})(t - \zeta)^{\alpha-1}d\zeta, \\
C_n(t) = \frac{(1-\alpha)}{B(\alpha)}\mathcal{G}_2(t, C_{n-1}) + \frac{\alpha}{B(\alpha)\Gamma(\alpha)}\int_0^t \mathcal{G}_2(\zeta, C_{n-1})(t - \zeta)^{\alpha-1}d\zeta, \\
A_n(t) = \frac{(1-\alpha)}{B(\alpha)}\mathcal{G}_3(t, A_{n-1}) + \frac{\alpha}{B(\alpha)\Gamma(\alpha)}\int_0^t \mathcal{G}_3(\zeta, A_{n-1})(t - \zeta)^{\alpha-1}d\zeta,
\end{cases}
\tag{39}
$$

and

$$S(0) = S_0(t), C(0) = C_0(t) \text{ and } A(0) = A_0(t). \tag{40}$$

The successive difference between the terms presented as

$$
\begin{cases}
\phi_{1n}(t) = S_n(t) - S_{n-1}(t) \\
\quad = \frac{(1-\alpha)}{B(\alpha)} \left(\mathcal{G}_1(t, S_{n-1}) - \mathcal{G}_1(t, S_{n-2}) \right) + \frac{\alpha}{B(\alpha)\Gamma(\alpha)} \int\limits_0^t \mathcal{G}_1(\zeta, S_{n-1})(t - \zeta)^{\alpha-1} d\zeta, \\
\phi_{2n}(t) = C_n(t) - C_{n-1}(t) \\
\quad = \frac{(1-\alpha)}{B(\alpha)} \left(\mathcal{G}_2(t, C_{n-1}) - \mathcal{G}_2(t, C_{n-2}) \right) + \frac{\alpha}{B(\alpha)\Gamma(\alpha)} \int\limits_0^t \mathcal{G}_2(\zeta, C_{n-1})(t - \zeta)^{\alpha-1} d\zeta, \\
\phi_{3n}(t) = A_n(t) - A_{n-1}(t) \\
\quad = \frac{(1-\alpha)}{B(\alpha)} \left(\mathcal{G}_3(t, A_{n-1}) - \mathcal{G}_3(t, A_{n-2}) \right) + \frac{\alpha}{B(\alpha)\Gamma(\alpha)} \int\limits_0^t \mathcal{G}_3(\zeta, A_{n-1})(t - \zeta)^{\alpha-1} d\zeta.
\end{cases}
\tag{41}
$$

Notice that

$$
\begin{cases}
S_n(t) = \sum\limits_{i=1}^n \phi_{1i}(t), \\
C_n(t) = \sum\limits_{i=1}^n \phi_{2i}(t), \\
A_n(t) = \sum\limits_{i=1}^n \phi_{3i}(t).
\end{cases}
\tag{42}
$$

By using Eq. (39) and applying the norm on the first term of Eq. (42), we have

$$\|\phi_{1n}(t)\| \le \frac{(1-\alpha)}{B(\alpha)} \eta_1 \left\| \phi_{1(n-1)}(t) \right\| + \frac{\alpha}{B(\alpha)\Gamma(\alpha)} \eta_1 \int\limits_0^t \left\| \phi_{1(n-1)}(\zeta) \right\| d\zeta. \tag{43}$$

Similarly, we have

$$
\begin{cases}
\|\phi_{2n}(t)\| \le \frac{(1-\alpha)}{B(\alpha)} \eta_2 \left\| \phi_{2(n-1)}(t) \right\| + \frac{\alpha}{B(\alpha)\Gamma(\alpha)} \eta_2 \int\limits_0^t \left\| \phi_{2(n-1)}(\zeta) \right\| d\zeta, \\
\|\phi_{3n}(t)\| \le \frac{(1-\alpha)}{B(\alpha)} \eta_3 \left\| \phi_{3(n-1)}(t) \right\| + \frac{\alpha}{B(\alpha)\Gamma(\alpha)} \eta_3 \int\limits_0^t \left\| \phi_{3(n-1)}(\zeta) \right\| d\zeta.
\end{cases}
\tag{44}
$$

Next using the above result, we have following results.

Theorem 4. The solution for Eq. (27) will exist and unique if we have t_0 then

$$\frac{(1-\alpha)}{B(\alpha)} \eta_i + \frac{\alpha}{B(\alpha)\Gamma(\alpha)} \eta_i < 1,$$

for $i = 1, 2$ and 3.

Proof. Let $S(t), C(t)$ and $A(t)$ be the bounded functions admitting the Lipschitz condition. Now, we have by Eqs. (43) and (45)

$$\|\phi_{1i}(t)\| \leq \|S_n(0)\| \left[\frac{(1-\alpha)}{\mathcal{B}(\alpha)} \eta_1 + \frac{\alpha}{\mathcal{B}(\alpha)\Gamma(\alpha)} \eta_1 \right]^n,$$

$$\|\phi_{2i}(t)\| \leq \|C_n(0)\| \left[\frac{(1-\alpha)}{\mathcal{B}(\alpha)} \eta_2 + \frac{\alpha}{\mathcal{B}(\alpha)\Gamma(\alpha)} \eta_2 \right]^n, \qquad (45)$$

$$\|\phi_{3i}(t)\| \leq \|A_n(0)\| \left[\frac{(1-\alpha)}{\mathcal{B}(\alpha)} \eta_3 + \frac{\alpha}{\mathcal{B}(\alpha)\Gamma(\alpha)} \eta_3 \right]^n.$$

This proves the continuity as well as existence. Now, we consider showing Eq. (46) is a solution for the system (27)

$$S(t) - S(0) = S_n(t) - \mathcal{K}_{1n}(t),$$
$$C(t) - C(0) = C_n(t) - \mathcal{K}_{2n}(t), \qquad (46)$$
$$A(t) - A(0) = A_n(t) - \mathcal{K}_{3n}(t).$$

To achieve the required result, we consider

$$\|\mathcal{K}_{1n}(t)\| = \left\| \frac{(1-\alpha)}{\mathcal{B}(\alpha)} (\mathcal{G}_1(t,S) - \mathcal{G}_1(t,S_{n-1})) \right.$$

$$\left. + \frac{\alpha}{\mathcal{B}(\alpha)\Gamma(\alpha)} \int_0^t (t-\zeta)^{\mu-1}(\mathcal{G}_1(\zeta,S) - \mathcal{G}_1(\zeta,S_{n-1}))d\zeta \right\|$$

$$\leq \frac{(1-\alpha)}{\mathcal{B}(\alpha)} \|(\mathcal{G}_1(t,S) - \mathcal{G}_1(t,S_{n-1}))\| \qquad (47)$$

$$+ \frac{\alpha}{\mathcal{B}(\alpha)\Gamma(\alpha)} \int_0^t \|(\mathcal{G}_1(\zeta,S) - \mathcal{G}_1(\zeta,S_{n-1}))\|d\zeta$$

$$\leq \frac{(1-\alpha)}{\mathcal{B}(\alpha)} \eta_1 \|S - S_{n-1}\| + \frac{\alpha}{\mathcal{B}(\alpha)\Gamma(\alpha)} \eta_1 \|S - S_{n-1}\| t.$$

In the same way at t_0, we can obtain

$$\|\mathcal{K}_{1n}(t)\| \leq \left(\frac{(1-\alpha)}{\mathcal{B}(\alpha)} + \frac{\alpha t_0}{\mathcal{B}(\alpha)\Gamma(\alpha)} \right)^{n+1} \eta_1^{n+1} M. \qquad (48)$$

We can see that form Eq. (49), when n approaches to ∞, $\|\mathcal{K}_{1n}(t)\|$ tends to 0. We can verify similarly for $\|\mathcal{K}_{2n}(t)\|$ and $\|\mathcal{K}_{3n}(t)\|$.

Now, we present the uniqueness. Suppose $S^*(t), C^*(t)$ and $A^*(t)$ be the set of other solutions, then we have

$$S(t) - S^*(t) = \frac{(1-\alpha)}{\mathcal{B}(\alpha)} (\mathcal{G}_1(t,S) - \mathcal{G}_1(t,S^*))$$

$$+ \frac{\alpha}{\mathcal{B}(\alpha)\Gamma(\alpha)} \int_0^t (\mathcal{G}_1(\zeta,S) - \mathcal{G}_1(\zeta,S^*))d\zeta. \qquad (49)$$

By employing norm on Eq. (51), we get

$$\|S(t) - S^*(t)\| = \left\| \frac{(1-\alpha)}{\mathcal{B}(\alpha)} (\mathcal{G}_1(t,S) - \mathcal{G}_1(t,S^*)) + \frac{\alpha}{\mathcal{B}(\alpha)\Gamma(\alpha)} \int_0^t (\mathcal{G}_1(\zeta,S) - \mathcal{G}_1(\zeta,S^*)) d\zeta \right\|$$

$$\leq \frac{(1-\alpha)}{\mathcal{B}(\alpha)} \eta_1 \|S(t) - S^*(t)\| + \frac{\alpha}{\mathcal{B}(\alpha)\Gamma(\alpha)} \eta_1 t \|S(t) - S^*(t)\|.$$

$$(50)$$

On simplification

$$\|S(t) - S^*(t)\| \left(1 - \frac{(1-\alpha)}{\mathcal{B}(\alpha)} \eta_1 - \frac{\alpha}{\mathcal{B}(\alpha)\Gamma(\alpha)} \eta_1 t \right) \leq 0. \qquad (51)$$

From the above condition, it is clear that $S(t) = S^*(t)$, if

$$\left(1 - \frac{(1-\alpha)}{\mathcal{B}(\alpha)} \eta_1 - \frac{\alpha}{\mathcal{B}(\alpha)\Gamma(\alpha)} \eta_1 t \right) \geq 0. \qquad (52)$$

Therefore Eq. (52) proves our result.

6 Numerical Results and Discussion

Here, we illustrated the nature of q-HATM solution for different α. The initial conditions for the proposed model is defined as

$$S(0) = S_0(t) = N, C(0) = C_0(t) = 0, A(0) = A_0(t) = 0.$$

where N is the total number of receptors and which is 4.4×10^4No./cell. In order to capture the behaviour, the value of the parameters cited in Eq. (2) are considered as follows

$$\delta = 0.0005 M^{-1} s^{-1}, \beta = 142.8 s^{-1}, \varepsilon = 1 unit/mL, \lambda = 0.12 s^{-1}.$$

The nature of results obtained by q-HATM for a considered model with different α is dissipated in Fig. 1 with different fractional order. To analyse the behaviour of archived results associated with $\hbar$, the $\hbar$-curves are plotted for distinct α is captured in Fig. 2. These help us to adjust and control the convergence region of the obtained results. For a suitable $\hbar$, the obtained results rapidly tend to an analytical solution. Moreover, in the plots the convergence region is denoted by the horizontal line. The captured figures show the degree of freedom and more simulating consequences about the hired model with different arbitrary order and also it signifies the novelty of the fractional operator employed. Further, from all plots one can observer that the projected solution procedure is and very effective and more accurate to examine the considered nonlinear problem.

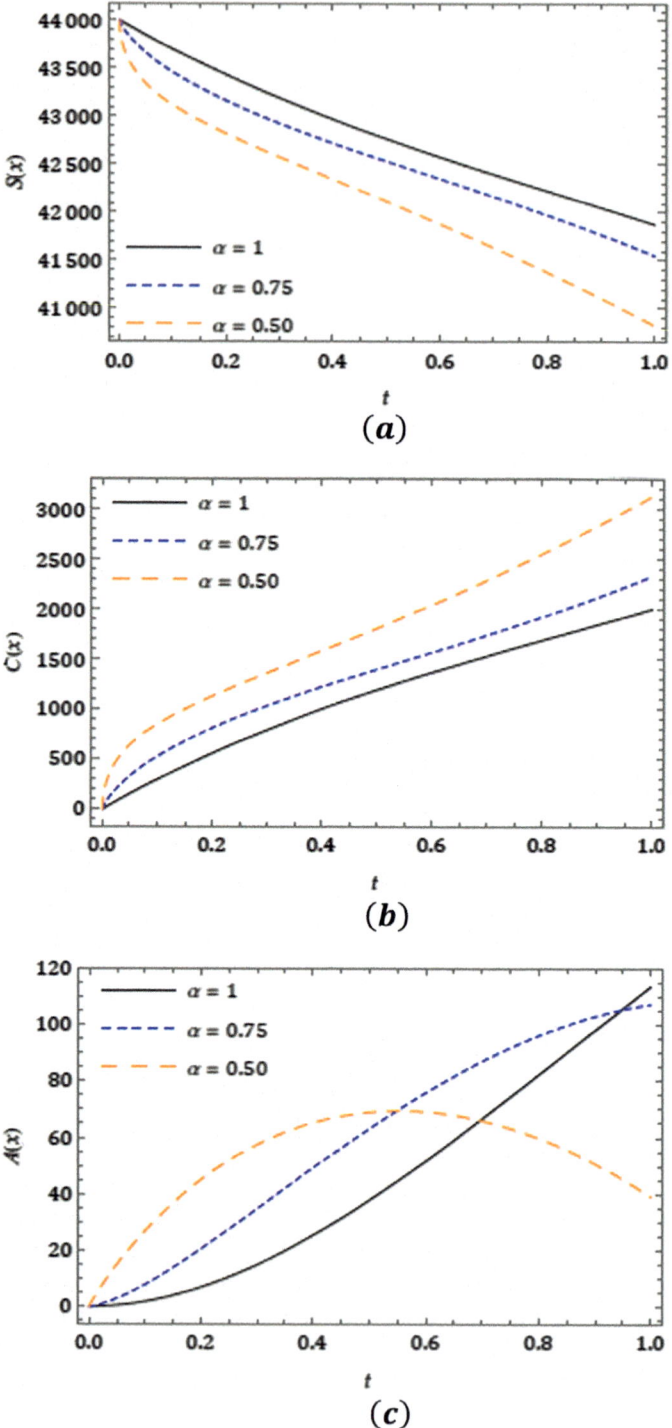

Fig. 1. Behaviour of the obtained results for (*a*) $S(t)$, (*b*) $C(t)$ and (*c*) $A(t)$ with different α at $n = 1$ and $\hbar = -1$.

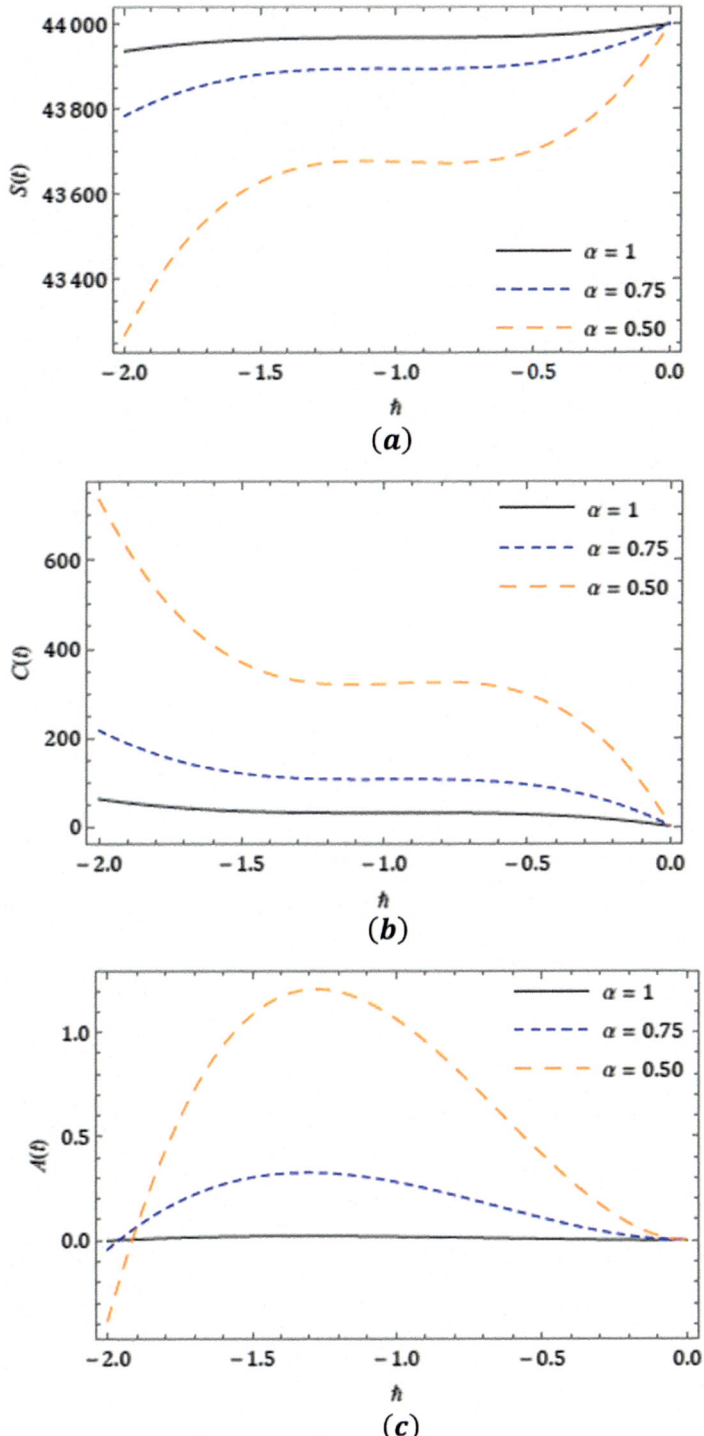

Fig. 2. $\hbar$-curves for (a) $S(t)$, (b) $C(t)$ and (c) $A(t)$ with distinct α at $t = 0.01$ and $n = 1$.

7   Conclusion

The q-HATM is employed efficiently in the present framework to find the solution for the system of equation with arbitrary order and illustrating the model of TRA mechanism in calcium signalling. Since, generalized Mittag-Leffler function is hired to define fractional-order AB integrals and derivatives, these operators help us to capture more simulating consequences and also it incorporate most essential behaviours of the models, and hence the current study exemplifies the effeteness of the projected derivative. Further, for the obtained results we presented the existence and uniqueness within the frame of fixed point hypothesis. As associated to consequences available in the literature, the results obtained by the help of projected method are more stimulating. The graphical representations show the dependence of the considered nonlinear model on parameters offered by the considered scheme and fractional order, and also it exemplifies the degree of freedom when we incorporate the fractional operator in the systems. We can be observed by the present study, the projected model is remarkably associated with the time instant and time history-based consequences, and which can be efficiently examined by the help of fractional calculus. Lastly, we can conclude that the present study can aid the researchers to analyse the nature system corresponded to very useful and interesting and consequences.

References

1. Berridge, M.J.: Inositol trisphosphate and diacylglycerol: two interacting second messengers. Annu. Rev. Biochem. **56**(1), 159–193 (1987)
2. Carafoli, E.: Intracellular calcium homeostasis. Annu. Rev. Biochem. **56**(1), 395–433 (1987)
3. Lenoci, L., Duvernay, M., Satchell, S., Benedetto, E.D., Hamm, H.E.: Mathematical model of PAR1-mediated activation of human platelets. Mol. BioSyst. **7**(4), 1129–1137 (2011)
4. Wiesner, T.F., Berk, B.C., Nerem, R.M.: A mathematical model of cytosolic calcium dynamics in human umbilical vein endothelial cells. Am. J. Physiol. Cell Physiol. **270**(5), 1556–1569 (1996)
5. Agarwal, R., Purohit, S.D.: A mathematical fractional model with nonsingular kernel for thrombin receptor activation in calcium signalling. Math. Meth. Appl. Sci. **42**(8), 7160–7171 (2019). https://doi.org/10.1002/mma.5822
6. Liouville, J.: Memoire surquelques questions de geometrieet de mecanique, et sur un nouveau genre de calcul pour resoudreces questions. J. Ecole. Polytech. **13**, 1–69 (1832)
7. Riemann, G.F.B.: Versuch Einer Allgemeinen Auffassung der Integration und Differentiation. Gesammelte Mathematische Werke, Leipzig (1896)
8. Caputo, M.: Elasticita e Dissipazione. Zanichelli, Bologna (1969)
9. Miller, K.S., Ross, B.: An introduction to fractional calculus and fractional differential equations. A Wiley, New York (1993)
10. Podlubny, I.: Fractional Differential Equations. Academic Press, New York (1999)
11. Kilbas, A.A., Srivastava, H.M., Trujillo, J.J.: Theory and applications of fractional differential equations. Elsevier, Amsterdam (2006)
12. Baleanu, D., Guvenc, Z.B., Tenreiro Machado, J.A.: New trends in nanotechnology and fractional calculus applications. Springer, New York (2010)

13. Esen, A., Sulaiman, T.A., Bulut, H., Baskonus, H.M.: Optical solitons and other solutions to the conformable space-time fractional Fokas-Lenells equation. Optik **167**, 150–156 (2018)
14. Veeresha, P., Prakasha, D.G.: An efficient technique for two-dimensional fractional order biological population model. Int. J. Model. Simul. Sci. Comput. **11**(1), 2050005 (2020). https://doi.org/10.1142/s1793962320500051
15. Baleanu, D., Wu, G.C., Zeng, S.D.: Chaos analysis and asymptotic stability of generalized Caputo fractional differential equations. Chaos, Solitons Fractals **102**, 99–105 (2017)
16. Veeresha, P., Prakasha, D.G.: New numerical surfaces to the mathematical model of cancer chemotherapy effect in Caputo fractional derivatives. Chaos: An Interdisciplinary J. Nonlinear Sci. **29**(1), 013119 (2019). https://doi.org/10.1063/1.5074099
17. Baskonus, H.M., Sulaiman, T.A., Bulut, H.: On the new wave behavior to the Klein-Gordon-Zakharov equations in plasma physics. Indian J. Phys. **93**(3), 393–399 (2019)
18. Prakasha, D.G., Veeresha, P.: Analysis of Lakes pollution model with Mittag-Leffler kernel. J. Ocean Eng. Sci., pp. 1–13 (2020), https://doi.org/10.1016/j.joes.2020.01.004
19. Gao, W., et al.: Novel dynamic structures of 2019-nCoV with nonlocal operator via powerful computational technique. Biology **9**(5), 107 (2020). https://doi.org/10.3390/biology9050107
20. Caputo, M., Fabrizio, M.: A new definition of fractional derivative without singular kernel. Progress Fract. Diff. Appl. **1**(2), 73–85 (2015)
21. Goswami, A., Sushila, J., Singh, D., Baleanu, D.: A new analysis of fractional fish farm model associated with Mittag-Leffler type kernel. Int. J. Biomath. **13**(02), 2050010 (2019)
22. Veeresha, P., Prakasha, D.G., Kumar, D., Baleanu, D., Singh, J.: An efficient computational technique for fractional model of generalized Hirota-Satsuma coupled Korteweg–de Vries and coupled modified Korteweg–de Vries equations. J. Comput. Nonlinear Dynam., 15 (7) (2020) https://doi.org/10.1115/1.4046898
23. Singh, J., Kumar, D.: Numerical computation of fractional Kersten-Krasil'shchik coupled KdV-mKdV system arising in multi-component plasmas. AIMS Math. **5**(3), 2346–2368 (2020)
24. Veeresha, P., Prakasha, D.G., Kumar, D.: Fractional SIR epidemic model of childhood disease with Mittag-Leffler memory. In: Fractional Calculus in Medical and Health Science. pp. 229–248 (2020)
25. Atangana, A., Baleanu, D.: New fractional derivatives with non-local and non-singular kernel theory and application to heat transfer model. Thermal Sci. **20**, 763–769 (2016)
26. Liao, S.J.: Homotopy analysis method and its applications in mathematics. J. Basic Sci. Eng. **5**(2), 111–125 (1997)
27. Liao, S.J.: Homotopy analysis method: a new analytic method for nonlinear problems. Appl. Math. Mech. **19**, 957–962 (1998)
28. Singh, J., Kumar, D., Swroop, R.: Numerical solution of time- and space-fractional coupled Burgers' equations via homotopy algorithm. Alexandria Eng. J. **55**(2), 1753–1763 (2016)
29. Srivastava, H.M., Kumar, D., Singh, J.: An efficient analytical technique for fractional model of vibration equation. Appl. Math. Model. **45**, 192–204 (2017)
30. Veeresha, P., Prakasha, D.G., Baskonus, H.M., Yel, G.: An efficient analytical approach for fractional Lakshmanan–Porsezian–Daniel model. Math. Meth. Appl. Sci. **43**(7), 4136–4155 (2020)
31. Bulut, H., Kumar, D., Singh, J., Swroop, R., Baskonus, H.M.: Analytic study for a fractional model of HIV infection of CD4 + T lymphocyte cells. Math. Nat. Sci. **2**(1), 33–43 (2018)
32. Veeresha, P., Prakasha, D.G.: Solution for fractional Zakharov-Kuznetsov equations by using two reliable techniques. Chinese J. Phys. **60**, 313–330 (2019)
33. Kumar, D., Agarwal, R.P., Singh, J.: A modified numerical scheme and convergence analysis for fractional model of Lienard's equation. J. Comput. Appl. Math. **399**, 405–413 (2018)

34. Veeresha, P., Prakasha, D.G.: Solution for fractional generalized Zakharov equations with Mittag-Leffler function. Results Eng. **5**, 1–12 (2020). https://doi.org/10.1016/j.rineng.2019. 100085
35. Veeresha, P., Prakasha, D.G., Baskonus, H.M.: Novel simulations to the time-fractional Fisher's equation. Math. Sci. **13**(1), 33–42 (2019). https://doi.org/10.1007/s40096-019-0276-6
36. Veeresha, P., Prakasha, D.G., Baskonus, H.M., Singh, J.: Fractional approach for equation describing the water transport in unsaturated porous media with Mittag-Leffler kernel. Front. Phys. **7**(193), 1–11 (2019). https://doi.org/10.3389/fphy.2019.00193
37. Singh, J., Kumar, D., Hammouch, Z., Atangana, A.: A fractional epidemiological model for computer viruses pertaining to a new fractional derivative. Appl. Math. Comput. **316**, 504–515 (2018)
38. Prakasha, D.G., Malagi, M.S., Veeresha, P.: New approach for fractional Schrödinger-Boussinesq equations with Mittag-Leffler kernel. Math. Meth. Appl. Sci., (2020) https://doi.org/10.1002/mma.6635
39. Gao, W., et al.: Iterative method applied to the fractional nonlinear systems arising in thermoelasticity with Mittag-Leffler kernel. Fractals (2020) https://doi.org/10.1142/s0218348x2040040x
40. Atangana, A., Alkahtani, B.T.: Analysis of the Keller-Segel model with a fractional derivative without singular kernel. Entropy **17**, 4439–4453 (2015)
41. Atangana, A., Alkahtani, B.T.: Analysis of non- homogenous heat model with new trend of derivative with fractional order. Chaos, Solitons Fractals **89**, 566–571 (2016)

Stability Analysis of Bifurcated Limit Cycles in a Labor Force Evolution Model

Sanaa ElFadily[1]($\boxtimes$), Najib Khalid[2], and Abdelilah Kaddar[3]

[1] LERMA Laboratory, Mohammadia School of Engineering, Mohammed V University in Rabat, Rabat, Morocco
elfadilysanaa@gmail.com
[2] LERMA Laboratory, National High School of Mines, Rabat, Morocco
najibkhalid@gmail.com
[3] Labsip Laboratory, National School of Applied Sciences, El Jadida, Chouaib Doukkali University, El Jadida, Morocco
a.kaddar@yahoo.fr

Abstract. This article focuses on the fluctuations observed in the labor markets. We divide the total population into three categories: employed, unemployed and inactive, then we describe the entry-exit flows between these different categories by two delay differential equations. Our contribution is to compute an indicator for determining the behavior of the model variables, in a neighborhood of the critical delay. Our findings show that the model can undergo a Hopf bifurcation and the bifurcated limit cycles is stable (or unstable), according to the crossing direction of critical delay.

Keywords: Differential equations · Delay · Periodic solutions · Hopf bifurcation · Limit cycles · Labor market · Employed persons · Unemployed

1 Introduction

The study of economic fluctuations and cycles has attracted much attention in macroeconomic theory and applied mathematics. On one hand, the resulting mathematical models have been involved solving many problems related to the explanation, identification and measurement or estimation of these phenomena, and on the other hand, the obtained results align well with proposed results in econometric studies [6,7,10,12,14]. Within this framework, we developed a mathematical model to study the fluctuations observed empirically in the three aggregates of labor markets namely, employment, unemployment and inactivity. The idea is to divide the population into three distinct categories of individuals: the employed, the unemployed and the inactive and describe the flows between these different categories by the following system (For more details, see Fig. 1):

$$
\begin{cases}
\frac{dL}{dt} = \gamma U(t) - (s + \alpha + m)L(t), \\
\frac{dU}{dt} = \rho(1 - \frac{L(t-r)+U(t-r)}{N_c})L(t-r) + sL(t) - (m + \gamma + \beta)U(t),
\end{cases}
\tag{1}
$$

© Springer Nature Switzerland AG 2021
Z. Hammouch et al. (Eds.): SM2A 2019, LNNS 168, pp. 61–77, 2021.
https://doi.org/10.1007/978-3-030-62299-2_5

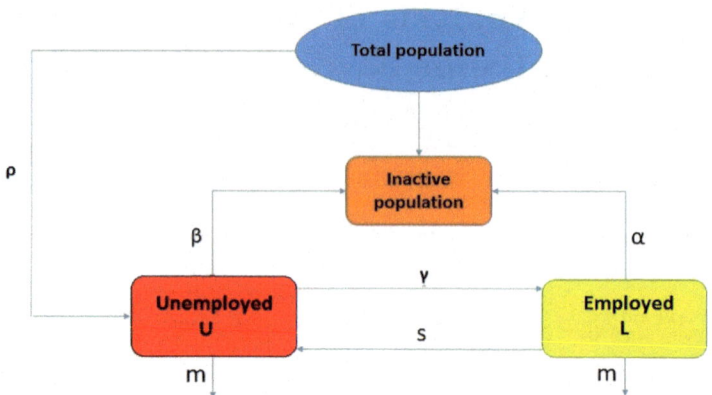

Fig. 1. Labor market flows

with the initial condition:

$$(L(\xi), U(\xi)) = (\varphi_1(\xi), \varphi_2(\xi)), \quad \forall \xi \in [-r, 0], \tag{2}$$

where the variable L is the number of the employed population, U is the number of the unemployed population, γ denotes the employment level, s indicates the job loss rate, α is the rate of workers who have withdrawn from labor market due to retirement or disability, β is the rate of unemployed people who are no longer able to work, m is the mortality rate, ρ is the maximum population growth rate, N_c is the maximum load capacity, r is the time needed for a new person who has found a job to contribute to the reproductive process and $\varphi_i \in \mathcal{C}([-r, 0], \mathcal{R}^+)$, $i = 1, 2$. Here $\mathcal{C}([-r, 0], \mathbb{R}^+)$ is the Banach space of continuous functions from the interval $[-r, 0]$ to the set of positive real numbers $\mathbb{R}^+$.

Model (1) is composed of two differential equations which model the inflows-outflows in the three categories of the total population (category U of the unemployed, category L of the employed and category I of inactive people). On the one hand, the second equation translates the feeding of category U by people who have reached working age and who are looking for a job. This flow is indicated by a logistic growth rate ρ. People in this category transform to category I with a disability rate β, or category L with a recruitment rate γ or exit with a death rate m. On the other hand, the first equation describes the evolution of category L by the difference between the newly employed, noted γU and those leaving this category by job loss, sL, by disability αL or by death, mL.

There have been several attempts in this area. We cite for example the model of labor force evolution, proposed by Farkas (in 1995 [16]), Only a handful of studies have been found to examine its dynamics systematically. Papers [9,15,20] study the local and global stability of labor force evolution model using linearization technique and Lyapunov method. The resulting numerical results usually give the local and global stability. In empirical studies, however, the observed data of the employed and the unemployed have oscillatory behavior.

In this work, we prove the existence of a Hopf bifurcation point and we also study the direction and stability of the periodic branches (limit cycles) that evolve around this point.

The study of the oscillatory behavior of dynamical systems can be done by fixed point methods [8,17] or by Hopf bifurcation theorem [13,25]. In the latter case, several researchers have proposed different techniques to investigate the behavior of dynamical systems in the neighborhoods of the critical delays. The first is to look for a normal form in a central manifold [5,22,23]. This method entails a long calculation. The second is the singular perturbation approach. This technique is preferable for its computational efficiency (multiple-scale analysis [4,21], Krylov-Bogoliubov-Mitropolsky method [2], Poincaré-Lindstedt method [1,19], harmonic balance method [3,18] and pseudo-oscillator analysis [27]). In this work we chose to work with the Kuznetsov method [11,26]. The method only requires a computation of the first Lyapunov coefficient to determine the behavior in the neighborhoods of the critical delays.

This work is structured as follows. In Sect. 2, we study the local stability and the Hopf bifurcation of the nontrivial equilibrium position of the System (1). In Sect. 3, we first give the essential calculations of the central manifold and the reduction of our model to a normal form. Then, we use the Kuzentsov method to determine the direction and stability of the periodic orbit resulting from the Hopf bifurcation. Numerical simulations are given in Sect. 4 to support the main aspects of our study. Finally, in Sect. 5, we summarize the main findings, our conclusion, the gaps we encountered and some perspectives on this study.

2   Hopf Bifurcation Analysis

2.1   Equilibria

In the following, we study the existence of equilibrium points for the system (1).

Proposition 1. *If* $\dfrac{\gamma\rho}{m(\gamma + s + \alpha + \alpha_2 + m) + \gamma\alpha + \beta(s + \alpha)} > 1$, *then system (1) admits two equilibria:* $E_0 = (0,0)$ *and a unique positive equilibrium* $E_* = (L_*, U_*)$, *where*

$$U_* = \frac{(s + \alpha + m)L_*}{\gamma}$$

and

$$L_* = \frac{\gamma N_c}{\gamma + s + \alpha + m}\left(1 - \frac{m(\gamma + s + \alpha + \beta + m) + \gamma\alpha + \beta(s + \alpha)}{\gamma\rho}\right).$$

Proof. Suppose (U, L) is an equilibrium point, that is,

$$\begin{cases} \gamma U - (s + \alpha + m)L = 0, \\ \rho(1 - \frac{L+U}{N_c})L + sL - (\gamma + \beta + m)U = 0. \end{cases} \tag{3}$$

It is clear that $(0,0)$ is a solution of the system (3). This gives that E_0 is a trivial equilibrium of system (1). Moreover, if (U, L) is an equilibrium such that $U > 0$ and $L > 0$ then we have

$$
\begin{cases}
U = \frac{(s+\alpha+m)L}{\gamma}, \\
\rho(1 - \frac{(\gamma+s+\alpha+m)L}{\gamma N_c}) - \frac{(s+\alpha+m)(\gamma+\beta+m)}{\gamma} = 0.
\end{cases} \tag{4}
$$

Under the condition $\frac{\gamma\rho}{m(\gamma+s+\alpha+\beta+m)+\gamma\alpha+\beta(s+\alpha)} > 1$, the system of linear equations (4) has a unique non-trivial solution $E_* = (L_*, U_*)$, where

$$
U_* = \frac{(s+\alpha+m)L_*}{\gamma}
$$

and

$$
L_* = \frac{\gamma N_c}{(\gamma+s+\alpha+m)}(1 - \frac{m(\gamma+s+\alpha+\beta+m)+\gamma\alpha+\beta(s+\alpha)}{\gamma\rho}).
$$

This completes the proof.

2.2   Local Stability

The Linearized system of Eqs. (1) at the positive equilibrium E_* is

$$
\begin{cases}
\frac{dx}{dt} = -(s+\alpha+m)x + \gamma y, \\
\frac{dy}{dt} = sx + (\frac{m(\gamma+s+\alpha+\beta+m)+\gamma\alpha+\beta(s+\alpha)}{\gamma} - \frac{\rho L_*}{N_c})x_r - (\gamma+\beta+m)y - \frac{\rho L_*}{N_c}y_r.
\end{cases} \tag{5}
$$

For System (5) the characteristic equation is:

$$
\lambda^2 + \theta_1\lambda + \theta_2\lambda e^{-\lambda r} + \theta_3 + \theta_4 e^{-\lambda r} = 0, \tag{6}
$$

where

$$
\theta_1 = \gamma + s + \alpha + \beta + 2m, \quad \theta_2 = \frac{\rho L_*}{N_c}
$$

$$
\theta_3 = m(\gamma + s + \alpha + \beta + m) + \gamma\alpha + \beta(s + \alpha)
$$

and

$$
\theta_4 = \rho(\gamma + s + \alpha + m)\frac{L_*}{N_c} - (m(\gamma + s + \alpha + \beta + m) + \gamma\alpha + \beta(s + \alpha)).
$$

Using the Routh-Hurwitz criterion and Kuang's results [25] for Eq. (6), we prove the following results:

Proposition 2. *If $r = 0$, then the positive equilibrium E_* is locally asymptotically stable.*

Proof. When $r = 0$, the Eq. (6) becomes

$$\lambda^2 + (\gamma + s + \alpha + \beta + 2m + \frac{\rho L_*}{N_c})\lambda + \rho(\gamma + s + \alpha + m)\frac{L_*}{N_c} = 0, \qquad (7)$$

Since $\gamma + s + \alpha + \beta + 2m + \frac{\rho L_*}{N_c} > 0$ and $\rho(\gamma + s + \alpha + m)\frac{L_*}{N_c} > 0$, then, by the Routh-Hurwitz criterion, all the roots of Eq. (6) have non-negative real parts, and therefore the positive equilibrium E_* is locally asymptotically stable.

Let (H_1): $\frac{\gamma\rho}{m(\gamma+s+\alpha+\beta+m)} > 3$.

Proposition 3. *If* (H_1) *is valid. Then there exists* $r_0 > 0$ *such that,*

(i) for $0 \le r < r_0$, E_* *is locally asymptotically stable;*
(ii) for $r > r_0$, E_* *is unstable;*
(iii) for $r = r_0$, *Eq. (5) admits two purely imaginary roots;*

where

$$r_0 = \frac{1}{\omega_0}\arccos(\frac{-\theta_1\theta_2\omega_0^2 - \theta_4(\theta_3 - \omega_0^2)}{\theta_2\omega_0^2 + \theta_4^2}),$$

and

$$\omega_0 = \sqrt{\frac{1}{2}\{(\theta_2^2 + 2\theta_3 - \theta_1^2) + \sqrt{(\theta_2^2 + 2\theta_3 - \theta_1^2)^2 + 4(\theta_3^2 - \theta_4^2)}\}}.$$

Proof. If $\frac{\gamma\rho}{m(\gamma+s+\alpha+\beta+m)} \ge 3$, then

$$\theta_3^2 - \theta_4^2 = \Lambda_1 \times \Lambda_2 \times \Lambda_3 < 0,$$

where

$$\Lambda_1 = [m(\gamma + s + \alpha + \beta + m) + \gamma\alpha + \beta(s + \alpha)]^2,$$

$$\Lambda_2 = \frac{\gamma\rho}{m(\gamma + s + \alpha + \beta + m) + \gamma\alpha + \beta(s + \alpha)} - 1$$

and

$$\Lambda_3 = 3 - \frac{\gamma\rho}{m(\gamma + s + \alpha + \beta + m) + \gamma\alpha + \beta(s + \alpha)}.$$

Consequently, Eq. (6) has only one purely imaginary solution,

$$i\omega_0 = i\sqrt{\frac{1}{2}\{(\theta_2^2 + 2\theta_3 - \theta_1^2) + \sqrt{(\theta_2^2 + 2\theta_1^2)^2 + 4(\theta_3^2 - \theta_4^2)}\}},$$

with $\omega_0 > 0$. By Theorem 2.7 in ([25], p. 77), we conclude that there exists $r_0 > 0$ which satisfies the three statements (i), (ii) and (iii) of Proposition 3.

2.3   Local Hopf Bifurcation

From (iii) of Proposition 3, we have proved that Eq. (5) has a pair of purely imaginary roots $\pm i\omega_0$, $\omega_0 > 0$, when the delay crosses the critical value r_0. In the following result we show the birth of the Hopf bifurcation, in a small vicinity of r_0.

Theorem 1. *Under hypothesis* (H_1), *the system (1) loses its stability through a Hopf bifurcation when* $r = r_0$, *i.e., a limit cycle appears out of the equilibrium* E_*.

Proof. From Proposition 3, the characteristic equation (6) has a pair of imaginary roots $\pm i\omega_0$ at $r = r_0$. It's easy to show that this root is simple. Thus it suffices to show that

$$\frac{dRe(\lambda)}{dr}(r_0) > 0$$

(see, for example [13]).
We have:

$$Sign\frac{dRe(\lambda)}{dr}|r_0 = Sign\{\theta_2^2 + 2\theta_3 - \theta_1^2 - 4(\theta_3^2 - \theta_4^2)\}.$$

After some calculations, we get:

$$\theta_2^2 + 2\theta_3 - \theta_1^2 - 4(\theta_3^2 - \theta_4^2) = \Gamma_1 - \Gamma_2(\Gamma_3 - 1)(3 - \Gamma_3), \tag{8}$$

where

$$\Gamma_1 = \frac{\gamma^2\rho^2(1 - m^2) + [\gamma - (\gamma + s + \alpha + \beta + m)^2]^2}{(\gamma + s + \alpha + \beta + m)^2},$$

$$\Gamma_2 = 4(m(\gamma + s + \alpha + \beta + m) + \gamma\alpha + \beta(s + \alpha))^2,$$

and

$$\Gamma_3 = \frac{\gamma\rho}{m(\gamma + s + \alpha + \beta + m) + \gamma\alpha + \beta(s + \alpha)}.$$

If the hypothesis (H_1) is verified, then $\Gamma_2(\Gamma_3 - 1)(3 - \Gamma_3) < 0$. Moreover, we have $0 < m < 1$. Consequently

$$\frac{dRe(\lambda)}{dr}(r_0) > 0.$$

3   Direction and Stability of the Hopf Bifurcation

In this section, we use Kuznetsov's method [24] to calculate an indicator of the direction and stability of the bifurcated branches (limit cycles) from E_*.

By the change of variables: $x(t) = L(rt) - L_*$, $y(t) = U(rt) - L_*$ and $r = r_0 + \varepsilon$, where $\varepsilon \in \mathbb{R}$ is the bifurcation parameter, the system (1) becomes

$$\begin{cases} \frac{dx}{dt} = (r_0 + \varepsilon)[a_1x + a_2y] \\ \frac{dy}{dt} = (r_0 + \varepsilon)[b_{10}x + b_{01}y + b'_{10}x_1 + b'_{11}(y_1 + x_1y_1 + x_1^2)] \end{cases} \tag{9}$$

with

$$x_1 := x(t-1), \quad y_1 := y(t-1),$$
$$a_1 = -(s + \alpha + m), \quad a_2 = \gamma,$$
$$b_{10} = s, \; b_{01} = -(\gamma + \beta + m),$$

$$b'_{10} = \left(\frac{m(\gamma + s + \alpha + \beta + m) + \gamma\alpha + \beta(s + \alpha)}{\gamma} - \frac{\rho L_*}{N_c} \right)$$

and

$$b'_{01} = -\frac{\rho L_*}{N_c}, \quad b'_{11} = -\frac{\rho}{N_c}.$$

Hence, system (9) is transformed into a functional differential equation in $\mathcal{C} := C([-1, 0], \mathbb{R}^2)$ as follows,

$$\dot{x}(t) = \mathcal{L}_\varepsilon(x_t) + f_\varepsilon(x_t), \tag{10}$$

where $x = (x, y)^T \in \mathcal{C}$, $x_t \in \mathcal{C}$ is defined by $x_t(\theta) = x(t + \theta)$ for any $\theta \in [-1, 0]$, $\varepsilon \in \mathbb{R}$ is the bifurcation parameter, $\mathcal{L}_\varepsilon : \mathcal{C} \to \mathbb{R}^2$, is a bounded linear operator and $f : \mathbb{R} \times \mathcal{C} \to \mathcal{C}$ is the nonlinear operator. $\mathcal{L}_\varepsilon$ and f are given respectively by:

$$\mathcal{L}_\varepsilon(\psi) := (r_0 + \varepsilon) \left(A_1(\psi(0)) + A_2(\psi(-1)) \right) \tag{11}$$

and

$$f_\varepsilon(\psi) = (r_0 + \varepsilon) \begin{pmatrix} 0 \\ b'_{11}\psi_1(-1)\psi_2(-1) + b'_{20}\psi_1^2(-1) \end{pmatrix}. \tag{12}$$

where

$$A_1 = \begin{pmatrix} -(s + \alpha + m) & \gamma \\ s & -(\gamma + \beta + m) \end{pmatrix}, \tag{13}$$

$$A_2 = \begin{pmatrix} 0 & 0 \\ \Lambda & \frac{-\rho L_*}{N_c} \end{pmatrix}, \tag{14}$$

with $\Lambda = \dfrac{m(\gamma + s + \alpha + \beta + m) + \gamma\alpha + \beta(s + \alpha)}{\gamma} - \dfrac{\rho L_*}{N_c}$.

Using the Riesz's representation (see [13]), we get

$$\mathcal{L} := \mathcal{L}_\varepsilon(\psi) = \int_{-1}^{0} d\eta(\theta, \varepsilon)\psi(\theta). \tag{15}$$

where,

$$\eta(\theta, \varepsilon) = (r_0 + \varepsilon) \left(A_1 \delta(\theta) + A_2 \delta(\theta + 1) \right) \tag{16}$$

The solution operator of Eq. (10) generates a $\mathcal{C}_0$-semigroup with the infinitesimal generator $\mathcal{A}_\varepsilon$ defined by

$$\mathcal{A}_\varepsilon \psi(\theta) = \begin{cases} \frac{d\psi}{d\theta}(\theta) & \text{for } \theta \in [-1, 0) \\ \int_{-1}^{0} d\eta(\theta, \varepsilon)\psi(\theta) & \text{for } \theta = 0 \end{cases} \tag{17}$$

We rewrite Eq. (10) as an abstract ordinary differential equation

$$\frac{dx_t}{dt} = \mathcal{A}_\varepsilon(x_t) + R_\varepsilon(x_t), \tag{18}$$

with the nonlinear term

$$R_\varepsilon(x_t) = \begin{cases} 0 \text{ for } \theta \in [-1,0), \\ f_\varepsilon(x_t) \text{ for } \theta = 0. \end{cases} \tag{19}$$

We denote by $\mathcal{A}^*$ the adjoint operator of $\mathcal{A}_\varepsilon$

$$\mathcal{A}^*\psi(s) = \begin{cases} -\frac{d\psi}{ds}(s), \text{ for } s \in (0,1] \\ \int_{-1}^0 \psi^T(-s)d\eta(s), \text{ for } s = 0, \end{cases} \tag{20}$$

where η^T is the transposed matrix of η.

In order to normalize the eigenvectors of operator A and A*, we define the bilinear form

$$<\psi, \phi> = \bar{\psi}(0)\phi(0) - \int_{-1}^0 \int_0^\theta \bar{\psi}(\xi - \theta)d\eta(\theta, 0)\phi(\xi)d\xi,$$

where $\phi \in \mathcal{C}$ and $\psi \in \mathcal{C}^* = C([0,1), (\mathbb{R}^{2*}))$.

Assume that $\mathcal{L}$ has two eigenvalues on the imaginary axis. Let $p(\theta)$ and $p^*(s)$ are eigenvectors of $\mathcal{A}_0$ and $\mathcal{A}^*$. In order to determine the Poincare normal form of operator A, we needs to calculate the eigenvector $p(\theta)$ and $p^*(s)$ corresponding to $i\omega_0 r_0$ and $-i\omega_0 r_0$, respectively, with $<p^*, p> = 1$ and $<p^*, \bar{p}> = 0$. Let P be the generalized eigenspace spanned by $p(\theta)$ and $\bar{p}(\theta)$ defined as

$$P = \{zp + \bar{z}\bar{p}, z \in \mathbb{C}\}.$$

Then the orthogonal complement of P in $\mathcal{C}$ is

$$Q = \{\psi \in \mathcal{C}, <p, \psi> = 0, <\bar{p}^*, \psi> = 0\}.$$

Therefore, we get a decomposition of $\mathcal{C}$ as follows

$$\mathcal{C} = P \oplus Q. \tag{21}$$

A straightforward calculation gives

$$p(\theta) = (p_1, 1)^T e^{i\theta\omega_0 r_0}$$

and

$$p^*(s) = \kappa(p_2, 1)^T e^{is\omega_0 r_0},$$

where $p_1 = \frac{i\omega_0 - b_{01} - b'_{01}}{b_{10} + b'_{10}}$, and $p_2 = \frac{-(i\omega_0 + b_{01} + b_{01})}{a_2}$.

Using the normalization condition $<p^*, p> = 1$, we get

$$\kappa = \frac{a_2(b'_{10} + b_{10})}{\Upsilon},$$

where

$$\Upsilon = a_2(b'_{10} + b_{10})(1 + b_{01}r_0 e^{i\omega_0 r_0}) + (i\omega_0 + b_{01} + b'_{01})((i\omega_0 + b_{01} + b'_{01}) - a_2 b_{10}).$$

From (21), the state variable x_t of Eq. (10) could be decomposed by

$$\begin{aligned} x_t &= \Phi z + w(z, \bar{z}, \theta) \\ &= -zp(\theta) - \bar{z}\bar{p}(\theta) + w(z, \bar{z}, \theta) \end{aligned} \tag{22}$$

where $w(z, \bar{z}, \theta) \in Q$. On the center manifold at $r = r_0$, we define

$$z(t) = <p^*, x_t>,$$

$$w(z, \bar{z}, \theta) = x_t(\theta) - Re\{z(t)p(\theta)\}, \quad w(z, \bar{z}) = w(z, \bar{z}, \theta).$$

Then

$$\begin{aligned} \dot{z}(t) &= <p^*, \dot{u}_t> \\ &= <p^*, \mathcal{A}_0 x_t + R_0 x_t> \\ &= <\mathcal{A}^* p^*, x_t> + <p^*, R_0 x_t> \end{aligned} \tag{23}$$

On the invariant manifold, system (1) can be written as

$$\dot{z}(t) = i\omega_0 r_0 z(t) + g(z, \bar{z}) \tag{24}$$

where

$$f_0(z, \bar{z}) = f(0, w(z, \bar{z}) + Re(z(t)p(\theta)) \tag{25}$$

$$\begin{aligned} g(z, \bar{z}) &= p^*(0)f_0(z, \bar{z}) \\ &= g_{20}\frac{z^2}{2} + g_{11}z\bar{z} + g_{02}\frac{\bar{z}^2}{2} + \cdots \end{aligned} \tag{26}$$

and

$$w(z, \bar{z}) = w_{20}(\theta)\frac{z^2}{2} + w_{11}(\theta)z\bar{z} + w_{02}(\theta)\frac{\bar{z}^2}{2} + \cdots$$

where z and $\bar{z}$ are local coordinates for center manifold in the direction of p and $\bar{p}^*$.

Thus, from (10) and (24), we have

$$\dot{w} = \dot{u}_t - p\dot{z} - \bar{p}\dot{\bar{z}},$$

which leads to

$$\dot{w} = \mathcal{A}_0 w + H(z, \bar{z}, \theta), \tag{27}$$

where

$$H(z, \bar{z}, \theta) = H_{20}(\theta)\frac{z^2}{2} + H_{11}(\theta)\frac{z\bar{z}}{2} + H_{02}(\theta)\frac{\bar{z}^2}{2} + \cdots \tag{28}$$

By expanding (28) and identifying its coefficients, we get

$$\begin{aligned} H_{20} &= -(\mathcal{A}_0 - 2i\omega_0 r_0)w_{20}(\theta), \\ H_{11} &= -\mathcal{A}_0 w_{11}(\theta), \\ H_{02} &= -(\mathcal{A}_0 + 2i\omega_0 r_0)w_{02}(\theta). \end{aligned} \tag{29}$$

By

$$\dot{x}_t = w(z, \bar{z}) + zp(\theta) + \bar{z}\bar{p}(\theta)$$

and

$$q(\theta) = (1, p_1)^T e^{i\theta\omega_0 r_0},$$

we get

$$x_1(t) = p_1 z + \bar{p_1}\bar{z} + \frac{1}{2}w_{20}(0)z^2 + w_{11}(0)z\bar{z} + \frac{1}{2}w_{02}(0)\bar{z}^2 + \dots$$

$$x_1(t-1) = p_1 z e^{-i\theta\omega_0 r_0} + \bar{p_1}\bar{z}e^{i\theta\omega_0 r_0} + \frac{1}{2}w_{20}(-1)z^2 + w_{11}(-1)z\bar{z} + \frac{1}{2}w_{02}(0)\bar{z}^2 + \dots$$

$$x_2(t) = z + \bar{z} + \frac{1}{2}w_{20}(0)z^2 + w_{11}(0)z\bar{z} + \frac{1}{2}w_{02}(0)\bar{z}^2 + \dots \tag{30}$$

$$x_2(t-1) = z e^{-i\theta\omega_0 r_0} + \bar{z}e^{i\theta\omega_0 r_0} + \frac{1}{2}w_{20}(-1)z^2 + \frac{1}{2}w_{11}(-1)z\bar{z} + \frac{1}{2}w_{02}(-1)\bar{z}^2 + \dots$$

Comparing with (27), we have the coefficients of (26):

$$g_{20} = -2r_0\bar{\kappa}\frac{\rho}{N_c}p_1(p_1 + 1)e^{2ir_0\omega_0},$$

$$g_{02} = -2\frac{\rho\bar{\kappa}r_0}{N_c}\bar{p_1}(\bar{p_1} + 1)e^{-2ir_0\omega_0},$$

$$g_{11} = -\frac{\rho\bar{\kappa}r_0}{N_c}(2\bar{p_1}p_1 + \bar{p_1} + p_1), \tag{31}$$

$$g_{21} = -\frac{\rho\bar{\kappa}r_0}{N_c}\left(e^{-i\omega_0 r_0}w_{111}(-1)(2p_1 + 1) + \frac{1}{2}\bar{p_1}e^{-i\omega_0 r_0}w_{220}(-1)\right.$$
$$\left. + p_1 e^{i\omega_0 r_0}w_{211}(-1) + (\frac{1}{2} + \bar{p_1})e^{-i\omega_0 r_0}w_{120}(-1)\right).$$

Next, we calculate $w_{11}(\theta)$, $w_{20}(\theta)$ and g_{21}.
For $\theta \in [-1, 0)$, we have

$$H(z, \bar{z}, \theta) = \bar{p}^*(0)f_0 p(0) - p^*(0)\bar{f_0}\bar{p}(0)$$
$$= -g(z, \bar{z})p(\theta) - \bar{g}(z, \bar{z})\bar{p}(\theta) \tag{32}$$
$$= -(g_{20}p(\theta) + \bar{g}_{02}\bar{p}(\theta))\frac{z^2}{2} - (g_{11}p(\theta) + \bar{g}_{11}\bar{p}(\theta))z\bar{z} + \dots$$

Using formula (27), we find

$$H_{20}(\theta) = -(g_{20}p(\theta) + \bar{g}_{02}\bar{p}(\theta)), \tag{33}$$

$$H_{11}(\theta) = -(g_{11}p(\theta) + \bar{g}_{11}\bar{p}(\theta)). \tag{34}$$

Substituting (34) into (25) and (33) into (25), respectively, we get

$$\dot{w}_{20} = 2ir_0\omega_0 w_{20}(\theta) + g_{20}p(\theta) + \bar{g}_{02}\bar{p}(\theta). \tag{35}$$

So

$$w_{20} = \frac{ig_{20}e^{ir_0\omega_0\theta}}{r_0\omega_0}p(0) + \frac{i\bar{g}_{02}e^{-ir_0\omega_0\theta}}{3r_0\omega_0}\bar{p}(0) + E_1 e^{2ir_0\omega_0\theta}, \tag{36}$$

$$w_{11} = \frac{-ig_{11}e^{ir_0\omega_0\theta}}{r_0\omega_0}p(0) + \frac{i\bar{g}_{11}e^{-ir_0\omega_0\theta}}{r_0\omega_0}\bar{p}(0) + E_2. \tag{37}$$

In the sequel, we determine E_1 and E_2.

By (27) and the operator $\mathcal{A}_0$, we have

$$\int_{-1}^{0} d\eta(\theta)w_{20}(\theta) = 2i\omega_0 r_0 w_{20}(0) - H_{20}(0), \tag{38}$$

$$\int_{-1}^{0} d\eta(\theta)w_{11}(\theta) = -H_{11}(0). \tag{39}$$

From (27) and (28), we obtain

$$H_{20}(\theta) = -(g_{20}q(0) + \bar{g}_{02}\bar{p}(0)) + \frac{\rho r_0 e^{2ir_0\omega_0}}{N_c}\begin{pmatrix} 0 \\ p_1(p_1 + 1) \end{pmatrix}, \tag{40}$$

$$H_{11}(\theta) = -(g_{11}q(0) + \bar{g}_{11}\bar{p}(0)) + \frac{\rho r_0}{N_c}\begin{pmatrix} 0 \\ 2\bar{p}_1 p_1 + \bar{p}_1 + p_1 \end{pmatrix}, \tag{41}$$

substituting (36) and (40) into (38) , and noticing that

$$\left(i\omega_0 r_0 I - \int_{-1}^{0} d\eta(\theta)e^{ir_0\omega_0\theta}\right)p(0) = 0, \tag{42}$$

$$\left(-i\omega_0 r_0 - \int_{-1}^{0} d\eta(\theta)e^{-ir_0\omega_0\theta}\right)\bar{p}(0) = 0, \tag{43}$$

we get

$$\left(2i\omega_0 r_0 - \int_{-1}^{0} d\eta(\theta)e^{2ir_0\omega_0\theta}\right)E_1 = -\frac{\rho r_0 e^{2ir_0\omega_0}}{N_c}\begin{pmatrix} 0 \\ p_1(p_1 + 1) \end{pmatrix}, \tag{44}$$

that is

$$\begin{pmatrix} a_1 - 2ir_0\omega_0 & a_2 \\ b_{10}e^{-2ir_0\omega_0} + b'_{10} & b_{01} + b'_{01}e^{-2ir_0\omega_0} - 2ir_0\omega_0 \end{pmatrix} E_1 = -\frac{\rho r_0 e^{2ir_0\omega_0}}{N_c}\begin{pmatrix} 0 \\ p_1(p_1 + 1) \end{pmatrix}, \tag{45}$$

where $E_1 = (E_{1_1}, E_{1_2})^T$, with

$$E_{1_1} = \frac{\rho r_0 a_2 p_1(p_1 + 1)e^{2ir_0\omega_0}}{N_c((a_1 - 2i\omega_0)(b_{01} + b'_{01}e^{-2ir_0\omega_0} - 2i\omega_0) - a_2(b_{10} + b'_{10}e^{-2ir_0\omega_0})},$$

and

$$E_{1_2} = -\frac{\rho r_0(a_1 - 2i\omega_0)E_{1_1}}{a_2 N_c}.$$

Similarly, substituting (37) and (41) into (39), we get

$$\int_{-1}^{0} d\eta(\theta) E_2 = -\frac{\rho r_0}{N_c} \begin{pmatrix} 0 \\ 2\bar{p}_1 p_1 + \bar{p}_1 + p_1 \end{pmatrix}, \tag{46}$$

that is

$$\begin{pmatrix} a_1 & a_2 \\ b_{10} + b'_{10} & b_{01} + b'_{01} \end{pmatrix} E_2 = -\frac{\rho r_0}{N_c} \begin{pmatrix} 0 \\ 2\bar{p}_1 p_1 + \bar{p}_1 + p_1 \end{pmatrix}, \tag{47}$$

where $E_2 = (E_{2_1}, E_{2_2})^T$, with

$$E_{2_1} = \frac{\rho r_0 a_2 p_1 (p_1 + 1)(a_1(b_{01} + b'_{01}) - a_2(b_{10} + b'_{10}))}{N_c},$$

and

$$E_{2_2} = -\frac{\rho r_0 a_1 E_{1_1}}{a_2 N_c}.$$

To give our main result, we recall the definition of an indicator of direction and stability of limit cycles.

Definition 1. The first Lyapunov coefficient is given by [24]

$$l_1(r) = \frac{Re(c_1)}{\omega r} + Re(\lambda)\frac{Im(c_1)}{\omega^2 r^2},$$

where

$$c_1 = \frac{g_{21}}{2} + \frac{|g_{11}|^2}{\lambda} + \frac{|g_{02}|^2}{2(2\lambda - \bar{\lambda})} + \frac{g_{20}g_{11}(2\lambda + \bar{\lambda})}{2|\lambda|^2}.$$

For $r = r_0$, we have $\lambda = \lambda_0 = i\omega_0$, and consequently, we obtain the following result.

Theorem 2. [24] Suppose that hypothesis (H_1) holds. Then, for λ in neighborhood of λ_0, the Eq. (10) is locally topologically equivalent to the following equation:

$$\dot{z} = (\sigma + i)z + sign(l_1(r_0))z \mid z \mid^2 + O(\mid z \mid^4), \tag{48}$$

with $\sigma = \frac{Re(\lambda)}{Im(\lambda)}|r_0$.

Theorem 3. [24] Suppose that hypothesis (H_1) holds. Then

(a) if $l_1(r_0) < 0$, then a stable limit cycle appears out of the equilibrium E_*, for $r > r_0$ (supercritical Hopf bifurcation).
(b) if $l_1(r_0) > 0$, then an unstable limit cycle appears out of the equilibrium E_*, for $r < r_0$ (sub-critical Hopf bifurcation).

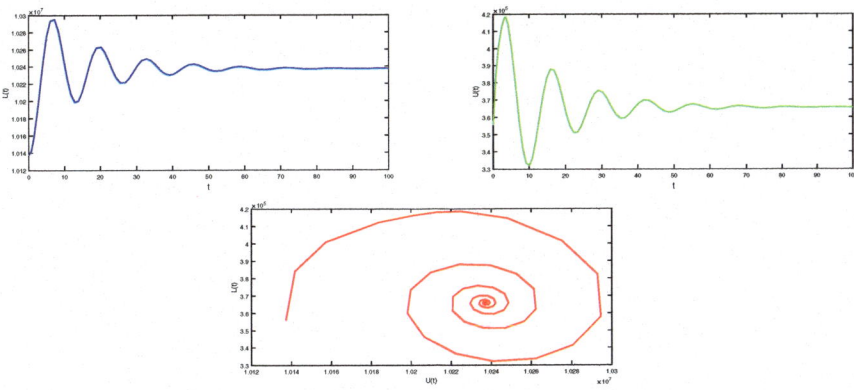

Fig. 2. Stable solutions of Model (1) for a delay, r smaller than the critical value, r_0: $r = 3.1068$ and $r_0 = 4.1068$

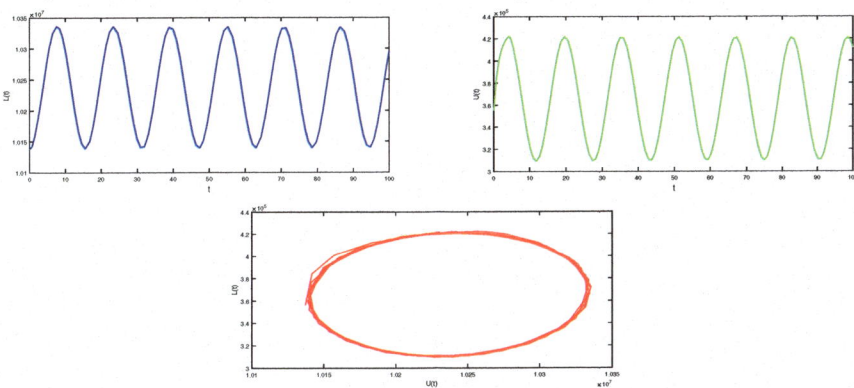

Fig. 3. Periodic solutions have bifurcated from the positive equilibrium of Model (1) for a delay closer to the critical value, $r_0 = 4.1068$

4   Numerical Simulations

4.1   Qualitative Behavior of Solutions

We consider the following hypothetical numerical parameters:

$$\gamma = 0.7, \ s = 0.01, \ \rho = 0.4, \ m = 0.005, \ \alpha = 0.01, \ \beta = 0.03, \ N_c = 11000000.$$

The positive equilibrium $E* = (1.0137 \times 10^7, 3.5561 \times 10^5)$. The first Lyapunov coefficient $l_1(r_0) = 1.106447511 \times 10^{-7}$, then the subcritical Hopf bifurcation exist and an unstable limit cycle appears out of $E *$.

According to Fig. 2, Fig. 3 and Fig. 4, we observe three oscillatory regimes on the labor market: convergent oscillations towards the equilibrium (see, Fig. 2),

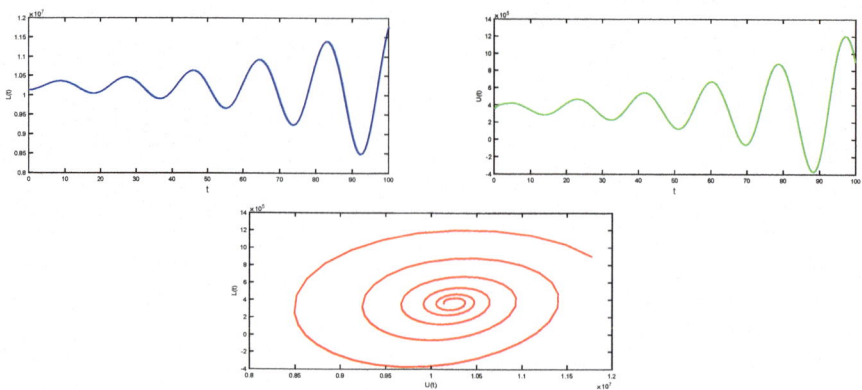

Fig. 4. Unstable solutions of Model (1) for a delay, r greater than the critical value: $r = 5.1068$ and $r_0 = 4.1068$

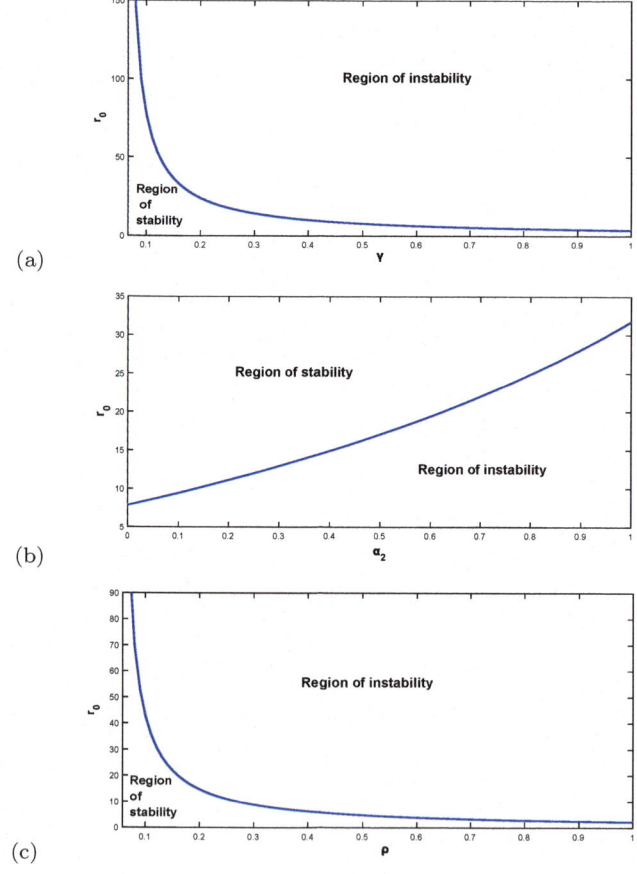

Fig. 5. The variation curve of the critical delay, r_0 as a function of: (a) employment level, γ, (b) the job loss rate, s and (c) the maximum population growth rate ρ

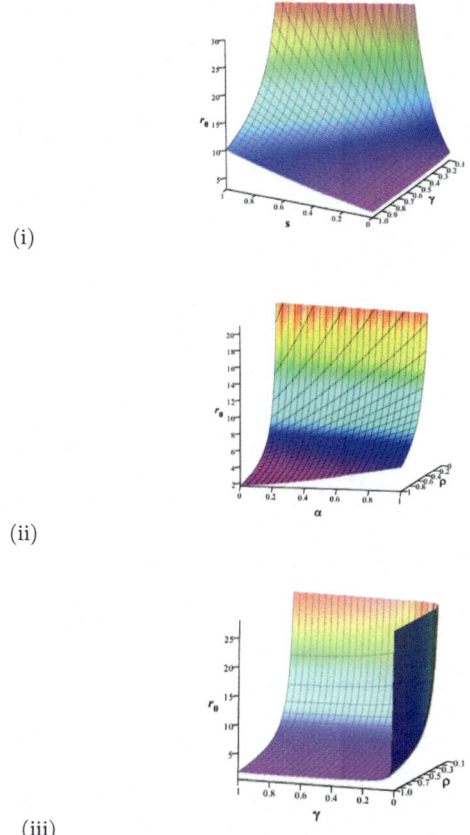

(i)

(ii)

(iii)

Fig. 6. The effect of the simultaneous variation of two parameters on the critical delay, r_0: (i) (β, ρ), (ii) (ρ, γ) and (iii) (γ, s)

periodic oscillations (see, Fig. 3) or divergent oscillations (see, Fig. 4). In summary, the number of the employed persons and the number of the unemployed oscillate around the labor market equilibrium, under the effect of the delay.

4.2   Effect of Parameters on Critical Delay

In this section, we examine the effect of the parameters on the critical delay. First, we vary one parameter and find that the critical value is a monotonic function, see Fig. 5. Next, we vary two parameters simultaneously and find a similar result, Fig. 6.

5   Conclusion

In this document, we have proposed a delayed labor model. We have studied the effect of lagging on job market fluctuations. From this analysis, we concluded that this time lag can destabilize the Model (1) via the Hopf bifurcation phenomenon. Using the Kuznetsov method [11], we have also shown that the proposed model undergo a subcritical Hopf bifurcation and that the bifurcated limit cycles are unstable, in the vicinity of the critical lag. These results can help to control the functioning of the labor market by rationalizing the reproduction process. The difficulties we encountered are essentially related to the non-linearity of our model and the presence of the temporal deviation. In order to develop our conclusions, we plan in our next work to study the effect of two delays.

References

1. Casal, A., Freedman, M.: A Poincaré-Lindstedt approach to bifurcation problems for differential-delay equations. IEEE Trans. Autom. Control **25**(5), 967–973 (1980)
2. Nayfeh, A.H.: Perturbation Methods. Shanghai Publishing House of Science and Technology, Shanghai (1984)
3. Nayfeh, A.H., Chin, C.M., Pratt, J.R.: Perturbation methods in nonlinear dynamics: applications to machining dynamics. ASME J. Manuf. Sci. Eng. **119**(4A), 485–493 (1997)
4. Nayfeh, A.H.: Order reduction of retarded nonlinear systems-the method of multiple scales versus center-manifold. Nonlinear Dyn. **51**, 483–500 (2008)
5. Hassard, B.D., Kazarinoff, N.D., Wan, Y.H.: Theory and Applications of Hopf Bifurcation. Cambridge University Press, Cambridge (1981)
6. Doz, C., Petronevich, A.: Dating Business Cycle Turning Points for the French Economy: A MS-DFM approach. In: Dynamic Factor Models. Advances in Econometrics, vol. 35, pp. 481–538. Emerald Publishing Ltd. (2016)
7. Koopmans, C.T.: The Econometric Approach to Business Fluctuations. The American Economic Review, Papers and Proceedings of the Sixty-first Annual Meeting of the American Economic Association, vol. 39, no. 3, pp. 64–72, May, 1949
8. Franco, D., Liz, E., Torres, P.J.: Existence of periodic solutions for functional equations with periodic delay. Indian J. Pure Appl. Math. **38**(3), 143–152 (2007)
9. Riad, D., Hattaf, K., Yousfi, N.: Dynamics of Capital-labour Model with Hattaf-Yousfi functional response. British J. Math. Comput. Sci. **18**(5), 1–7 (2016)
10. Diebold, F.X., Rudebusch, G.: Measuring business cycles: a modern perspective. Rev. Econ. Stat. **78**(1), 67–77 (1996)
11. Wang, H.P., Li, J., Zhang, K.: Non-resonant response, bifurcation and oscillation suppression of a non-autonomous system with delayed position feedback control. Nonlinear Dyn. **51**, 447–464 (2008)
12. Blatt, J.M.: On the econometric approach to business-cycle analysis. Oxford Economic Papers, New Series, vol. 30, no. 2, pp. 292–300 (1978)
13. Hale, J.K., Verduyn Lunel, S.M.: Introduction to Functional Differential Equations. Springer, New York (1993). https://doi.org/10.1007/978-1-4612-4342-7
14. Tinbergen, J.: Econometric business cycle research. Rev. Econ. Stud. **7**(2), 73–90 (1940)
15. Balázsi, L., Kiss, K.: Cross-diffusion modeling in macroeconomics. Differ. Equ. Dyn. Syst. **23**, 147–166 (2015)
16. Farkas, M.: On the distribution of capital and labour in a closed economy. SEA Bull. Math. **19**(2), 27–36 (1995)

17. Hua, N.: The fixed point theory and the existence of the periodic solution on a nonlinear differential equation. J. Appl. Math. **2018**, 1–11 (2018). Article ID 6725989
18. MacDonald, N.: Harmonic balance in delay-differential equations. J. Sound Vib. **186**(4), 649–656 (1995)
19. Rand, R., Verdugo, A.: Hopf bifurcation formula for first order differential-delay equations Commun. Nonlinear Sci. Numer. Simul. **12**(6), 859–864 (2007)
20. Aly, S.: Spatial Inhomogenity due to Turing instability in a capital-labour market. Appl. Math. **3**(2), 172–176 (2012)
21. Das, S.L., Chatterjee, A.: Multiple scales without center manifold reduction for delay differential equations near Hopf bifurcations. Nonlinear Dyn. **30**, 323–335 (2002)
22. Faria, T., Magalhaes, L.T.: Normal forms for retarded functional differential equations and applications to Bogdanov-Takens singularity. J. Differ. Equ. **122**, 201–224 (1995)
23. Faria, T., Magalhaes, L.T.: Normal forms for retarded functional differential equations with parameters and applications to Hopf bifurcation. J. Differ. Equ. **122**, 181–200 (1995)
24. Kuznetsov, Y.A.: Elements of applied bifurcation theory. J. Appl. Math. Sci. **112** (1998). 2nd edn. Springer, New York, USA
25. Kuang, Y.: Delay Differential Equations with Applications in Population Dynamics. Academic Press, Boston (1993)
26. Wang, Z.H.: An iteration method for calculating the periodic solution of time-delay systems after a Hopf bifurcation. Nonlinear Dyn. **53**, 1–11 (2008)
27. Wang, Z.H., Hu, H.Y.: Pesudo-oscillator analysis of scalar nonlinear time-delay systems near a Hopf bifurcation. Internat. J. Bifur. Chaos. **17**(8), 2805–2814 (2007)

Existence and Uniqueness Results of Fractional Differential Equations with Fuzzy Data

Atimad Harir$^{(\boxtimes)}$, Said Melliani, and Lalla Saadia Chadli

Laboratory of Applied Mathematics and Scientific Computing, Sultan Moulay
Slimane University, Beni Mellal, Morocco
atimad.harir@gmail.com, s.melliani@usms.ma, sa.chadli@yahoo.fr

Abstract. In this paper, we are going to study the existence and uniqueness solutions of fractional differential equations with fuzzy data, involving the fuzzy fractional differential operators of the order $\gamma \in R_+$. The
aid method of successive approximation is provided with adequate conditions for the existence and uniqueness solution. Examples are given to
explain the theory obtained.

1 Introduction

Fractional differential equations (FDEs) is a generalization and integration of
ordinary differential equations into arbitrary non-integer orders. This is com-
monly and effectively used to explain many phenomena that occur in specific
scientific fields and engineering. Indeed, many applications can be found in vis-
coelasticity, electrochemistry, power, porous media, electromagnetic, etc. (see
[1–3]). And they got a lot of attention. For the most recent work on the existence and uniqueness of solutions of initial and boundary value problems for
fractional differential equations, we list [4,5,10,23,24].

Agarwal et al. [6] have taken an initiative to incorporate the idea of a solution
for fuzzy fractional differential equations in order $\gamma > 0$ to get a more practical
model than (FDEs). This contribution has inspired other writers to draw some
results about the solution's existence and uniqueness. (see [8,10,15,16,20–23,25].
In this paper, we will study the fuzzy fractional differential equation

$$\begin{cases} D^\gamma y(t) = F(t, y(t)), \ t \in [0, a], \ \gamma \in R_+, \\ D^j y(t)|_{t=0} = y_j(0) \ , \ j = 0, 1, 2, \dots, \ k = [\gamma], \end{cases} \tag{1}$$

where $F : [0, a] \times R_\mathscr{F} \to R_\mathscr{F}$ is continuous, we shall consider this equation
with some appropriate initial condition for a given equation $y_j(0) \in R_\mathscr{F}$ ($R_\mathscr{F}$ be
the set of fuzzy real numbers [20,21]) and D^γ is the fuzzy fractional differential
operator (γ be a positive real number with $j = [\gamma]$ ($[\gamma]$ is the smallest integer
greater than or equal to γ)). For earlier works concerning the crisp problem 1, the
first author studied it in [11,12] when $F \in [0, a] \times C([0, a]) \to C([0, a])$ ($C([0, a])$

© Springer Nature Switzerland AG 2021
Z. Hammouch et al. (Eds.): SM2A 2019, LNNS 168, pp. 78–90, 2021.
https://doi.org/10.1007/978-3-030-62299-2_6

is the set of continuous functions defined on $[0, a]$) and [13] when $F \in [0, a] \times X \to X$, ($X$ is the Banach space). Here we generalize this work for fuzzy set $\mathrm{R}_{\mathscr{F}}$.

The paper is structured as follows: In Sect. 2, we remember some basic knowledge of fuzzy calculus. Several basic principles and properties of fuzzy fractional calculus are introduced in Sect. 3 and in Sect. 4 we prove some results on the existence and uniqueness of solutions of fuzzy fractional differential equations. We denote to some examples, finally.

2 Preliminaries

We now recall some definitions needed in throughout the paper. Let us denote by $\mathrm{R}_{\mathscr{F}}$ the class of fuzzy subsets of the real axis $y : \mathrm{R} \to [0, 1]$ satisfying the following properties:

(i) y is normal: there exists $x_0 \in \mathrm{R}$ with $y(x_0) = 1$,
(ii) y is convex fuzzy set: for all $x, t \in \mathrm{R}$ and $0 < \lambda \le 1$, it holds that

$$y(\lambda x + (1 - \lambda)t) \ge \min\{y(x), y(t)\},$$

(iii) y is upper semicontinuous: for any $x_0 \in \mathrm{R}$, it holds that

$$y(x_0) \ge \lim_{x \to x_0} y(x) ,$$

(iv) $[y]^0 = cl\{x \in \mathrm{R} | y(x) > 0\}$ is compact.

Then $\mathrm{R}_{\mathscr{F}}$ is called the space of fuzzy numbers see [27]. Obviously, $\mathrm{R} \subset \mathrm{R}_{\mathscr{F}}$. If y is a fuzzy set, we define $[y]^\alpha = \{x \in \mathrm{R} | y(x) \ge \alpha\}$ the α-level (cut) sets of y, with $0 < \alpha \le 1$. Also, if $y \in \mathrm{R}_{\mathscr{F}}$ then α-cut of y denoted by $[y]^\alpha = [y_1^\alpha, y_2^\alpha]$.

Lemma 1. *See ([14]) Let $y, z : \mathrm{R}_{\mathscr{F}} \to [0, 1]$ be the fuzzy sets. Then $y = z$ if and only if $[y]^\alpha = [z]^\alpha$ for all $\alpha \in [0, 1]$.*

For $y, z \in \mathrm{R}_{\mathscr{F}}$ and $\lambda \in \mathrm{R}$ the sum $y + z$ and the product λy are defined by

$$[y + z]^\alpha = [y_1^\alpha + z_1^\alpha, y_2^\alpha + z_2^\alpha],$$

$$[\lambda y]^\alpha = \lambda[y]^\alpha = \begin{cases} [\lambda y_1^\alpha, \lambda y_2^\alpha], & \lambda \ge 0; \\ [\lambda y_2^\alpha, \lambda y_1^\alpha], & \lambda < 0, \end{cases}$$

$\forall \alpha \in [0, 1]$. Additionally if we denote $\hat{0} = \chi_{\{0\}}$, then $\hat{0} \in \mathrm{R}_{\mathscr{F}}$ is a neutral element with respert to $+$.

Let $d : \mathrm{R_F} \times \mathrm{R}_{\mathscr{F}} \to \mathrm{R} + \cup \{0\}$ by the following equation:

$$d(y, z) = \sup_{\alpha \in [0, 1]} d_H([y]^\alpha, \ [z]^\alpha), for \ all y, z \in \mathrm{R}_{\mathscr{F}},$$

where d_H is the Hausdorff metric defined as:

$$d_H([y]^\alpha, \ [z]^\alpha) - \max\{|y_1^\alpha - z_1^\alpha|, \ |y_2^\alpha - z_2^\alpha$$

The following properties are well-known see [26]:

$$d(y + w, z + w) = d(y, z) \quad and \quad d(y, z) = d(z, y), \quad \forall\, y, z, w \in R_{\mathscr{F}},$$
$$d(ky, kz) = |k|d(y, z), \quad \forall k \in R, \; y, z \in R_{\mathscr{F}} \tag{2}$$
$$d(y + z, w + e) \le d(y, w) + d(z, e), \quad \forall\, y, z, w, e \in R_{\mathscr{F}},$$

and $(R_{\mathscr{F}}, d)$ is a complete metric space.

Remark 1. We denote by $C([0, a], R_{\mathscr{F}})$ the space of all continuous fuzzy functions on $[0, a]$ and is a complete metric space with respect to the metric

$$h(y, z) = \sup_{t \in [0,a]} d\big(y(t), z(t)\big).$$

We denote by $L^1([0, a], R_{\mathscr{F}})$ the space of all fuzzy functions $F : [0, a] \to R_{\mathscr{F}}$ which are Lebesgue integrable on the bounded interval $[0, a]$.

Definition 1. The mapping $y : [0, a] \to R_{\mathscr{F}}$ for some interval $[0, a]$ is called a fuzzy process. Therefore, its α-level set can be written as follows:

$$[y(t)]^\alpha = [y_1^\alpha(t), y_2^\alpha(t)], \; t \in [0, a], \; \alpha \in [0, 1].$$

Theorem 1. *[8] Let* $y : [0, a] \to R_{\mathscr{F}}$ *be Seikkala differentiable and denote* $[y(t)]^\alpha = [y_1^\alpha(t), y_2^\alpha(t)]$. *Then, the boundary function* $y_1^\alpha(t)$ *and* $y_2^\alpha(t)$ *are differentiable and*

$$[y'(t)]^\alpha = [(y_1^\alpha)'(t), \; (y_2^\alpha)'(t)], \; \alpha \in [0, 1].$$

Definition 2. [9] *Let* $y : [0, a] \to R_{\mathscr{F}}$. *The fuzzy integral, denoted by* $\int_b^c y(t)dt, b, c \in [0, a]$, *is defined levelwise by the following equation:*

$$\left[\int_b^c y(t)dt \right]^\alpha = \left[\int_b^c y_1^\alpha(t)dt, \; \int_b^c y_2^\alpha(t)dt \right],$$

for all $0 \le \alpha \le 1$. In [9], if $y : [0, a] \to R_{\mathscr{F}}$ is continuous, it is fuzzy integrable.

Theorem 2. *[7] If* $y \in R_{\mathscr{F}}$, *then the following properties hold:*

(i) $[y]^{\alpha_2} \subset [y]^{\alpha_1}$, *if* $0 \le \alpha_1 \le \alpha_2 \le 1$;

(ii) $\{\alpha_k\} \subset [0, 1]$ *is a nondecreasing sequence which converges to* α *then*

$$[y]^\alpha = \bigcap_{k \ge 1} [y]^{\alpha_k}.$$

Conversely if $A_\alpha = \{[y_1^\alpha, y_2^\alpha]; \alpha \in (0, 1]\}$ *is a family of closed real intervals verifying* (i) *and* (ii), *then* $\{A_\alpha\}$ *defined a fuzzy number* $y \in R_{\mathscr{F}}$ *such that* $[y]^\alpha = A_\alpha$.

3   Fuzzy Fractional Integral and Fuzzy Fractional Derivative

Let $\gamma \in R_+$ and $y : [0, a] \to R_{\mathscr{F}}$ be such that $[y(t)]^\alpha = [y_1^\alpha(t), y_2^\alpha(t)]$ for all $t \in [0, a]$. Suppose that $y_1^\alpha, y_2^\alpha \in C([0, a], R) \cap L^1([0, a], R)$ for all $\alpha \in [0, 1]$ and let

$$A_\alpha := \frac{1}{\Gamma(\gamma)} \Big[\int_0^t (t - s)^{\gamma - 1} y_1^\alpha(s) ds, \int_0^t (t - s)^{\gamma - 1} y_2^\alpha(s) ds \Big],$$

$$:= [\Psi_\gamma(t) * y_1^\alpha(t), \ \Psi_\gamma(t) * y_2^\alpha(t)]. \tag{3}$$

Lemma 2. *See ([10]) The family* $\{A_\alpha; \alpha \in [0, 1]\}$ *given by 3, defined a fuzzy number* $y \in R_{\mathscr{F}}$ *such that* $[y]^\alpha = A_\alpha$.

Now for any positive real number $\gamma > 0$, we define

$$\Psi_\gamma(t) = \begin{cases} \frac{t^{\gamma - 1}}{\Gamma(\gamma)}, & t > 0, \\ 0, & t \le 0, \end{cases}$$

and

$$\Psi_{-\gamma}(t) = \Psi_{1+k-\gamma}(t) * \delta^{1+k}(t), \quad k = [\gamma],$$
$$\Psi_{-n}(t) = \delta^n(t), \quad n = 0, 1, 2, \ldots.$$

with the property $\Psi_\gamma(t) * \Psi_p(t) = \Psi_{\gamma+p}(t)$ for $p > 0$, where $\delta^n(t)$ is the nth derivative of the delta function and Γ is the gamma function (for the properties of $\Psi_\gamma(t)$ see [17, 18]).

Definition 3. Let $y \in C([0, a], R_{\mathscr{F}}) \cap L^1([0, a], R_{\mathscr{F}})$. The fuzzy fractional primitive of order $\gamma > 0$ of y, is defined by

$$I^\gamma y(t) = \frac{1}{\Gamma(\gamma)} \int_0^t (t - s)^{\gamma - 1} y(s) ds,$$

by

$$\big[I^\gamma y(t) \big]^\alpha = \frac{1}{\Gamma(\gamma)} \Big[\int_0^t (t - s)^{\gamma - 1} y_1^\alpha(s) ds, \int_0^t (t - s)^{\gamma - 1} y_2^\alpha(s) ds \Big], \tag{4}$$
$$= [y_1^\alpha(t) * \Psi_\gamma(t), y_2^\alpha(t) * \Psi_\gamma(t)],$$

For $\gamma = 1$, we obtain $I^1 y(t) = \int_0^t y(s) ds, \ t \in [0, a]$, that is, the integral operator. Also, Subsequent properties are evident.

(i) $I^\gamma(\lambda y)(t) = \lambda I^\gamma(y)(t)$ for each constant $\lambda \in R_{\mathscr{F}}$,
(ii) $I^\gamma(y + z)(t) = I^\gamma(y)(t) + I^\gamma(z)(t)$.

Proposition 1. *[10] If $y \in C([0,a], R_{\mathscr{F}}) \cap L^1([0,a], R_{\mathscr{F}})$ and $p, \gamma > 0$, then we have*

$$I^p I^\gamma y = I^{p+\gamma} y.$$

Definition 4. Let $y \in C^{1+k}([0,a], R_{\mathscr{F}}) \cap L^1([0,a], R_{\mathscr{F}})$ be a given function such that $[y]^\alpha = [y_1^\alpha, y_2^\alpha]$ for all $t \in [0,a]$ and $\alpha \in [0,1]$ the fuzzy fractional differential operator is defined

$$D^\gamma y(t) = \frac{1}{\Gamma(1+k-\gamma)} \int_0^t (t-s)^{k-\gamma} D^{1+k} y(s) ds, \tag{5}$$

$$= D^{1+k} y(t) * \Psi_{1+k-\gamma}(t),$$

by

$$\left[D^\gamma y(t) \right]^\alpha = \frac{1}{\Gamma(1+k-\gamma)} \left[\int_0^t (t-s)^{k-\gamma} D^{1+k} y_1^\alpha(s) ds, \int_0^t (t-s)^{k-\gamma} D^{1+k} y_2^\alpha(s) ds \right]$$

$$= \left[D^{1+k} y_1^\alpha(t) * \Psi_{1+k-\gamma}(t), D^{1+k} y_2^\alpha(t) * \Psi_{1+k-\gamma}(t) \right].$$

For $k = 0$, we obtain

$$\left[D^\gamma y(t) \right]^\alpha = \frac{1}{\Gamma(1-\gamma)} \left[\int_0^t (t-s)^{-\gamma} \frac{d}{ds} y_1^\alpha(t) ds, \int_0^t (t-s)^{-\gamma} \frac{d}{ds} y_2^\alpha(t) ds \right],$$

provided that the equation defines a fuzzy number $D^\gamma y(t) \in R_{\mathscr{F}}$. In fact $\left[D^\gamma y(t) \right]^\alpha = [D^\gamma y_1^\alpha(t), D^\gamma y_2^\alpha(t)]$ for all $t \in [0,a]$ and $\alpha \in [0,1]$.

4   Existence and Uniqueness of the Fuzzy Solution

We now consider the fuzzy fractional differential equation

$$\begin{cases} D^\gamma y(t) = F(t, y(t)), & t \in [0,a], \\ D^j y(t)|_{t=0} = y_j(0) \in R_{\mathscr{F}}, & j = 0, 1, 2, \ldots, k. \end{cases} \tag{6}$$

where $\gamma \in R_+$ and $F \in [0,a] \times R_{\mathscr{F}} \to R_{\mathscr{F}}$ is a continuous function on $(0,a] \times R_{\mathscr{F}}$. We call $y : [0,a] \to R_{\mathscr{F}}$ a fuzzy solution of 6, if

$$D^\gamma y_1^\alpha(t) = f_1(t, y(t)), \quad D^j y_1^\alpha(t)|_{t=0} = y_{1j}^\alpha(0)$$

$$D^\gamma y_2^\alpha(t) = f_2(t, y(t)), \quad D^j y_1^\alpha(t)|_{t=0} = y_{2j}^\alpha(0) \tag{7}$$

for $t \in [0,a]$ and $0 < \alpha \leq 1$, where

$$\left[F(t,y) \right]^\alpha = \left[f_1(t,y), f_2(t,y) \right]$$

$$= \left[\min\{F(t,x) : x \in [y_1^\alpha, y_2^\alpha]\}, \max\{F(t,x) : x \in [y_1^\alpha, y_2^\alpha]\} \right].$$

If we can solve it (uniquely), we have only to verify that the intervals $[y_1^\alpha(t), y_2^\alpha(t)]$, for all $\alpha \in (0,1]$, define a fuzzy number $y(t) \in R_{\mathscr{F}}$.

Definition 5. A mapping $y : [0, a] \rightarrow R_{\mathscr{F}}$ is a solution to the problem 6 if it is continuous and satisfies the integral equation

$$y(t) = \sum_{j=0}^{k} \frac{t^j}{j!} y_j(0) + \frac{1}{\Gamma(\gamma)} \int_0^t (t-s)^{\gamma-1} F(s, y(s)) ds. \qquad (8)$$

According to the method of successive approximation, let us consider the sequence $\{y_n(t)\}$ such that $y_0 : [0, a] \rightarrow R_{\mathscr{F}}$ be continuous,

$$y_n(t) = \sum_{j=0}^{k} \frac{t^j}{j!} y_j(0) + \frac{1}{\Gamma(\gamma)} \int_0^t (t-s)^{\gamma-1} F(s, y_{n-1}(s)) ds, \qquad (9)$$

where $n = 1, 2, 3, \ldots$.

Now we are proving the following theorem on equivalence.

Theorem 3. *Let* $F : [0, a] \times R_{\mathscr{F}} \rightarrow R_{\mathscr{F}}$ *be continuous on* $[0, a] \times R_{\mathscr{F}}$. *And suppose* $\exists \eta > 0$, *such that*

$$d\big(F(t, y(t)), F(t, z(t))\big) \leq \eta d(y(t), z(t)), \qquad (10)$$

for every $y(t)$, $z(t) \in R_{\mathscr{F}}$, $t \in [0, a]$. *If* $\left| \dfrac{\eta a^\gamma}{\Gamma(\gamma+1)} \right| < 1$ *then the problem 6 has a unique solution* $y(t) \in C\big([0, a], R_{\mathscr{F}}\big)$.

Proof. By using the definition 4, we can write 6 in the form

$$\left[D^{1+k} y_1^\alpha(t) * \Psi_{1+k-\gamma}(t), D^{1+k} y_2^\alpha(t) * \Psi_{1+k-\gamma}(t) \right] = \left[F(t, y) \right]^\alpha,$$

from lemma 1

$$D^{1+k} y_1^\alpha(t) * \Psi_{1+k-\gamma}(t) = F(t, y_1^\alpha, y_2^\alpha),$$
$$D^{1+k} y_2^\alpha(t) * \Psi_{1+k-\gamma}(t) = F(t, y_1^\alpha, y_2^\alpha), \qquad (11)$$

where $\overline{F} = (f_1, f_2)$, operating with the convolution of $\Psi_\gamma(t)$, we get

$$D^{1+k} y_1^\alpha(t) * \Psi_{1+k}(t) = \overline{F}(t, y_1^\alpha(t), y_2^\alpha(t)) * \Psi_\gamma(t),$$
$$D^{1+k} y_2^\alpha(t) * \Psi_{1+k}(t) = \overline{F}(t, y_1^\alpha(t), y_2^\alpha(t)) * \Psi_\gamma(t),$$

and taking into consideration the initial values 6 by choosing $y_n(0) = [y_{1n}^\alpha(0), y_{2n}^\alpha(0)]$ we obtain

$$y(t) = \sum_{j=0}^{k} \frac{t^j}{j!} y_j(0) + \frac{1}{\Gamma(\gamma)} \int_0^t (t-s)^{\gamma-1} \overline{F}(s, y(s)) ds, \qquad (12)$$

where $y(t) = (y_1^\alpha(t), y_2^\alpha(t))$ to 7 for all $\alpha \in [0, 1]$.

We will prove that the intervals $[y_1^\alpha(t), y_2^\alpha(t)]$, for $0 < \alpha \leq 1$, define a fuzzy number. $y(t) \in R_{\mathscr{F}}$ for each $t \geq 0$; Means that y is a fuzzy solution to 6.

The successive approximation $y_0 \in \mathbb{R}_{\mathscr{F}}$,

$$y_n(t) = \sum_{j=0}^{k} \frac{t^j}{j!} y_j(0) + \frac{1}{\Gamma(\gamma)} \int_0^t (t-s)^{\gamma-1} \overline{F}(s, y_{n-1}(s)) ds,$$

where $n = 1, 2, 3..., $. And the integral is the fuzzy integral, define a sequence of fuzzy numbers $y_n(t) \in \mathbb{R}_{\mathscr{F}}$. Let us show that there exists a fuzzy set-valued mapping $y : [0, a] \to \mathbb{R}_{\mathscr{F}}$ such that $d(y_n(t), y(t)) \to 0$ uniformly on $t \in [0, a]$ as $n \to \infty$.

Let $t \in [0, a]$, from 9, it follows that, for $n = 1$

$$y_1(t) = \sum_{j=0}^{k} \frac{t^j}{j!} y_j(0) + \frac{1}{\Gamma(\gamma)} \int_0^t (t-s)^{\gamma-1} \overline{F}(s, y_0(s)) ds, \tag{13}$$

and for $n = 2$ from 9

$$y_2(t) = \sum_{j=0}^{k} \frac{t^j}{j!} y_j(0) + \frac{1}{\Gamma(\gamma)} \int_0^t (t-s)^{\gamma-1} \overline{F}(s, y_1(s)) ds. \tag{14}$$

From 13 and 14, we have

$$d_H\left([y_2(t)]^\alpha, [y_1(t)]^\alpha\right) = d_H\left(\left[\frac{1}{\Gamma(\gamma)} \int_0^t (t-s)^{\gamma-1} F(s, y_1(s)) ds\right]^\alpha, \left[\frac{1}{\Gamma(\gamma)} \int_0^t (t-s)^{\gamma-1} F(s, y_0(s)) ds\right]^\alpha\right)$$
$$\leq \frac{1}{\Gamma(\gamma)} \int_0^t (t-s)^{\gamma-1} d_H\left(\left[F(s, y_1(s))\right]^\alpha, \left[F(s, y_0(s))\right]^\alpha\right) ds, \tag{15}$$

for any $\alpha \in [0, 1]$.

According to the condition 10 and using proprieties 2, we get

$$d(y_2(t), y_1(t)) \leq \frac{\eta}{\Gamma(\gamma)} \int_0^t (t-s)^{\gamma-1} d(y_1(s), y_0(s)) ds$$
$$\leq \frac{\eta}{\Gamma(\gamma)} \int_0^t (t-s)^{\gamma-1} \sup_{s \in [0,a]} d(y_1(s), y_0(s)) ds \tag{16}$$

Now, we can apply 16 to get

$$d(y_2(t), y_1(t)) \leq \frac{\eta a^\gamma}{\Gamma(\gamma+1)} h(y_1, y_0). \tag{17}$$

Starting from 16 and 17, we assume that

$$d(y_n(t), y_{n-1}(t)) \leq \left(\frac{\eta a^\gamma}{\Gamma(\gamma+1)}\right)^{n-1} h(y_1, y_0), \tag{18}$$

and we will show that inequality holds for $d(y_{n+1}(t), y_n(t))$.

Indeed, from 9 and condition 10, so

$$d_H\left([y_{n+1}(t)]^\alpha, [y_n(t)]^\alpha\right) = d_H\left(\left[\frac{1}{\Gamma(\gamma)}\int_0^t (t-s)^{\gamma-1}F(s,y_n(s))ds\right]^\alpha,\right.$$

$$\left.\left[\frac{1}{\Gamma(\gamma)}\int_0^t (t-s)^{\gamma-1}F(s,y_{n-1}(s))ds\right]^\alpha\right)$$

$$\leq \frac{1}{\Gamma(\gamma)}\int_0^t (t-s)^{\gamma-1}d_H\left(\left[F(s,y_n(s))\right]^\alpha,\right.$$

$$\left.\left[f(s,y_{n-1}(s))\right]^\alpha\right)ds, \tag{19}$$

for any $\alpha \in [0,1]$. And by properties 2, we obtain

$$d\left(y_{n+1}(t), y_n(t)\right) \leq \frac{\eta}{\Gamma(\gamma)}\int_0^t (t-s)^{\gamma-1}d(y_n(s), y_{n-1}(s))ds$$

$$\leq \frac{\eta}{\Gamma(\gamma)}\int_0^t (t-s)^{\gamma-1}\left(\frac{\eta a^\gamma}{\Gamma(\gamma+1)}\right)^{n-1}h(y_1, y_0)ds \tag{20}$$

$$\leq \frac{\eta}{\Gamma(\gamma)}\left(\frac{\eta a^\gamma}{\Gamma(\gamma+1)}\right)^{n-1}h(y_1, y_0)\int_0^t (t-s)^{\gamma-1}ds.$$

Considering 18 we have

$$d(y_{n+1}(t), y_n(t)) \leq \left(\frac{\eta a^\gamma}{\Gamma(\gamma+1)}\right)^n h(y_1, y_0). \tag{21}$$

Consequently, inequality 18 holds for n = 1, 2, ... We can also write

$$d\left(y_n(t), y_{n-1}(t)\right) \leq \left(\frac{\eta a^\gamma}{\Gamma(\gamma+1)}\right)^{n-1}h(y_1, y_0). \tag{22}$$

From 22, with to the convergence, it follows that the suite having the general term $\left(\frac{\eta a^\gamma}{\Gamma(\gamma+1)}\right)^{n-1} \to 0$, so $d(y_n(t), y_{n-1}(t)) \to 0$ uniformly on $0 \leq t \leq a$ as $n \to \infty$.

Hence, there exists a fuzzy set-valued mapping $y : [0,a] \to \mathbb{R}_\mathscr{F}$ such that $d\left(y_n(t), y(t)\right) \to 0$ uniformly on $0 \leq t \leq a$ as $n \to \infty$.

From 10 and by 2, we get

$$d\left(F(t, y_n(t)), F(t, y(t))\right) \leq \eta d\left(y_n(t), y(t)\right) \to 0, \tag{23}$$

uniformly on $0 \leq t \leq a$ as $n \to \infty$.

With 23 into account, from 9, we obtain, for $n \to \infty$

$$y(t) = \sum_{j=0}^k \frac{t^j}{j!}y_j(0) + \frac{1}{\Gamma(\gamma)}\int_0^t (t-s)^{\gamma-1}F\left(s, y(s)\right)ds, \tag{24}$$

by the convergence of sequence 9, the end points of $[y_n(t)]^\alpha$ converge to $y_1^\alpha(t)$ and

$y_2^\alpha(t)$ respectively. Therefore at least one continuous solution exists 6. Now, we prove that this solution is unique that, is from

$$z(t) = \sum_{j=0}^{k} \frac{t^j}{j!} y_j(0) + \frac{1}{\Gamma(\gamma)} \int_0^t (t-s)^{\gamma-1} F(s, z(s)) ds, \tag{25}$$

it follows that $d(y(t), z(t)) \equiv 0$ Indeed, from 9 and 25, we obtain

$$d_H \left([z(t)]^\alpha, [y_n(t)]^\alpha \right) = d_H \left(\left[\frac{1}{\Gamma(\gamma)} \int_0^t (t-s)^{\gamma-1} F(s, z(s)) ds \right]^\alpha, \right.$$

$$\left. \left[\frac{1}{\Gamma(\gamma)} \int_0^t (t-s)^{\gamma-1} F(s, y_{n-1}(s)) ds \right]^\alpha \right)$$

$$\leq \frac{1}{\Gamma(\gamma)} \int_0^t (t-s)^{\gamma-1} d_H \left(\left[F(s, z(s)) \right]^\alpha, \left[F(s, y_{n-1}(s)) \right]^\alpha \right) ds, \tag{26}$$

for any $\alpha \in [0, 1]$. And by 2, we obtain

$$d(z(t), y_n(t)) \leq \frac{\eta}{\Gamma(\gamma)} \int_0^t (t-s)^{\gamma-1} d(z(s), y_{n-1}(s)) ds, \quad n = 1, 2, \ldots, \tag{27}$$

but $\sup_{t \in [0,a]} d(z(t), y_0(t)) < \infty$ being a solution of 25. It follows from 27 that

$$d(z(t), y_1(t)) \leq \eta \frac{a^\gamma}{\Gamma(\gamma+1)} h(z, y_0), \quad t \in [0, a]. \tag{28}$$

Assume that

$$d(z(t), y_n(t)) \leq \left(\eta \frac{a^\gamma}{\Gamma(\gamma+1)} \right)^n h(z, y_0), \quad t \in [0, a]. \tag{29}$$

From

$$d(z(t), y_{n+1}(t)) \leq \frac{\eta}{\Gamma(\gamma)} \int_0^t (t-s)^{\gamma-1} d(z(s), y_n(s)) ds, \quad t \in [0, a], \tag{30}$$

and 29, one obtains

$$d(z(t), y_{n+1}(t)) \leq \left(\eta \frac{a^\gamma}{\Gamma(\gamma+1)} \right)^{n+1} h(z, y_0), \quad t \in [0, a]. \tag{31}$$

Consequently, (29) holds for any n, therefore we have

$$d(z(t), y_n(t)) = d(y(t), y_n(t)) \to 0 \tag{32}$$

on $t \in [0, a]$ as $n \to \infty$. This proves the uniqueness of the solution for 6.

Now write $\gamma = 1 + k - p$, $0 < p < 1$, $k = [\gamma]$ and consider the problem

$$\begin{cases} D^{1+k} z(t) = F(t, z(t)), & t \in [0, a], \\ D^j z(t)|_{t=0} = z_j(0) \in R_{\mathscr{F}}, j = 0, 1, 2, \ldots, k. \end{cases} \tag{33}$$

Then, we get the following result.

Theorem 4. *Let $F \in [0, a] \times R_\mathscr{F} \to R_\mathscr{F}$ be continuous on $[0, a] \times R_\mathscr{F}$. and satisfy the Lipschitz condition 10. If $p \to 0$ (i.e $\gamma \to 1 + k$) then the solution of 6 coincides with the solution of 33.*

Proof. Suppose that $y(t)$ is a solution of 6 and $z(t)$ is a solution of 33, then by the equivalence between 6 and the integral Eq. 8 and matching equivalence between 33 and the integral equation

$$z(t) = \sum_{j=0}^{k} \frac{t^j}{j!} y_j(0) + \frac{1}{\Gamma(1+k)} \int_0^t (t-s)^k F(s, z(s)) ds, \tag{34}$$

we have

$$d(y(t), z(t)) \leq d\left(\frac{1}{\Gamma(\gamma)} \int_0^t (t-s)^{\gamma-1} F(s, y(s)) ds, \frac{1}{\Gamma(1+k)} \int_0^t (t-s)^k F(s, z(s)) ds \right)$$

$$= d\left(\frac{1}{\Gamma(\gamma)} \int_0^t (t-s)^{\gamma-1} F(s, y(s)) ds + \frac{1}{\Gamma(\gamma)} \int_0^t (t-s)^{\gamma-1} F(s, z(s)) ds, \right.$$

$$\left. \frac{1}{\Gamma(1+k)} \int_0^t (t-s)^k F(s, z(s)) ds + \frac{1}{\Gamma(\gamma)} \int_0^t (t-s)^{\gamma-1} F(s, z(s)) ds \right)$$

$$\leq \frac{1}{\Gamma(\gamma)} d\left(\int_0^t (t-s)^{\gamma-1} F(s, y(s)) ds, \int_0^t (t-s)^{\gamma-1} F(s, z(s)) ds \right)$$

$$+ \frac{1}{\Gamma(1+k)} d\left(\int_0^t (t-s)^k F(s, z(s)) ds + \frac{\Gamma(1+k)}{\Gamma(\gamma)} \int_0^t (t-s)^{\gamma-1} F(s, z(s)) ds \right)$$

$$\leq \frac{1}{\Gamma(\gamma)} \int_0^t (t-s)^{\gamma-1} d(F(s, y(s)), F(s, z(s))) ds$$

$$+ \frac{1}{\Gamma(1+k)} \int_0^t (t-s)^k (d(F(s, z(s)), \hat{0}) |1 - \frac{\Gamma(1+k)}{\Gamma(\gamma)} (t-s)^{\gamma-1-k}|) ds.$$

Therefore

$$h(y, z) \leq \frac{\eta a^\gamma}{\Gamma(\gamma+1)} h(y, z) + \kappa_p, \tag{35}$$

and hence

$$h(y, z) \leq \frac{\kappa_p}{\left(1 - \frac{\eta a^\gamma}{\Gamma(\gamma+1)} \right)}, \tag{36}$$

where

$$\kappa_p = \frac{1}{\Gamma(1+k)} \int_0^t (t-s)^k \left(d(F(s, z(s)), 0) |1 - \frac{\Gamma(1+k)}{\Gamma(\gamma)} (t-s)^p| \right) ds. \tag{37}$$

Now, since

$$(t-s)^k d(F(s, z(s)), \hat{0}) |1 - \frac{\Gamma(1+k)}{\Gamma(1+k-p)} (t-s)^p| \leq (t-s)^k d(F(s, z(s)), \hat{0})$$

and

$$(t-s)^k d\left(F(s, z(s)), \hat{0} \right) |1 - \frac{\Gamma(1+k)}{\Gamma(1+k-p)} (t-s)^p| \to 0 \text{ as } p \to 0,$$

It follows from a theorem dominated by Lebesgue [19] that $\kappa_p \to 0$ as $p \to 0$ which proves that $h(y,z) \to 0$ as $\gamma \to 1+k$

$$\lim_{\gamma \to 1+k} y(t) = z(t) \ in \ C\Big([0,a], \mathrm{R}_{\mathscr{F}}\Big).$$

Remark 2. If the assumptions of Theorem 3 are satisfied, then

$$\lim_{\gamma \to 1+k} D^\gamma y(t) = D^{1+k} z(t) \ in \ C\Big([0,a], \mathrm{R}_{\mathscr{F}}\Big).$$

From 6 and 33 we have

$$\lim_{\gamma \to 1+k} d(D^\gamma y(t), D^{1+k} z(t)) = \lim_{\gamma \to 1+k} d\Big(F(t,y(t)), F(t,z(t))\Big)$$
$$\leq \eta \lim_{\gamma \to 1+k} d\big(y(t), z(t)\big),$$

then

$$\lim_{\gamma \to 1+k} h(D^\gamma y, D^{1+k} z) \leq \eta \lim_{\gamma \to 1+k} h(y,z) = 0$$

which proves the result.

5   Examples

In order to illustrate the previous results, we give here two examples.

Example 1. Let $t \in [0,a]$, so the function $F(t,y(t)) = t + y(t)$ is continuous on $[0,a] \times \mathrm{R}_{\mathscr{F}}$ and Lipschitzians

$$d\Big(F(t,y(t)), F(t,z(t))\Big) \leq d(y(t), z(t)),$$

for all $y, z \in \mathrm{R}_{\mathscr{F}}$ and $t \in [0,a]$ it follows that
$d(F(t,y(t)), F(t,z(t))) \leq \eta d(y(t), z(t))$, *with* $\eta = 1$
Hence, we can apply our theorems to the initial value problem

$$\begin{cases} D^\gamma y(t) = t + y(t), & t \in [0,a], \ \gamma \in \mathrm{R}_+, \\ D^j y(t)|_{t=0} = y_j(0) \in \mathrm{R}_{\mathscr{F}}, & j = 0,1,2\ldots, k. \end{cases}$$

Example 2. Let $t \in [0,a]$, then the function $F(t,z(t)) = z^2(t)$ is continuous on $[0,a] \times \mathrm{R}_{\mathscr{F}}$ and Lipschitzians

$$\begin{aligned}
d_H\Big([F(t,z(t))]^\alpha, [F(t,y(t))]^\alpha\Big) &= d_H\left(F(t, [z(t)]^\alpha), F(t, [y(t)]^\alpha)\right) \\
&= d_H\Big([z^2(t)]^\alpha, [y^2(t)]^\alpha\Big) \\
&= \max\Big\{\Big|\big(z_1^\alpha(t)\big)^2 - (y_1^\alpha(t))^2\Big|, \ \Big|\big(z_2^\alpha(t)\big)^2 - (y_2^\alpha(t))^2\Big|\Big\} \\
&\leq \max\Big\{\Big|\big(z_1^\alpha(t) - y_1^\alpha(t)\big)\big(z_1^\alpha(t) + y_1^\alpha(t)\big)\Big|, \\
&\qquad \Big|\big(z_2^\alpha(t) - y_2^\alpha(t)\big)\big(z_2^\alpha(t) + y_2^\alpha(t)\big)\Big|\Big\} \\
&\leq \max\Big\{\Big|\big(z_2^\alpha(t) + y_2^\alpha(t)\big)\Big|\big(\big|(z_1^\alpha(t) - y_1^\alpha(t))\big|, \ \big|(z_2^\alpha(t) - y_2^\alpha(t))\big|\big)\Big\} \\
&\leq \Big|\big(z_2^\alpha(t) + y_2^\alpha(t)\big)\Big| d_H\Big([z(t)]^\alpha, [y(t)]^\alpha\Big)
\end{aligned}$$

$$d\Big(F(t, z(t)), F(t, y(t))\Big) \leq \sup_{\alpha \in [0,1]} \Big|\Big(z_2^\alpha(t) + y_2^\alpha(t)\Big)\Big| d(z(t), y(t)),$$

for all $z, y \in \mathbb{R}_{\mathscr{F}}$ and $t \in [0, a]$ it follows that $d\big(F(t, z(t)), F(t, y(t))\big) \leq \eta d\big(z(t), y(t)\big)$

with $\eta = \sup\limits_{\alpha \in [0,1]} \big|(z_2^\alpha(t) + y_2^\alpha(t))\big|$.

Hence, using our results to the initial value problem

$$\begin{cases} D^\gamma z(t) = z^2(t), & t \in [0, a], \ \gamma \in \mathbb{R}_+, \\ D^j z(t)|_{t=0} = z_j(0) \in \mathbb{R}_{\mathscr{F}}, & j = 0, 1, 2, \ldots, k. \end{cases}$$

References

1. Lin, W.: Global existence theory and chaos control of fractional differential equations. J. Math. Anal. Appl. **332**, 709–726 (2007)
2. Kilbas, A.A., Srivastava, H.M., Trujillo, J.J.: Theory, and Applications of Fractional Differential Equations. Elsevier, Amsterdam (2006)
3. Diethelm, K., Freed, A.D.: On the solution of nonlinear fractional order differential equations used in the modelling of viscoplasticity. In: Keil, F., Mackens, W., Voss, H. (eds.) Scientific Computing in Chemical Engineering II-Computational Fluid Dynamics and Molecular Properties, pp. 217–224. SpringerVerlag, Heidelberg (1999)
4. Lakshmikantham, V.: Theory of fractional functional differential equations. Nonlinear Anal. **69**, 3337–3343 (2008)
5. Lakshmikantham, V., Vatsala, A.S.: Basic theory of fractional differential equations. Nonlinear Anal. **69**, 2677–2682 (2008)
6. Agarwal, R.P., Lakshmikantham, V., Nieto, J.J.: On the concept of solution for fractional differential equations with uncertainty. Nonlinear Anal. Theor. Meth. Appl. **72**(6), 2859–2862 (2010)
7. Negoita, C.V., Ralescu, D.A.: Applications of Fuzzy Sets to System Analysis. Birkhauser, Basel (1975)
8. Kaleva, O.: A note on fuzzy differential equations. Nonlinear Anal. **64**, 895–900 (2006)
9. Song, S., Guo, L., Feng, C.: Global existence of solutions to fuzzy differential equations. Fuzzy Set. Syst. **115**, 371–376 (2000)
10. Harir, A., Melliani, S., Chadli, L.S.: Fuzzy fractional evolution equations and fuzzy solution operators. Adv. Fuzzy Syst. **2019**(5734190), 10 (2019). https://doi.org/10.1155/2019/5734190
11. EI-Sayed, A.M.A.: Fractional differential equations. Kyungpook Math. J. **28**(2), 119–122 (1988)
12. EI-Sayed, A.M.A.: On the fractional differential equations: On the fractional differential equations. Appl. Math. Comput. **49**, 2–3 (1992)
13. El-Sayed, A.M.A., Ibrahim, A.G.: Multivalued fractional differential equations. Appl. Math. Comput. **68**, 15–25 (1995)
14. Goo, H.Y., Park, J.S.: On the continuity of the Zadeh extensions. J. Chungcheong Math. Soc. **20**(4), 525–533 (2007)
15. Chadli, L.S., Harir, A., Melliani, S.: Solutions of fuzzy heat-like equations by variational iterative method. Ann. Fuzzy Math. Inform. **10**(1), 29–44 (2015)

16. Harir, A., Melliani, S., Chadli, L.S.: Fuzzy generalized conformable fractional derivative. Adv. Fuzzy Syst. **2020**(954975), 7 (2019). https://doi.org/10.1155/2020/1954975
17. Gelfand, I.M., Shilvoe, G.E.: Generalized Functions, vol. 1, Moscow (1958)
18. Shilove, G.E.: Generalized functions and partial differential equations, in Mathematics and its Applications, Science Publishers, Inc. (1968)
19. Curtain, R.F., Priehard, A.J.: Functional Analysis in Modern Applied Mathematics. Academic Press, Cambridge (1977)
20. Diamond, P., Kloeden, P.E.: Metric Spaces of Fuzzy Sets. World Scientific, Singapore (1994)
21. Lakshmikantham, V., Mohapatra, R.N.: Theory of Fuzzy Differential Equations and Applications. Taylor and Francis, London (2003)
22. Arshad, S., Lupulescu, V.: Fractional differential equation with the fuzzy initial condition. Electron. J. Differ. Equ. **2011**(34), 1–8 (2011)
23. Lupulescu, V., O'Regan, D.: A new derivative concept for set-valued and fuzzy-valued functions. differential and integral calculus in quasilinear metric spaces. Fuzzy Sets Syst. (2020)
24. Shuqin, Z.: Monotone iterative method for initial value problem involving Riemann-Liouville fractional derivatives. Nonlinear Anal. **71**, 2087–2093 (2009)
25. Bede, B., Gal, S.G.: Generalizations of the differentiability of fuzzy-number-valued functions with applications to fuzzy differential equations. Fuzzy Sets Syst. **151**, 581–599 (2005)
26. Wu, C., Gong, Z.: On Henstock integral of fuzzy-number-valued functions I. Fuzzy Sets Syst. **120**, 523–532 (2001)
27. Dubois, D., Prade, H.: Fuzzy numbers: an overview. In: Bezdek, J. (ed.) Analysis of Fuzzy Information, pp. 112–148. CRC Press, Boca Raton (1987)

Approximate Efficient Solutions of Nonsmooth Vector Optimization Problems via Approximate Vector Variational Inequalities

Mohsine Jennane[1] and El Mostafa Kalmoun[2(✉)]

[1] Department of Mathematics, Dhar El Mehrez,
Sidi Mohamed Ben Abdellah University, Fes, Morocco
mohsine.jennane@usmba.ac.ma
[2] Department of Mathematics, Statistics and Physics,
College of Arts and Sciences, Qatar University, Doha, Qatar
ekalmoun@qu.edu.qa

Abstract. In this work, we demonstrate the connection between the solutions of approximate vector variational inequalities and approximate efficient solutions of corresponding nonsmooth vector optimization problems via generalized approximate invex functions. The underlying variational inequalities are stated under the Clarke's generalized Jacobian.

1   Introduction

Various significant applications in engineering and economics can only be stated as a multiobjective optimization problem [1]. Nowadays, the connection of these problems to vector variational inequalities is well-established for differentiable convex functions [2]. In particular, results in this direction were developed under various assumptions of generalized convexity [3–7] and nonsmooth invexity [8–11]. On the other hand, relationships between a vector variational inequality and a nonsmooth vector optimization problem (NVOP) were established under the generalized approximate convexity assumption [12–14].

This paper is devoted to the case of NVOP involving generalized approximate invex multiobjective functions, which we have introduced in [15]. Our aim is to use approximate vector variational inequalities (AVVIs) of Stampacchia and Minty type in terms of Clarke's generalized Jacobian to characterize approximate efficient solutions. It is worth mentioning that, as generalized approximate invexity is an extension of generalized approximate convexity, the results obtained in our work are improvements and generalizations of the main results in [14].

The paper is organized as follows: in Sect. 2, we give some preliminary definitions, notation, and auxiliary results. In Sect. 3, we introduce the concept of approximate efficiency for NVOPs, and derive their relationships to AVVIs using the assumption of approximate invex functions. In Sect. 4, we give an example to illustrate our main results. Finally, we conclude our paper in Sect. 5.

© Springer Nature Switzerland AG 2021
Z. Hammouch et al. (Eds.): SM2A 2019, LNNS 168, pp. 91–101, 2021.
https://doi.org/10.1007/978-3-030-62299-2_7

2   Preliminaries

Let $\mathbb{R}^n$ be the n-dimensional Euclidean spaces, $S \subseteq \mathbb{R}^n$ be a given nonempty set and $C \subseteq \mathbb{R}^m$ be a solid pointed convex cone. We use the following partial ordering relations:

$$u \geq_C v \Leftrightarrow u - v \in C;$$

$$u >_C v \Leftrightarrow u - v \in int C.$$

Definition 1 ([16]). Let $F : S \rightarrow \mathbb{R}^m$ be a vector-valued function. F is locally Lipschitz if for each $w \in S$ there is $k > 0$ and $\rho > 0$ such that, for all $u, v \in B(w; \rho)$

$$\|F(u) - F(v)\| \leq k\|u - v\|.$$

Throughout this paper, we let $F := (F_1, ..., F_m) : S \rightarrow \mathbb{R}^m$ be a locally lipschitz function, $\theta : S \times S \rightarrow \mathbb{R}^n$ be a mapping and $\tau >_C 0$ be a vector.

Definition 2 ([16]). The Clarke's generalized Jacobian of F at $u \in S$ is given by

$$\partial F(u) = conv\{ \lim_{i \rightarrow +\infty} JF(u^{(i)}) : u^{(i)} \rightarrow u, u^{(i)} \in D\},$$

where $conv$ denotes the convex hull, $JF(u^{(i)})$ indicates the Jacobian of F at $u^{(i)}$, and D is the differentiability set of F.

We note that the Clarke's generalized Jacobian is not equal to the cartesian product of the components' Clarke subdifferentials. Nevertheless, one has

$$\partial F(u) \subseteq \partial F_1(u) \times ... \times \partial F_m(u).$$

Note also that $\partial(-F)(u) = -\partial F(u)$.

We recall some definitions given in [15] which are a generalization of the concepts of generalized approximate convexity provided in [12,14,17].

Definition 3. F is called approximate $(\theta, \tau)-$invex ($A(\theta, \tau)I$) at $w \in S$, if there is $\rho > 0$ satisfying

$$F(u) - F(v) \geq_C A_v \theta(u, v) - \tau\|\theta(u, v)\|, \quad \text{for each} \quad u, v \in B(w, \rho), A_v \in \partial F(v).$$

If F is $A(\theta, \tau)I$ at each $w \in S$, we say that F is $A(\theta, \tau)I$ on S.

Taking $\theta(u, v) = u - v$, approximate invexity reduces to approximate convexity [18]. The counter-example given in [15, Example 2.2] shows that approximate invexity is still more general.

Definition 4. • F is approximate pseudo $(\theta, \tau)-$invex of type 1 ($AP(\theta, \tau)I$-1) at $w \in S$ if there is $\rho > 0$ such that, whenever $u, v \in B(w, \rho)$ and if

$$F(u) - F(v) <_C -\tau\|\theta(u, v)\|,$$

then

$$A_v \theta(u, v) <_C 0 \text{ for each } A_v \in \partial F(v).$$

- F is approximate pseudo $(\theta, \tau)-$invex of type 2 (AP(θ, τ)I-2) at $w \in S$ if there is $\rho > 0$ such that, whenever $u, v \in B(w, \rho)$ and if

$$F(u) - F(v) <_C 0,$$

then

$$A_v \theta(u, v) + \tau \|\theta(u, v)\| <_C 0 \text{ for all } A_v \in \partial F(v).$$

Proposition 1. *If F is AP(θ, τ)I-2 at $w \in S$, then F is AP(θ, τ)I-1 at w.*

Proof. Assume that there is $\overline{\rho} > 0$ satisfying for each $u, v \in B(w, \overline{\rho})$

$$F(u) - F(v) <_C -\tau \|\theta(u, v)\|,$$

then

$$F(u) - F(v) <_C 0.$$

Since F is AP(θ, τ)I-2 at w, then there is $\rho > 0$, $\rho < \overline{\rho}$, satisfying for each $u, v \in B(w, \rho)$

$$A_v \theta(u, v) + \tau \|\theta(u, v)\| <_C 0 \text{ for each } A_v \in \partial F(v),$$

which further implies that

$$A_v \theta(u, v) <_C 0 \text{ for each } A_v \in \partial F(v).$$

Hence F is AP(θ, τ)I-1 at $w \in S$.

Definition 5. • F is approximate quasi $(\theta, \tau)-$invex of type 1 (AQ(θ, τ)I-1) at $w \in S$ if there is $\rho > 0$ such that for each $u, v \in B(w, \rho)$

$$A_v \theta(u, v) - \tau \|\theta(u, v)\| >_C 0, \quad \text{for some } A_v \in \partial F(v),$$

implies

$$F(u) >_C F(v).$$

- F is approximate quasi $(\theta, \tau)-$invex of type 2 (AQ(θ, τ)I-2) at $w \in S$ if there is $\rho > 0$ such that, for each $u, v \in B(w, \rho)$

$$A_v \theta(u, v) >_C 0, \quad \text{for some } A_v \in \partial F(v),$$

implies

$$F(u) - F(v) >_C \tau \|\theta(u, v)\|.$$

The next proposition can be easily proven.

Proposition 2. *If F is AQ(θ, τ)I-2 at $v \in S$, then F is AQ(θ, τ)I-1 at v.*

Remark 1. • A(θ, τ)I $\Rightarrow \left[\text{AP}(\theta, \tau)\text{I-1 and AQ}(\theta, \tau)\text{I-1} \right]$.

- There is no relation between AP(θ, τ)I-2 and AQ(θ, τ)I-2 and A(θ, τ)I (for examples, see [14]).

Now, we consider the following NVOP:

$$(NVOP) \quad \min F(u) := (F_1(u), \cdots, F_m(u)) \text{ subject to } u \in S,$$

where each $F_i : S \to \mathbb{R}$ are real-valued functions for any $i \in \{1, \cdots, m\}$.

Definition 6. Let $\zeta \in S$.

(i) ζ is an efficient solution of (NVOP) iff there is no vector $u \in S$ such that

$$F(u) \leq_C F(\zeta).$$

(ii) ζ is an τ-approximate efficient solution (τ-AES) of (NVOP) iff there is no $\rho > 0$ such that, for each $u \in B(\zeta; \rho) \setminus \{\zeta\}$

$$F(u) - F(\zeta) \leq_C -\tau \|\theta(u, \zeta)\|.$$

3   Relationships Between NVOP and AVVI

Consider the following AVVI of Stampacchia and Minty type in terms of Clarke subdifferentials as follows:

(ASVVI). To find $\zeta \in S$ such that, there is no $\rho > 0$ satisfying for each $u \in B(\zeta, \rho)$ and $A_\zeta \in \partial F(\zeta)$

$$A_\zeta \theta(u, \zeta) \leq_C -\tau \|\theta(u, \zeta)\|.$$

(AMVVI). To find $\zeta \in S$ such that, there is no $\rho > 0$ satisfying for each $u \in B(\zeta, \rho)$ and $A_u \in \partial F(u)$

$$A_u \theta(u, \zeta) \leq_C -\tau \|\theta(u, \zeta)\|.$$

The following theorems describe relations between AVVI and NVOP.

Theorem 1. Let F be A(θ, τ)I at $\zeta \in S$. If ζ solves (ASVVI) w.r.t. τ, then ζ is a 2τ-AES of (NVOP).

Proof. Assume ζ fails to be a 2τ-AES of (NVOP). It means that there is $\overline{\rho} > 0$ satisfying for each $u \in B(\zeta, \overline{\rho})$

$$F(u) - F(\zeta) \leq_C -2\tau \|\theta(u, \zeta)\|. \tag{1}$$

As F is A(θ, τ)I at ζ, it follows that there is $\widetilde{\rho} > 0$, satisfying

$$F(u) - F(\zeta) \geq_C A_\zeta \theta(u, \zeta) - \tau \|\theta(u, \zeta)\| \quad \forall\, u \in B(\zeta, \widetilde{\rho}), \; A_\zeta \in \partial F(\zeta).$$

By using (1) and the definition of approximate $(\theta, \tau)-$ invexity, and by taking $\rho := min(\overline{\rho}, \widetilde{\rho})$, we obtain

$$A_\zeta \theta(u, \zeta) - \tau\|\theta(u, \zeta)\| \leq_C -2\tau\|\theta(u, \zeta)\|.$$

Hence

$$A_\zeta \theta(u, \zeta) \leq_C -\tau\|\theta(u, \zeta)\|.$$

This means ζ does not solve (ASVVI) w.r.t τ.

Theorem 2. *Let* $-F$ *be* $A(\theta, \tau)I$ *at* $\zeta \in S$. *If* $\zeta \in S$ *is a* τ-*AES for (NVOP), then* ζ *solves (ASVVI) w.r.t* 2τ.

Proof. Assume ζ fails to be a solution of (ASVVI) w.r.t 2τ. It means that there is $\overline{\rho} > 0$ such that, for each $u \in B(\zeta, \overline{\rho}), A_\zeta \in \partial F(\zeta)$, we have

$$A_\zeta \theta(u, \zeta) \leq_C -2\tau\|\theta(u, \zeta)\|.$$

Then

$$-A_\zeta \theta(u, \zeta) \geq_C 2\tau\|\theta(u, \zeta)\|. \tag{2}$$

By $\partial(-F)(\zeta) = -\partial F(\zeta)$ we deduce that $-A_\zeta \in \partial(-F)(\zeta)$.
As $-F$ is $A(\theta, \tau)I$ at ζ, it yields that there is $\widetilde{\rho} > 0$ satisfying

$$(-F)(u) - (-F)(\zeta) \geq_C -A_\zeta \theta(u, \zeta) - \tau\|\theta(u, \zeta)\| \qquad \forall u \in B(\zeta, \widetilde{\rho}).$$

By using (2) and by taking $\rho := min(\overline{\rho}, \widetilde{\rho})$, we obtain

$$-F(u) + F(\zeta) + \tau\|\theta(u, \zeta)\| \geq_C -A_\zeta \theta(u, \zeta) \geq_C 2\tau\|\theta(u, \zeta)\| \qquad \forall u \in B(\zeta, \rho) \setminus \{\zeta\},$$

which implies

$$F(u) - F(\zeta) \leq_C -\tau\|\theta(u, \zeta)\|.$$

Therefore ζ cannot be a τ-AES of (NVOP).

Theorem 3. *Let* F *be* $A(\theta, \tau)I$ *at* $\zeta \in S$ *and* $\theta(u, \zeta) + \theta(\zeta, u) = 0$ *for any* $u \in S$. *If* ζ *solves (AMVVI) w.r.t* τ, *then* ζ *is a* 2τ-*AES of (NVOP).*

Proof. Assume ζ fails to be a 2τ-AES of (NVOP). It means that there is $\overline{\rho} > 0$ satisfying for each $u \in B(\zeta, \overline{\rho})$

$$F(u) - F(\zeta) \leq_C -2\tau\|\theta(u, \zeta)\|. \tag{3}$$

As $-F$ is $A(\theta, \tau)I$ at ζ, it yields that there is $\widetilde{\rho} > 0$ satisfying

$$(-F)(\zeta) - (-F)(u) \geq_C A_v \theta(\zeta, u) - \tau\|\theta(\zeta, u)\| \qquad \forall u \in B(\zeta, \widetilde{\rho}), A_v \in \partial(-F)(u),$$

then

$$F(u) - F(\zeta) \geq_C A_v \theta(\zeta, u) - \tau\|\theta(\zeta, u)\|.$$

By using (3) and by taking $\rho := min(\overline{\rho}, \widetilde{\rho})$, we obtain

$$A_v\theta(\zeta, u) - \tau\|\theta(\zeta, u)\| \leq_C -2\tau\|\theta(u, \zeta)\| \qquad \forall u \in B(\zeta, \rho) \setminus \{\zeta\}.$$

From $\partial(-F)(u) = -\partial F(u)$, there is $A_u = -A_v \in \partial F(u)$. Consequently, using $\theta(u, \zeta) + \theta(\zeta, u) = 0$ together with the above inequality, we deduce

$$A_u\theta(u, \zeta) \leq_C -\tau\|\theta(u, \zeta)\|.$$

This means ζ does not solve (AMVVI) w.r.t τ.

Theorem 4. *Let* $-F$ *be* $A(\theta, \tau)I$ *at* $\zeta \in S$ *and* $\theta(u, \zeta) + \theta(\zeta, u) = 0$ *for all* $u \in S$. *If* $\zeta \in S$ *is a* τ-*AES for (NVOP), then* ζ *solves (AMVVI) w.r.t* 2τ.

Proof. Assume ζ fails to be a solution of (AMVVI) w.r.t 2τ. Thus, there is $\overline{\rho} > 0$ satisfying for any $u \in B(\zeta, \overline{\rho})$, $A_u \in \partial F(u)$

$$A_u\theta(u, \zeta) \leq_C -2\tau\|\theta(u, \zeta)\|. \tag{4}$$

As F is $A(\theta, \tau)I$ at ζ, it yields that there is $\widetilde{\rho} > 0$, such that

$$F(\zeta) - F(u) \geq_C A_u\theta(\zeta, u) - \tau\|\theta(\zeta, u)\| \quad \forall u \in B(\zeta, \widetilde{\rho}), A_u \in \partial F(u).$$

Since $\theta(\zeta, u) = -\theta(u, \zeta)$, then

$$F(u) - F(\zeta) - \tau\|\theta(u, \zeta)\| \leq_C A_u\theta(\zeta, u).$$

By using (3) and by taking $\rho := min(\overline{\rho}, \widetilde{\rho})$, we obtain

$$F(u) - F(\zeta) \leq_C -\tau\|\theta(u, \zeta)\|.$$

We conclude that ζ cannot be a τ-AES of (NVOP).

Theorem 5. *Let* F *be* $AP(\theta, \tau)I$-2 *at* $\zeta \in S$. *If* ζ *solves (ASVVI) w.r.t.* τ, *then* ζ *is a* τ-*AES of (NVOP).*

Proof. Assume ζ fails to be a τ-AES of (NVOP). It means that there is $\overline{\rho} > 0$ satisfying for all $u \in B(\zeta, \overline{\rho})$

$$F(u) - F(\zeta) \leq_C -\tau\|\theta(u, \zeta)\| <_C 0. \tag{5}$$

As F is $AP(\theta, \tau)I$-2 at ζ, it yields that there is $\widetilde{\rho} > 0$, such that, whenever $u \in B(\zeta, \widetilde{\rho})$

$$F(u) - F(\zeta) <_C 0 \Rightarrow A_\zeta\theta(u, \zeta) <_C -\tau\|\theta(u, \zeta)\|, \quad \forall A_\zeta \in \partial F(\zeta).$$

By using (5) and the definition of approximate quasi (θ, τ)-invexity type 2, and by taking $\rho := min(\overline{\rho}, \widetilde{\rho})$, we obtain

$$A_\zeta\theta(u, \zeta) \leq_C -\tau\|\theta(u, \zeta)\|.$$

This means ζ does not solve (ASVVI) w.r.t. τ.

Theorem 6. *Let $-F$ be $AQ(\theta, \tau)I$-2 at $\zeta \in S$. If ζ is a τ-AES of (NVOP), then ζ solves (ASVVI) w.r.t. τ.*

Proof. Assume ζ fails to be a solution of (ASVVI) w.r.t. τ, then, there is $\rho > 0$ satisfying for each $A_\zeta \in \partial F(\zeta)$ and $u \in B(\zeta, \overline{\rho})$

$$A_\zeta \theta(u, \zeta) \leq_C -\tau \|\theta(u, \zeta)\|.$$

Then

$$-A_\zeta \theta(u, \zeta) \geq_C \tau \|\theta(u, \zeta)\| >_C 0. \tag{6}$$

As $\partial(-F)(\zeta) = -\partial F(\zeta)$ it yields that $-A_\zeta \in \partial(-F)(\zeta)$.
Since $-F$ is $AQ(\theta, \tau)I$-2 at ζ, it follows that there is $\widetilde{\rho} > 0$ such that, whenever $u \in B(\zeta, \widetilde{\rho})$

$$-A_\zeta \theta(u, \zeta) >_C 0 \Rightarrow -F(u) - (-F(\zeta)) >_C \tau \|\theta(u, \zeta)\|.$$

By using (6) and the definition of approximate pseudo (θ, τ)–invexity type 2, and by taking $\rho := \min(\overline{\rho}, \widetilde{\rho})$, we get

$$F(u) - F(\zeta) \leq_C -\tau \|\theta(u, \zeta)\|.$$

Consequently ζ cannot be a τ-AES of (NVOP).

The following corollary can be deduced from Theorems 5 and 6.

Corollary 1. *Let F be $AP(\theta, \tau)I$-2 at $\zeta \in S$ and $-F$ be $AQ(\theta, \tau)I$-2 at ζ. ζ is a τ-AES of (NVOP) if and only if ζ solves (ASVVI) w.r.t. τ.*

Theorem 7. *Let F be $AQ(\theta, \tau)I$-2 at ζ and $\theta(u, \zeta) + \theta(\zeta, u) = 0$, $\forall u \in S$. If ζ is a τ-AES of (NVOP), then ζ solves (AMVVI) w.r.t. τ.*

Proof. Assume ζ fails to be a solution of (AMVVI) w.r.t. τ. Then, there is $\overline{\rho} > 0$ satisfying for each $A_u \in \partial F(u)$ and $u \in B(\zeta, \overline{\rho})$

$$A_u \theta(u, \zeta) \leq_C -\tau \|\theta(u, \zeta)\|.$$

From $\theta(u, \zeta) + \theta(\zeta, u) = 0$, we obtain

$$A_u \theta(\zeta, u) \geq_C \tau \|\theta(\zeta, u)\| >_C 0. \tag{7}$$

As F is $AQ(\theta, \tau)I$-2 at ζ, it yields that, there is $\widetilde{\rho} > 0$ such that, whenever $u \in B(\zeta, \widetilde{\rho})$

$$A_u \theta(\zeta, u) >_C 0 \Rightarrow F(\zeta) - F(u) >_C \tau \|\theta(\zeta, u)\|.$$

By using (7) and the definition of approximate quasi (θ, τ)–invexity type 2, and by taking $\rho := \min(\overline{\rho}, \widetilde{\rho})$, we deduce

$$F(u) - F(\zeta) \leq_C -\tau \|\theta(u, \zeta)\|.$$

This means that ζ is not a τ-AES of (NVOP).

Theorem 8. *Let* $-F$ *be* $AP(\theta,\tau)I$-2 *at* ζ *and* $\theta(u,\zeta) + \theta(\zeta,u) = 0$, $\forall u \in S$. *If* $\zeta \in S$ *solves (AMVVI) w.r.t.* τ, *then* ζ *is a* τ-*AES of (NVOP)*.

Proof. Assume ζ fails to be a τ-AES of (NVOP). It means that there is $\overline{\rho} > 0$ satisfying for any $u \in B(\zeta,\overline{\rho})$

$$F(u) - F(\zeta) \leq_C -\tau\|\theta(u,\zeta)\|.$$

Thus

$$-F(\zeta) - (-F)(u) \leq_C -\tau\|\theta(u,\zeta)\| <_C 0. \tag{8}$$

As $-F$ is $AP(\theta,\tau)I$-2 at ζ, it yields that there is $\widetilde{\rho} > 0$, such that, whenever $u \in B(\zeta,\widetilde{\rho})$

$$-F(\zeta) - (-F)(u) <_C 0 \Rightarrow A_v\theta(u,\zeta) <_C -\tau\|\theta(u,\zeta)\|, \quad \forall A_v \in \partial(-F)(u).$$

By using (8) and the definition of approximate pseudo (θ,τ)−invexity type 2, and by taking $\rho := \min(\overline{\rho},\widetilde{\rho})$, we obtain

$$A_v\theta(u,\zeta) \leq_C -\tau\|\theta(u,\zeta)\|, \quad \forall A_v \in \partial(-F)(u), \ u \in B(\zeta,\rho).$$

Using $\partial(-F)(u) = -\partial F(u)$, there is $A_u = -A_v \in \partial F(u)$, then we have

$$-A_u\theta(\zeta,u) \leq_C -\tau\|\theta(u,\zeta)\|.$$

Since $\theta(u,\zeta) + \theta(\zeta,u) = 0$, therefore,

$$A_u\theta(u,\zeta) \leq_C -\tau\|\theta(u,\zeta)\|.$$

This means ζ does not solve (AMVVI) w.r.t. τ.

The following corollary can be deduced from Theorems 7 and 8.

Corollary 2. *Let* F *be* $AQ(\theta,\tau)I$-2 *at* $\zeta \in S$ *and* $-F$ *be* $AP(\theta,\tau)I$-2 *at* ζ *and* $\theta(u,\zeta) + \theta(\zeta,u) = 0$, $\forall u \in S$. ζ *is a* τ-*AES of (NVOP) if and only if* ζ *solves (AMVVI) w.r.t.* τ.

4   Example

Consider the following NVOP as an example to illustrate the obtained results.

$$\min_{u \in S} F(u) = \begin{cases} u^2 + 3u, & u \geq 0 \\ -u^2 + 4u, & u < 0, \end{cases}$$

where $S = \mathbb{R}$, $C = \mathbb{R}^+$ and $\theta(u,v) = (u - v)^3$ for each $u, v \in S$.

The Clarke subdifferential of F at $u \in S$ is defined by

$$\partial F(u) = \begin{cases} 2u + 3, & u > 0; \\ [3, 4], & u = 0; \\ -2u + 4, & u < 0. \end{cases}$$

For $1 < \tau < 2$, there is $\rho = \frac{1}{2} > 0$ such that, for each $u, v \in B(\zeta, \rho)$, $\zeta = 0$, $A_v \in \partial F(v)$, we have

$$F(u) - F(v) = \begin{cases} (u - v)(u + v + 3) > 0, & \text{if } v > 0, u > 0, u - v > 0; \\ (u - v)(u + v + 3) < 0, & \text{if } v > 0, u > 0, u - v < 0; \\ -u^2 + 4u - v^2 - 3v < 0, & \text{if } v > 0, u \leq 0; \\ u^2 + 3u + v(v - 4) > 0, & \text{if } v < 0, u \geq 0; \\ (u - v)(4 - u - v) > 0, & \text{if } v < 0, u < 0, u - v > 0; \\ (u - v)(4 - u - v) < 0, & \text{if } v < 0, u < 0, u - v < 0; \\ u^2 + 3u > 0, & \text{if } v = 0, u > 0; \\ -u^2 + 4u < 0, & \text{if } v = 0, u < 0. \end{cases}$$

Also,

$$A_v \theta(u, v) + \tau \|\theta(u, v)\| = \begin{cases} (2v + 3 - \tau)(u - v)^3 < 0, & \text{if } v > 0, u > 0, u - v < 0; \\ (2v + 3 - \tau)(u - v)^3 < 0, & \text{if } v > 0, u \leq 0; \\ (-2v + 4 - \tau)(u - v)^3 < 0, & \text{if } v < 0, u < 0, u - v < 0; \\ ku^3 < 0, & \text{if } v = 0, u < 0, \end{cases}$$

where $k \in [3, 4]$. Hence, F is AP(θ, τ)I-2 at $\zeta = 0$.

Since for any $u > 0$, one has

$$A_\zeta \theta(u, \zeta) + \tau \|\theta(u, \zeta)\| = ku^3 + \tau u^3 > 0, \ k \in [2, 3].$$

Hence, there is no $\rho > 0$ satisfying for each $u \in B(\zeta, \rho)$ and $A_\zeta \in \partial F(\zeta)$

$$A_\zeta \theta(u, \zeta) \leq_C -\tau \|\theta(u, \zeta)\|.$$

Thus, $\zeta = 0$ solves (ASVVI) w.r.t. τ.

Finally, as F is AP(θ, τ)I-2 at $\zeta = 0$, then, from Theorem 5, $\zeta = 0$ should be a τ-AES of (NVOP). Indeed, for all $u > 0$ we have

$$F(u) - F(\zeta) + \tau \|\theta(u, \zeta)\| = u^2 + 3u + \tau u^3 > 0.$$

Hence, there is no $\rho > 0$ such that, for each $u \in B(\zeta; \rho) \setminus \{\zeta\}$

$$F(u) - F(\zeta) \leq_C -\tau \|\theta(u, \zeta)\|.$$

Therefore, $\zeta = 0$ is a τ-AES of (NVOP).

Remark 2. In the above example, the function $-F$ is AQ(θ, τ)I-2 at $\zeta = 0$ and $\theta(u, \zeta) + \theta(\zeta, u) = 0$, $\forall u \in S$. We can easily show that it verifies the conditions of Theorem 6.

5   Conclusions

We have shown the relationships between AVVI in terms of Clarke's generalized Jacobian and NVOP using the concepts of approximate efficiency and generalized approximate invexity. Our work improves that of Gupta and Mishra [14] with respect to two aspects:

- If the generalized approximate invexity assumption is replaced by generalized approximate convexity assumption, then the proof arguments remain the same. Consequently, our theorems are more general since the concept of invexity includes that of convexity as a special case.
- In addition to necessary conditions of approximate efficient solutions of NVOP, we have also provided sufficient conditions using the generalized approximate invexity of $-F$.

Acknowledgments. The authors are most grateful to Dr. Lhoussain Elfadil for continued help throughout the preparation of this paper.

References

1. Eichfelder, G., Jahn, J.: Vector optimization problems and their solution concepts. Recent Developments in Vector Optimization, pp. 1–27. Springer, Berlin, Heidelberg (2012)
2. Giannessi, F.: On Minty variational principle. New Trends in Mathematical Programming, pp. 93–99. Kluwer Academic Publishers, Dordrecht (1998)
3. Yang, X.M., Yang, X.Q., Teo, K.L.: Some remarks on the Minty vector variational inequality. J. Optim. Theory Appl. **121**(1), 193–201 (2004)
4. Gang, X., Liu, S.: On Minty vector variational-like inequality. Comput. Maths. Appl. **56**, 311–323 (2008)
5. Fang, Y.P., Hu, R.: A nonsmooth version of Minty variational principle. Optimization **58**(4), 401–412 (2009)
6. Al-Homidan, S., Ansari, Q.H.: Generalized Minty vector variational-like inequalities and vector optimization problems. J. Optim. Theory Appl. **144**, 1–11 (2010)
7. Oveisiha, M., Zafarani, J.: Vector optimization problem and generalized convexity. J. Glob. Optim. **52**, 29–43 (2012)
8. Long, X.J., Peng, J.W., Wu, S.Y.: Generalized vector variational-like inequalities and nonsmooth vector optimization problems. Optimization **61**(9), 1075–1086 (2012)
9. Mishra, S.K., Wang, S.Y.: Vector variational-like inequalities and non-smooth vector optimization problems. Nonlinear Anal. Theory, Methods Appl. **64**(9), 1939–1945 (2006)
10. Yang, X.M., Yang, X.Q.: Vector variational-like inequalities with pseudoinvexity. Optimization **55**(1–2), 157–170 (2006)
11. Ansari, Q.H., Rezaei, M.: Generalized vector variational-like inequalities and vector optimization in Asplund spaces. Optimization **62**, 721–734 (2013)
12. Bhatia, D., Gupta, A., Arora, P.: Optimality via generalized approximate convexity and quasiefficiency. Optim. Lett. **7**, 127–135 (2013)
13. Mishra, S.K., Laha, V.: On minty variational principle for nonsmooth vector optimization problems with approximate convexity. Optim. Lett. **10**(3), 577–589 (2015)

14. Gupta, P., Mishra, S.K.: On Minty variational principle for nonsmooth vector optimization problems with generalized approximate convexity. Optimization **67**, 1157–1167 (2018)
15. Jennane, M., El Fadil, L., Kalmoun, E.M.: On local quasi efficient solutions for nonsmooth vector optimization. Croatian Oper. Res. Rev. **11**(1), 1–10 (2020)
16. Clarke, F.H.: Optimization and Nonsmooth Analysis. Wiley-Interscience, New York (1983)
17. Aslam Noor, M., Inayat Noor, K.: Some characterizations of strongly preinvex functions. J. Math. Anal. Appl. **316**, 697–706 (2006)
18. Ngai, H.V., Luc, D., Thera, M.: Approximate convex functions. J. Nonlinear Convex Anal. **1**, 155–176 (2000)

Existence of Entropy Solutions for Anisotropic Elliptic Nonlinear Problem in Weighted Sobolev Space

Adil Abbassi, Chakir Allalou, and Abderrazak Kassidi[(✉)]

Laboratory LMACS, Faculty of Science and Technology of Beni Mellal, Sultan Moulay Slimane University, Beni Mellal, Morocco
adil.abbassi@usms.ma, chakir.allalou@yahoo.fr, abderrazakassidi@gmail.com

Abstract. In this paper, we will study the existence of an entropy solution to the unilateral problem for a class of nonlinear anisotropic elliptic equation, with second term being an element of $L^1(\Omega)$. Our technical approach is based on a monotony method and the truncation techniques in the framework of the weighted anisotropic Sobolev space.

1 Introduction

The unilateral elliptic problems in weighted anisotropic Sobolev space have recently attracted the attention of many authors (see [5,8,12]), who used different methods to solve the question of the existence of solutions in the framework of weighted anisotropic Sobolev space (we refer to [1,2,12,13] for more details). One of the motivations for studying the unilateral elliptic problems comes from applications of mathematical modeling of physical and mechanical processes in anisotropic continuous medium.

The purpose of this paper is to study the unilateral problem for a class of nonlinear anisotropic elliptic equation of type:

$$\begin{cases} Au - div(\phi(u)) = f & \text{in } \Omega, \\ u = 0 & \text{on } \partial\Omega, \end{cases} \tag{1}$$

where $\Omega \subset \mathbb{R}^N$ $(N \geq 2)$ is a bounded open subset with smooth boundary $\partial\Omega$, $1 < p_1, \cdots, p_N < +\infty$, $\overrightarrow{p}$ and $\overrightarrow{w}$ are respectively the exponent and weight function vectors, which will be specified in the following. The term $\phi = (\phi_1, \cdots, \phi_N)$ belongs to $C^0(\mathbb{R}, \mathbb{R}^N)$, $Au = -div(a(x,u,\nabla u))$ is the Leray-Lions operator defined on $W_0^{1,\overrightarrow{p}}(\Omega, \overrightarrow{w})$, with $a(x,u,\nabla u)$ is a Carathéodory's function satisfying some hypotheses which will be stated later. Finally, we mention that the second member f belongs to $L^1(\Omega)$.

In the non weighted case $w_i \equiv 1$ for any $i \in \{1,...,N\}$, by using monotony method and the truncation techniques, the authors in [4] has established the existence of an entropy solutions for anisotropic elliptic unilateral problem like (1). we refer the reader to the papers [8,15] and the references therein. Moreover, Boccardo et al. [10] studied the existence of weak solutions for nonlinear elliptic problem with

© Springer Nature Switzerland AG 2021
Z. Hammouch et al. (Eds.): SM2A 2019, LNNS 168, pp. 102–122, 2021.
https://doi.org/10.1007/978-3-030-62299-2_8

$$Au = -\sum_{i=1}^{N} \frac{\partial}{\partial x_i} \left(\left| \frac{\partial u}{\partial x_i} \right|^{p_i-2} \frac{\partial u}{\partial x_i} \right), \text{ when } \phi_i(u) = 0 \text{ for } i = 1, \cdots, N \text{ and the right-hand side}$$

is a bounded Radon measure on Ω.

In general the function ϕ_i does not belongs to $L^1_{loc}(\Omega)$. Then, the problem (1) does not admit weak solution. To avoid this situation, we use entropy solutions in this paper, this concept of entropy solution was first proposed by Benilan et al. see [7].

Motivated by the above cited papers and the results in [4], we show the existence result for the anisotropic unilateral nonlinear elliptic problem related to the equation in the problem (1). Specifically, we show the existence result of an entropy solutions for the following unilateral anisotropic problem,

$$\begin{cases} u \geq \psi \text{ a.e. in} \Omega, T_k(u) \in W_0^{1,\vec{p}}(\Omega, \vec{\omega}), \\ \sum_{i=1}^{N} \int_{\Omega} a_i(x,u,\nabla u)\partial_i T_k(u-v)dx + \sum_{i=1}^{N} \int_{\Omega} \phi_i(u)\partial_i T_k(u-v)dx \\ \leq \int_{\Omega} f T_k(u-v)dx, \ \forall v \in K_\psi \cap L^\infty(\Omega), \ \forall k > 0, \end{cases} \quad (2)$$

in the convex class $K_\psi := \{ u \in W_0^{1,\vec{p}}(\Omega, \vec{\omega}), u \geq \psi \quad \text{a.e in } \Omega \}$, where ψ is a measurable function on Ω such that

$$\psi^+ \in W_0^{1,\vec{p}}(\Omega, \vec{\omega}) \cap L^\infty(\Omega). \quad (3)$$

Note that the uniqueness result being a rather delicate one, due to a counter-example by Serrin (see [16]), we also mention some works [11,14] for further remarks.

The paper is outlined as follow: In the next section, we will give a brief discussion of the weighted Lebesgue space and the weighted anisotropic Sobolev space. The Sect. 3 is dedicated to some necessary lemmas and basic assumptions of our problem. In the last section, we present the main result and proofs.

2 Preliminaries

In this section, we recall some basic properties of the weighted Lebesgue-Sobolev spaces needed to study problem (1), and we give the fundamental definitions and lemmas which will be used in the following pages.

Let Ω be a bounded open subset of $\mathbb{R}^N (N \geq 2)$ with smooth boundary $\partial\Omega$. Let $p_1, \ldots, p_N$ be N real numbers and $\vec{p} = \{p_1, \ldots, p_N\}$ be a vector of exponent, the following vector $\vec{w} = \{w_1, \ldots, w_N\}$ be a vector of weight functions, i.e., every component w_i is a measurable function which is positive a.e. in Ω. Further, we suppose in all our considerations that

(H_1) $w_i \in L^1_{loc}(\Omega)$ and $w_i^{\frac{-1}{p_i-1}} \in L^1_{loc}(\Omega)$.

for any $i = 1, \ldots, N$, we denote

$$\partial_i u = \frac{\partial u}{\partial x_i} \quad \text{for} \quad i = 1, \ldots, N,$$

$$p^- = \min\{p_1, \ldots, p_N\}, \quad p^+ = \max\{p_1, \ldots, p_N\}.$$

We define the weighted Lebesgue space $L^p(\Omega, \gamma)$ with weight γ in Ω as, the set of all measurable functions u on Ω.

we endow it

$$\|u\|_{L^p(\Omega,\gamma)} \equiv \|u\|_{p,\gamma} = \left(\int_\Omega |u|^p \gamma(x) dx \right)^{\frac{1}{p}} \qquad 1 \le p < \infty. \tag{4}$$

We denote by $W^{1,\vec{p}}(\Omega, \vec{w})$ the weighted anisotropic Sobolev space of all functions $u \in L^1_{loc}(\Omega)$ such that the derivatives $\partial_i u$ are in $L^{p_i}(\Omega, w_i)$ for any $i = 1, \ldots, N$.

This set of functions is a Banach space with respect to norm (see [12])

$$\|u\|_{1,\vec{p},\vec{w}} = \|u\|_{L^1(\Omega)} + \sum_{i=1}^{N} \|\partial_i u\|_{p_i, w_i}. \tag{5}$$

In the following to study the Dirichlet problem, we use the functional space $W_0^{1,\vec{p}}(\Omega, \vec{w})$ defined as the closure of $C_0^\infty(\Omega)$ in $W^{1,\vec{p}}(\Omega, \vec{w})$ with respect to the norm (5).

Let us remark that $C_0^\infty(\Omega)$ is dense in $W_0^{1,\vec{p}}(\Omega, \vec{w})$ and $\left(W_0^{1,\vec{p}}(\Omega, \vec{w}), \|\cdot\|_{1,\vec{p},\vec{w}} \right)$ is a reflexive Banach space, for all $i = 1, \ldots, N$ such that $1 < p_i < \infty$, (see [2] for more details).

We next recall that the dual of the weighted anisotropic Sobolev space $W_0^{1,\vec{p}}(\Omega, \vec{w})$ is equivalent to $W^{-1,\vec{p'}}(\Omega, \vec{w^*})$, where $\vec{p'}$ is the conjugate of $\vec{p}$, i.e. $p_i' = \dfrac{p_i}{p_i - 1}$ and $\vec{w^*} = \left\{ w_i^* = w_i^{1-p_i'}, i = 1, \ldots, N \right\}$.

Remark 1. suppose there is $s_i \in]\frac{N}{p_i}, +\infty[\cap [\frac{1}{p_i-1}, +\infty[$ such that

$$w_i^{-s_i} \in L^1(\Omega), \quad \text{for all } i = 1, \cdots, N. \tag{6}$$

Then, the expression

$$\|u\|_{W_0^{1,\vec{p}}(\Omega,\vec{w})} = \sum_{i=1}^{N} \|\partial_i u\|_{p_i, w_i} \tag{7}$$

is a norm defined on $W_0^{1,\vec{p}}(\Omega, \vec{w})$ which is equivalent to (5).

Note that (6) is stronger than the second integrability condition in (H_1).

Let us consider the following exponent vector $\vec{p_s} = \left\{ p_{s_i} = \dfrac{p_i s_i}{s_i + 1}, i = 1, \ldots, N \right\}$.

Lemma 1. *Suppose that (H_1) and (6) hold, we have*

- *If $p^- < N$, then $W_0^{1,\vec{p}}(\Omega, \vec{w}) \subset L^q(\Omega)$ for all $q \in [p^-, p^*[$, with $p^* = \dfrac{Np^-}{N-p^-}$.*
- *If $p^- = N$, then $W_0^{1,\vec{p}}(\Omega, \vec{w}) \subset L^q(\Omega)$ for all $q \in [p^-, +\infty[$.*

Furthermore, the above embeddings are compacts.

The proof of this lemma comes from the fact that the following embedding (see [9] for more details)

$$W_0^{1,\vec{p}}(\Omega, \vec{w}) \subset W_0^{1,\vec{p_s}}(\Omega) \subset W_0^{1,p^-}(\Omega)$$

We consider the space

$$\mathscr{T}_0^{1,\vec{p}}(\Omega, \vec{w}) := \left\{ u \text{ measurable in } \Omega, \ T_k(u) \in W_0^{1,\vec{p}}(\Omega, \vec{w}), \text{ for any } k > 0 \right\},$$

where

$$T_k(z) := \begin{cases} z & \text{if } |z| \leq k, \\ k\dfrac{z}{|z|} & \text{if } |z| > k. \end{cases}$$

3 Basic Assumptions and Notion of Solutions

In this section, we recall some useful technical lemmas to show our aim, and we give the assumptions of our problem.

We suppose that $a_i : \Omega \times \mathbb{R} \times \mathbb{R}^N \mapsto \mathbb{R}$ are Carathéodory functions, for $i = 1, \ldots, N$ which satisfies the following assumptions, for every ξ, $\xi' \in \mathbb{R}^N$, $\theta \in \mathbb{R}$ and a.e. in $x \in \Omega$,

$$a_i(x, \theta, \xi) \cdot \xi_i \geq \alpha w_i |\xi_i|^{p_i}, \tag{8}$$

$$|a_i(x, \theta, \xi)| \leq \beta w_i^{1/p_i} \left(R_i(x) + \sigma^{1/p_i'} |\theta|^{p_i/p_i'} + w_i^{1/p_i'} |\xi_i|^{p_i-1} \right), \tag{9}$$

$$(a_i(x, \theta, \xi) - a_i(x, \theta, \xi')) \cdot (\xi_i - \xi_i') > 0 \quad \text{for} \quad \xi_i \neq \xi_i', \tag{10}$$

where $R_i(\cdot)$ is a nonnegative function lying in $L^{p_i'}(\Omega)$ and $\alpha, \beta > 0$.

Moreover, we suppose that

$$\phi_i \in C^0(\mathbb{R}, \mathbb{R}) \quad \text{for} \quad i = 1, \ldots, N, \tag{11}$$

and

$$f \in L^1(\Omega). \tag{12}$$

Lemma 2. *[1] Let $g \in L^r(\Omega, \gamma)$ and $g_n \subset L^r(\Omega, \gamma)$ such that $\|g_n\|_{r,\gamma} \leq C$, $1 < r < \infty$, If $g_n(x) \to g(x)$ a.e. in Ω then $g_n \rightharpoonup g$ weakly in $L^r(\Omega, \gamma)$.*

Lemma 3. *[3] Suppose that (8)–(10) hold, let $(u_n)_n$ a sequence in $W_0^{1,\vec{p}}(\Omega, \vec{w})$ such that $u_n \rightharpoonup u$ weakly in $W_0^{1,\vec{p}}(\Omega, \vec{w})$ and*

$$\sum_{i=1}^N \int_\Omega (a_i(x, u_n, \nabla u_n) - a_i(x, u, \nabla u)) \partial_i(u_n - u) dx \to 0,$$

then $u_n \longrightarrow u$ strongly in $W_0^{1,\vec{p}}(\Omega, \vec{w})$.

Lemma 4. *Let* $(u_n)_n$ *be a sequence from* $W_0^{1,\overrightarrow{p}}(\Omega, \overrightarrow{w})$, *if* $u_n \rightharpoonup u$ *weakly in* $W_0^{1,\overrightarrow{p}}(\Omega, \overrightarrow{w})$. *Then* $T_k(u_n)$ *weakly converges to* $T_k(u)$ *in* $W_0^{1,\overrightarrow{p}}(\Omega, \overrightarrow{w})$.

Proof. We have $u_n \rightharpoonup u$ weakly in $W_0^{1,\overrightarrow{p}}(\Omega, \overrightarrow{w})$ and $W_0^{1,\overrightarrow{p}}(\Omega, \overrightarrow{w}) \hookrightarrow\hookrightarrow L^q(\Omega)$, we obtain $u_n \to u$ strongly in $L^q(\Omega)$ and a.e. in Ω, thus $T_k(u_n) \to T_k(u)$ a.e. in Ω.
On the other hand

$$\|T_k(u_n)\|_{W_0^{1,\overrightarrow{p}}(\Omega,\overrightarrow{w})} = \sum_{i=1}^{N} \|\partial_i T_k(u_n)\|_{p_i,w_i}$$
$$\leq \sum_{i=1}^{N} \left(\int_{\Omega} |T_k'(u_n)\partial_i u_n|^{p_i} w_i(x) dx \right)^{1/p_i}$$
$$\leq \sum_{i=1}^{N} \left(\int_{\Omega} |\partial_i u_n|^{p_i} w_i(x) dx \right)^{1/p_i} = \|u_n\|_{W_0^{1,\overrightarrow{p}}(\Omega,\overrightarrow{w})}$$

Thus $(T_k(u_n))_n$ is bounded in $W_0^{1,\overrightarrow{p}}(\Omega, \overrightarrow{w})$, consequently $T_k(u_n) \rightharpoonup T_k(u)$ weakly in $W_0^{1,\overrightarrow{p}}(\Omega, \overrightarrow{w})$. $\qquad\square$

Lemma 5. *[3] If* $u \in W_0^{1,\overrightarrow{p}}(\Omega, \overrightarrow{w})$ *then* $\sum_{i=1}^{N} \int_{\Omega} \partial_i u \, dx = 0$.

Proof. Since $u \in W_0^{1,\overrightarrow{p}}(\Omega, \overrightarrow{w})$ there exists $u_k \in C_0^\infty(\Omega)$ such that $u_k \to u$ strongly in $W_0^{1,\overrightarrow{p}}(\Omega, \overrightarrow{w})$

Moreover, since $u_k \in C_0^\infty(\Omega)$ by Green's Formula, we have

$$\sum_{i=1}^{N} \int_{\Omega} \partial_i u_k \, dx = \int_{\partial\Omega} u_k . \overrightarrow{n} \, ds = 0$$

Since $\partial_i u_k \to \partial_i u$ strongly in $L^{p_i}(\Omega, w_i)$ we have $\partial_i u_k \to \partial_i u$ strongly in $L^1(\Omega)$

Passing to the limit in (3), we conclude that $\sum_{i=1}^{N} \int_{\Omega} \partial_i u \, dx = 0$. $\qquad\square$

4   Main Results

In this section we state and show the main result of our article.

The definition of an entropy solution for problem (1) can be defined as follows.

Definition 1. *A function* $u \in \mathcal{T}_0^{1,\overrightarrow{p}}(\Omega, \overrightarrow{w})$ *such that* $u \geq \psi$ *a.e. in* Ω *is said to be an entropy solution for the unilateral problem (1), if*

$$\sum_{i=1}^{N} \int_{\Omega} [a_i(x, u, \nabla u) \partial_i T_k(u - \varphi) + \phi_i(u) \partial_i T_k(u - \varphi)] \, dx \leq \int_{\Omega} f T_k(u - \varphi) dx$$

for all $\varphi \in K_\psi \cap L^\infty(\Omega)$.

Theorem 1. *Under the Assumptions (8)–(12), then the problem (1) admits at least one entropy solution.*

Proof:

Step l: Approximate problems.

let us consider the following approximate problems

$$
\begin{cases}
u_n \in K_\psi \\
\displaystyle\sum_{i=1}^{N} \int_\Omega a_i(x, u_n, \nabla u_n) \partial_i(u_n - v)dx + \sum_{i=1}^{N} \int_\Omega \phi_i^n(u_n)\partial_i(u_n - v)dx \le \int_\Omega f_n(u_n - v)dx \\
\qquad \forall v \in K_\psi \quad \text{and} \ \forall k > 0,
\end{cases}
\tag{13}
$$

where $f_n = T_n(f)$ and $\phi_i^n(s) = \phi_i(T_n(s))$.

We define the operators Φ_n of K_ψ to $W_0^{-1,\vec{p}'}(\Omega, \vec{w}^*)$ by:

$$
\langle \Phi_n u, v \rangle = \sum_{i=1}^{N} \int_\Omega \phi_i(T_n(u))\, \partial_i v\, dx \quad \text{for all } u \in K_\psi \text{ and } v \in W_0^{1,\vec{p}}(\Omega, \vec{\omega}).
$$

Lemma 6. *The operator $B_n = A + \Phi_n$ is pseudomonotone. Furthermore, B_n is coercive in the following sense: there exists $v_0 \in K_\psi$ such that*

$$
\frac{\langle B_n v, v - v_0 \rangle}{\|v\|_{1,\vec{p},\vec{w}}} \longrightarrow +\infty \qquad \text{if} \quad \|v\|_{1,\vec{p},\vec{w}} \to +\infty \qquad \text{for} \quad v \in K_\psi.
$$

Proof. In light of the Hölder's type inequality, we get for every $u, v \in W_0^{1,\vec{p}}(\Omega, \vec{w})$,

$$
\begin{aligned}
|\langle \Phi_n u, v \rangle| &\le \sum_{i=1}^{N} \int_\Omega \phi_i(T_n(u)) \partial_i v\, w_i^{\frac{-1}{p_i}} w_i^{\frac{1}{p_i}} dx \\
&\le \sum_{i=1}^{N} \left(\int_\Omega |\phi_i(T_n(u)) w_i^{\frac{-1}{p_i}}|^{p_i'} dx \right)^{\frac{1}{p_i'}} \left(\int_\Omega |\partial_i v\, w_i^{\frac{1}{p_i}}|^{p_i} dx \right)^{\frac{1}{p_i}} \\
&\le \sum_{i=1}^{N} \left(\int_\Omega \sup_{|s| \le n} |\phi_i(s)|^{p_i'} w_i^{\frac{-p_i'}{p_i}} dx \right)^{\frac{1}{p_i'}} \left(\int_\Omega |\partial_i v|^{p_i} w_i dx \right)^{\frac{1}{p_i}} \\
&\le \sum_{i=1}^{N} \left(\int_\Omega (\sup_{|s| \le n} |\phi_i(s)| + 1)^{p_i'} w_i^{\frac{-1}{p_i-1}} dx \right)^{\frac{1}{p_i'}} \left(\int_\Omega |\partial_i v|^{p_i} w_i dx \right)^{\frac{1}{p_i}} \\
&\le \sum_{i=1}^{N} \left(\sup_{|s| \le n} |\phi_i(s)| + 1 \right) \left(\int_\Omega w_i^{\frac{-1}{p_i-1}} dx \right)^{\frac{1}{p_i'}} \left(\int_\Omega |\partial_i v|^{p_i} w_i dx \right)^{\frac{1}{p_i}} \\
&\le C(n) \|v\|_{W_0^{1,\vec{p}}(\Omega, \vec{w})},
\end{aligned}
$$

which implies that $\dfrac{|<\Phi_n u, v>|}{\|v\|_{1,\vec{p},\vec{w}}} \le C(n).$

Let $v_0 \in K_\psi$, thanks to Hölder's inequality and (9), by using the following continuous embeddings $W_0^{1,p_i}(\Omega, w_i) \hookrightarrow L^{p_i}(\Omega, w_i)$, we obtain

$$
\begin{aligned}
|<Av,v_0>| &\leq \sum_{i=1}^{N} \int_\Omega |a_i(x,v,\nabla v)\partial_i v_0 \, w_i^{\frac{-1}{p_i}} w_i^{\frac{1}{p_i}}| dx \\
&\leq \sum_{i=1}^{N} \left(\int_\Omega |a_i(x,v,\nabla v) w_i^{\frac{-1}{p_i}}|^{p_i'} dx \right)^{\frac{1}{p_i'}} \left(\int_\Omega |\partial_i v_0 \, w_i^{\frac{1}{p_i}}|^{p_i} dx \right)^{\frac{1}{p_i}} \\
&\leq \beta \sum_{i=1}^{N} \left(\int_\Omega R_i^{p_i'}(x) + \sigma |v|^{p_i} + |\partial_i v|^{p_i} w_i dx \right)^{\frac{1}{p_i'}} \left(\int_\Omega |\partial_i v_0|^{p_i} w_i dx \right)^{\frac{1}{p_i}} \\
&\leq \beta \sum_{i=1}^{N} \left(C_1 + C_2 \int_\Omega |\partial_i v|^{p_i} w_i dx + \int_\Omega |\partial_i v|^{p_i} w_i dx \right)^{\frac{1}{p_i'}} \left(\int_\Omega |\partial_i v_0|^{p_i} w_i dx \right)^{\frac{1}{p_i}} \\
&\leq \beta \sum_{i=1}^{N} C_1^{\frac{1}{p_i'}} \left(1 + \frac{C_2+1}{C_1} \sum_{i=1}^{N} \int_\Omega |\partial_i v|^{p_i} w_i dx \right)^{\frac{1}{p_i'}} \left(\int_\Omega |\partial_i v_0|^{p_i} w_i dx \right)^{\frac{1}{p_i}} \\
&\leq \beta C_4 \sum_{i=1}^{N} \left(1 + \frac{C_2+1}{C_1} \sum_{i=1}^{N} \int_\Omega |\partial_i v|^{p_i} w_i dx \right)^{\frac{1}{p_-}} \left(\int_\Omega |\partial_i v_0|^{p_i} w_i(x) dx \right)^{\frac{1}{p_i}} \\
&\leq \beta C_4 \sum_{i=1}^{N} \left(1 + C_3 \left(\sum_{i=1}^{N} \int_\Omega |\partial_i v|^{p_i} w_i dx \right)^{\frac{1}{p_-}} \right) \left(\int_\Omega |\partial_i v_0|^{p_i} w_i dx \right)^{\frac{1}{p_i}} \\
&\leq \beta C_4 \left(1 + C_3 \left(\sum_{i=1}^{N} \int_\Omega |\partial_i v|^{p_i} w_i dx \right)^{\frac{1}{p_-}} \right) \sum_{i=1}^{N} \left(\int_\Omega |\partial_i v_0|^{p_i} w_i dx \right)^{\frac{1}{p_i}} \\
&\leq \beta C_4 \left(1 + C_3 \left(\sum_{i=1}^{N} \int_\Omega |\partial_i v|^{p_i} w_i dx \right)^{\frac{1}{p_-}} \right) \|v_0\|_{W_0^{1,\vec{p}}(\Omega,\vec{w})}.
\end{aligned}
$$

Therefore

$$
\frac{|<Av,v-v_0>|}{\|v\|_{W_0^{1,\vec{p}}(\Omega,\vec{w})}} \geq \alpha \frac{\sum_{i=1}^{N} \int_\Omega |\partial_i v|^{p_i} w_i dx}{\|v\|_{W_0^{1,\vec{p}}(\Omega,\vec{w})}} - \frac{\beta C_4 \|v_0\|_{W_0^{1,\vec{p}}(\Omega,\vec{w})}}{\|v\|_{W_0^{1,\vec{p}}(\Omega,\vec{w})}}
$$
$$
- \frac{\beta C_4 C_3}{\|v\|_{W_0^{1,\vec{p}}(\Omega,\vec{w})}} \left(\sum_{i=1}^{N} \int_\Omega |\partial_i v|^{p_i} w_i dx \right)^{\frac{1}{p_-}} \|v_0\|_{W_0^{1,\vec{p}}(\Omega,\vec{w})}.
$$

Then,

$$
\frac{|<Av,v-v_0>|}{\|v\|_{W_0^{1,\vec{p}}(\Omega,\vec{w})}} \geq \alpha \frac{\sum_{i=1}^{N} \int_\Omega |\partial_i v|^{p_i} w_i dx}{\|v\|_{W_0^{1,\vec{p}}(\Omega,\vec{w})}} \left[1 - \frac{\beta}{\alpha} C_4 C_3 \left(\sum_{i=1}^{N} \int_\Omega |\partial_i v|^{p_i} dx \right)^{\frac{1}{p_-}-1} \right.
$$
$$
\left. \|v_0\|_{W_0^{1,\vec{p}}(\Omega,\vec{w})} \right] - \frac{\beta C_4 \|v_0\|_{W_0^{1,\vec{p}}(\Omega,\vec{w})}}{\|v\|_{W_0^{1,\vec{p}}(\Omega,\vec{w})}}.
$$

According to Jensen's inequality, we obtain

$$\|v\|_{W_0^{1,\vec{p}}(\Omega,\vec{w})}^{p_-^+} = \left(\sum_{i=1}^{N} \left(\int_{\Omega} |\partial_i v|^{p_i} w_i dx \right)^{\frac{1}{p_i}} \right)^{p_-^+}$$

$$\leq \left(\sum_{i=1}^{N} \left(\int_{\Omega} |\partial_i v|^{p_i} w_i dx \right)^{\frac{1}{p_-^+}} \right)^{p_-^+}$$

$$\leq C \sum_{i=1}^{N} \int_{\Omega} |\partial_i v|^{p_i} w_i dx,$$

where

$$p_-^+ = \begin{cases} p^- & \text{if } \|\partial_i v\|_{L^{p_i}(\Omega, w_i)} \geq 1 \\ p^+ & \text{if } \|\partial_i v\|_{L^{p_i}(\Omega, w_i)} < 1. \end{cases}$$

Then

$$\frac{\sum_{i=1}^{N} \int_{\Omega} |\partial_i v|^{p_i} w_i dx}{\|v\|_{W_0^{1,\vec{p}}(\Omega,\vec{w})}} \to +\infty \text{ and } \sum_{i=1}^{N} \int_{\Omega} |\partial_i v|^{p_i} w_i dx \to +\infty \text{ as } \|v\|_{W_0^{1,\vec{p}}(\Omega,\vec{w})} \to +\infty.$$

Using (4), we obtain $\dfrac{|<Av, v - v_0>|}{\|v\|_{W_0^{1,\vec{p}}(\Omega,\vec{w})}} \to +\infty$ as $\|v\|_{1,\vec{p},\vec{w}} \to +\infty$.

Since $\dfrac{<\Phi_n v, v>}{\|v\|_{W_0^{1,\vec{p}}(\Omega,\vec{w})}}$ and $\dfrac{<\Phi_n v, v_0>}{\|v\|_{W_0^{1,\vec{p}}(\Omega,\vec{w})}}$ are bounded, then we get

$$\frac{<B_n v, v, -v_0>}{\|v\|_{W_0^{1,\vec{p}}(\Omega,\vec{w})}} = \frac{<Av, v - v_0>}{\|v\|_{W_0^{1,\vec{p}}(\Omega,\vec{w})}} + \frac{<\Phi_n v, v, -v_0>}{\|v\|_{W_0^{1,\vec{p}}(\Omega,\vec{w})}} \to +\infty \text{ as } \|v\|_{W_0^{1,\vec{p}}(\Omega,\vec{w})} \to +\infty.$$

We conclude that $B_n = A + \Phi_n$ is coercive.

It remains to show that B_n is pseudomonotone.

Let $(u_k)_k$ be a sequence in $W_0^{1,\vec{p}}(\Omega,\vec{w})$ such that

$$\begin{cases} u_k \rightharpoonup u & \text{weakly in } W_0^{1,\vec{p}}(\Omega,\vec{w}) \\ B_n u_k \rightharpoonup \chi & \text{weakly in } W_0^{-1,\vec{p}'}(\Omega,\vec{w}^*) \\ \limsup_{k \to +\infty} <B_n u_k, u_k> \leq <\chi, u>. \end{cases}$$

We will show that $\chi = B_n u$ and $<B_n u_k, u_k> \longrightarrow <\chi, u>$ as $k \to +\infty$. Since $W_0^{1,\vec{p}}(\Omega,\vec{w}) \hookrightarrow\hookrightarrow L^{p^-}(\Omega)$, then $u_k \to u$ strongly in $L^{p^-}(\Omega)$ and a.e. in Ω for a subsequence denoted again $(u_k)_k$. Since $(u_k)_k$ is bounded in $W_0^{1,\vec{p}}(\Omega,\vec{w})$. By using (9) we have $(a_i(x, u_k, \nabla u_k))_k$ is bounded in $L^{p'_i}(\Omega, w_i^*)$, then there exists a function $\varphi_i \in L^{p'_i}(\Omega, w_i^*)$ such that

$$a_i(x, u_k, \nabla u_k) \rightharpoonup \varphi_i \quad \text{as} \quad k \to +\infty \tag{14}$$

Moreover, since $(\phi_i^n(u_k))_k$ is bounded in $L^{p_i'}(\Omega, w_i^*)$ and $\phi_i^n(u_k) \to \phi_i^n(u)$ a.e. in Ω, we obtain

$$\phi_i^n(u_k) \to \phi_i^n(u) \quad \text{strongly in} \quad L^{p_i'}(\Omega, w_i) \quad \text{as} \quad k \to +\infty. \tag{15}$$

For all $v \in W_0^{1,\vec{p}}(\Omega, \vec{w})$ combining (14) and (15), we have

$$
\begin{aligned}
< \chi, v > &= \lim_{k \to +\infty} < B_n u_k, v > \\
&= \lim_{k \to +\infty} \sum_{i=1}^N \int_\Omega a_i(x, u_k, \nabla u_k) \partial_i v dx + \lim_{k \to +\infty} \sum_{i=1}^N \int_\Omega \phi_i^n(u_k) \partial_i v dx \\
&= \sum_{i=1}^N \int_\Omega \varphi_i \partial_i v dx + \sum_{i=1}^N \int_\Omega \phi_i^n(u) \partial_i v dx.
\end{aligned}
$$

Hence, we obtain

$$
\begin{aligned}
\limsup_{k \to +\infty} < B_n u_k, u_k > &= \limsup_{k \to +\infty} \Big[\sum_{i=1}^N \int_\Omega a_i(x, u_k, \nabla u_k) \partial_i u_k dx + \sum_{i=1}^N \int_\Omega \phi_i^n(u_k) \partial_i u_k dx \Big] \\
&= \limsup_{k \to +\infty} \sum_{i=1}^N \int_\Omega a_i(x, u_k, \nabla u_k) \partial_i u_k dx + \sum_{i=1}^N \int_\Omega \phi_i^n(u) \partial_i u dx \\
&\leq \quad < \chi, u > \\
&= \sum_{i=1}^N \int_\Omega \varphi_i \partial_i u dx + \sum_{i=1}^N \int_\Omega \phi_i^n(u) \partial_i u dx
\end{aligned}
$$

as a result

$$\limsup_{k \to +\infty} \sum_{i=1}^N \int_\Omega a_i(x, u_k, \nabla u_k) \partial_i u_k dx \leq \sum_{i=1}^N \int_\Omega \varphi_i \partial_i u dx. \tag{16}$$

Using (10), we get $\sum_{i=1}^N \int_\Omega (a_i(x, u_k, \nabla u_k) - a_i(x, u_k, \nabla u))(\partial_i u_k - \partial_i u) dx > 0$. Then

$$
\begin{aligned}
\sum_{i=1}^N \int_\Omega a_i(x, u_k, \nabla u_k) \partial_i u_k dx \geq &- \sum_{i=1}^N \int_\Omega a_i(x, u_k, nablau) \partial_i u dx \\
&+ \sum_{i=1}^N \int_\Omega a_i(x, u_k, \nabla u_k) \partial_i u dx + \sum_{i=1}^N \int_\Omega a_i(x, u_k, \nabla u) \partial_i u_k dx.
\end{aligned}
$$

By (14), we have

$$\liminf_{k \to +\infty} \sum_{i=1}^N \int_\Omega a_i(x, u_k, \nabla u_k) \partial_i u_k dx \geq \sum_{i=1}^N \int_\Omega \varphi_i \partial_i u dx. \tag{17}$$

Using (16) and (17), we get

$$\lim_{k \to +\infty} \sum_{i=1}^N \int_\Omega a_i(x, u_k, \nabla u_k) \partial_i u_k dx = \sum_{i=1}^N \int_\Omega \varphi_i \partial_i u dx \tag{18}$$

$$\lim_{k\to+\infty} <B_n u_k, u_k> = \lim_{k\to+\infty} \sum_{i=1}^{N} \int_{\Omega} a_i(x, u_k, \nabla u_k)\partial_i u_k dx + \lim_{k\to+\infty} \sum_{i=1}^{N} \int_{\Omega} \phi_i^n(u_k)\partial_i u_k dx$$

$$= \sum_{i=1}^{N} \int_{\Omega} \varphi_i \partial_i u dx + \sum_{i=1}^{N} \int_{\Omega} \phi_i^n(u)\partial_i u dx$$

$$= <\chi, u>.$$

Moreover, since $a_i(x, u_k, \nabla u) \longrightarrow a_i(x, u, \nabla u)$ strongly in $L^{p_i'}(\Omega, w_i)$, by using (18) we have

$$\sum_{i=1}^{N} \int_{\Omega} (a_i(x, u_k, \nabla u_k) - a_i(x, u_k, \nabla u))(\partial_i u_k - \partial_i u)dx = 0.$$

Using Lemma 3, we obtain u_k converges to u strongly in $W_0^{1,\vec{p}}(\Omega, \vec{w})$ and a.e. in Ω, then $a_i(x, u_k, \nabla u)$ converges to $a_i(x, u, \nabla u)$ weakly in $L^{p_i'}(\Omega, w_i)$ and $\phi_i^n(u)$ converges to $\phi_i^n(u)$ strongly in $L^{p_i'}(\Omega, w_i)$. Then for all $v \in W_0^{1,\vec{p}}(\Omega, \vec{w})$, we get

$$<\chi, v> = \lim_{k\to+\infty} <B_n u_k, v>$$

$$= \lim_{k\to+\infty} \sum_{i=1}^{N} \int_{\Omega} a_i(x, u_k, \nabla u_k)\partial_i v dx + \lim_{k\to+\infty} \sum_{i=1}^{N} \int_{\Omega} \phi_i(u_k)\partial_i v dx$$

$$= \sum_{i=1}^{N} \int_{\Omega} a_i(x, u, \nabla u)\partial_i v dx + \sum_{i=1}^{N} \int_{\Omega} \phi_i(u)\partial_i v dx$$

$$= <B_n u, v>$$

Therefore $B_n u = \chi$.   □

Proposition 1. *Assume that (8)–(12) hold, then the problem (13) admits at least one solution.*

Proof. From Lemma 6 and Theorem 8.2 chapter 2 in [13], then the problem (13) admits at least one solution.   □

Step 2: A priori estimate.

Proposition 2. *Under the assumptions (8)–(12) and if u_n is a solution of the approximate problem (13). Then the following assertion is valid:*

$$\sum_{i=1}^{N} \int_{\Omega} |\partial_i T_k(u_n)|^{p_i} w_i dx \le C(k+1) \qquad \text{for all } k > 0,$$

where C is a constant.

Proof. Let $v = u_n - \eta T_k(u_n^+ - \psi^+)$ where $\eta \ge 0$. Since $v \in W_0^{1,\vec{p}}(\Omega, \vec{w})$ and for all η small enough, we get $v \in K_\psi$. We take v as test function in problem (13), we obtain

$$\sum_{i=1}^{N} \int_{\Omega} a_i(x, u_n, \nabla u_n)\partial_i T_k(u_n^+ - \psi^+)dx + \sum_{i=1}^{N} \int_{\Omega} \phi_i^n(u_n)\partial_i T_k(u_n^+ - \psi^+)dx$$

$$\le \int_{\Omega} f_n T_k(u_n^+ - \psi^+)dx.$$

As result

$$\sum_{i=1}^{N} \int_{\Omega} a_i(x, u_n, \nabla u_n) \partial_i T_k(u_n^+ - \psi^+) dx \le \int_{\Omega} f_n T_k(u_n^+ - \psi^+) dx$$
$$+ \sum_{i=1}^{N} \int_{\Omega} |\phi_i^n(u_n)| |\partial_i T_k(u_n^+ - \psi^+)| dx.$$

Since $\partial_i T_k(u_n^+ - \psi^+) = 0$ on the set $\{u_n^+ - \psi^+ > k\}$, we get

$$\sum_{i=1}^{N} \int_{\{u_n^+ - \psi^+ \le k\}} a_i(x, u_n, \nabla u_n) \partial_i(u_n^+ - \psi^+) dx \le \int_{\Omega} f_n T_k(u_n^+ - \psi^+) dx$$
$$+ \sum_{i=1}^{N} \int_{\{u_n^+ - \psi^+ \le k\}} |\phi_i^n(u_n)| |\partial_i(u_n^+ - \psi^+)| dx,$$

thus, we have

$$\sum_{i=1}^{N} \int_{\{u_n^+ - \psi^+ \le k\}} a_i(x, u_n^+, \nabla u_n^+) \partial_i u_n^+ dx \le \int_{\Omega} f_n T_k(u_n^+ - \psi^+) dx$$
$$+ \sum_{i=1}^{N} \int_{\{u_n^+ - \psi^+ \le k\}} |\phi_i^n(u_n)| |\partial_i u_n^+| w_i^{\frac{-1}{p_i}} w_i^{\frac{1}{p_i}} dx + \sum_{i=1}^{N} \int_{\{u_n^+ - \psi^+ \le k\}} |\phi_i^n(u_n)| |\partial_i \psi^+| dx$$
$$+ \sum_{i=1}^{N} \int_{\{u_n^+ - \psi^+ \le k\}} |a_i(x, u_n^+, \nabla u_n^+) \partial_i \psi^+| dx$$

According to Young's inequalities, we have for a positive constant λ

$$\sum_{i=1}^{N} \int_{\{u_n^+ - \psi^+ \le k\}} a_i(x, u_n^+, \nabla u_n^+) \partial_i u_n^+ dx \le \int_{\Omega} f_n T_k(u_n^+ - \psi^+) dx$$
$$+ C_1(\alpha) \sum_{i=1}^{N} \int_{\{u_n^+ - \psi^+ \le k\}} |\phi_i^n(T_{k+\|\psi\|_\infty}(u_n))|^{p_i'} w_i^{\frac{-1}{p_i-1}} dx + \frac{\alpha}{6} \sum_{i=1}^{N} \int_{\{u_n^+ - \psi^+ \le k\}} |\partial_i u_n^+|^{p_i} w_i dx$$
$$+ \sum_{i=1}^{N} \int_{\{u_n^+ - \psi^+ \le k\}} |\phi_i^n(T_{k+\|\psi\|_\infty}(u_n))| |\partial_i \psi^+| dx$$
$$+ \sum_{i=1}^{N} \frac{\lambda^{p_i'}}{p_i} \int_{\{u_n^+ - \psi^+ \le k\}} |a_i(x, u_n, \nabla u_n)|^{p_i'} w_i^{1-p_i'} dx + \sum_{i=1}^{N} \frac{1}{p_i \lambda^{p_i}} \int_{\{u_n^+ - \psi^+ \le k\}} |\partial_i \psi^+|^{p_i} w_i dx.$$

Using to (9) and taking $\lambda = \left(\dfrac{p_i' \alpha}{6\beta} \right)^{\frac{1}{p_i'}}$, we have

$$\sum_{i=1}^{N} \int_{\{u_n^+ - \psi^+ \leq k\}} a_i(x, u_n, \nabla u_n) \partial_i u_n^+ \, dx \leq \int_{\Omega} f_n T_k(u_n^+ - \psi^+) \, dx$$

$$+ C_1(\alpha) \sum_{i=1}^{N} \int_{\{u_n^+ - \psi^+ \leq k\}} |\phi_i^n(T_{k+\|\psi\|_\infty}(u_n))|^{p_i'} w_i^{\frac{-1}{p_i-1}} \, dx + \frac{\alpha}{6} \sum_{i=1}^{N} \int_{\{u_n^+ - \psi^+ \leq k\}} |\partial_i u_n^+|^{p_i} w_i \, dx$$

$$+ \sum_{i=1}^{N} \int_{\{u_n^+ - \psi^+ \leq k\}} |\phi_i^n(T_{k+\|\psi\|_\infty}(u_n))| |\partial_i \psi^+| \, dx + \sum_{i=1}^{N} \frac{\alpha}{6} \int_{\{u_n^+ - \psi^+ \leq k\}} R_i(x)|^{p_i'} \, dx$$

$$+ \sum_{i=1}^{N} \frac{\alpha}{6} \int_{\{u_n^+ - \psi^+ \leq k\}} |u_n^+|^{p_i} w_i \, dx + \sum_{i=1}^{N} \frac{\alpha}{6} \int_{\{u_n^+ - \psi^+ \leq k\}} |\partial_i u_n^+|^{p_i} w_i \, dx$$

$$+ \sum_{i=1}^{N} \frac{(6\beta)^{p_i-1}}{p_i(p_i'\alpha)^{p_i-1}} \int_{\{u_n^+ - \psi^+ \leq k\}} |\partial_i \psi^+|^{p_i} w_i \, dx.$$

Combining (3), (8), (9), (10) and (H_1), we have

$$\sum_{i=1}^{N} \int_{\{u_n^+ - \psi^+ \leq k\}} |\partial_i u_n^+|^{p_i} w_i \, dx \leq Ck + C' \tag{19}$$

As $\{x \in \Omega, u^+ \leq k\} \subset \{x \in \Omega, u^+ - \psi^+ \leq k + \|\psi^+\|_\infty\}$, then

$$\sum_{i=1}^{N} \int_{\Omega} |\partial_i T_k(u_n^+)|^{p_i} w_i \, dx = \sum_{i=1}^{N} \int_{\{u^+ \leq k\}} |\partial_i u_n^+|^{p_i} w_i \, dx \leq \sum_{i=1}^{N} \int_{\{u^+ - \psi^+ \leq k + \|\psi^+\|_\infty\}} |\partial_i u_n^+|^{p_i} w_i \, dx.$$

Hence, thanks to (19), we get

$$\sum_{i=1}^{N} \int_{\Omega} |\partial_i T_k(u_n^+)|^{p_i} w_i \, dx \leq (k + \|\psi^+\|_\infty) C + C' \quad \forall k > 0. \tag{20}$$

Similarly taking $v = u_n + T_k(u_n^-)$ as test function in approximate problem (13), we have

$$\sum_{i=1}^{N} \int_{\Omega} |\partial_i T_k(u_n)|^{p_i} w_i \, dx \leq C''(k+1). \tag{21}$$

By (20) and (21), we obtain

$$\sum_{i=1}^{N} \int_{\Omega} |\partial T_k(u_n)|^{p_i} w_i(x) \, dx \leq (k + \|\psi^+\|_\infty + 1) C' \quad \text{for all } k > 0.$$

$\square$

Step 3: Strong convergence of truncations.

Proposition 3. *If u_n is a solution of approximate problem* (13). *Then there is a function u and a subsequence of u_n such that*

$$T_k(u_n) \to T(u) \quad \text{strongly in} \quad W_0^{1,\vec{p}}(\Omega, \vec{w})$$

Proof. According to Proposition 2, we obtain

$$\|T_k(u_n)\|_{W_0^{1,\vec{p}}(\Omega, \vec{w})} \le C(k + \|\psi^+\|_\infty + 1)^{\frac{1}{p^-}}. \tag{22}$$

Firstly, we shall demonstrate that $(u_n)_n$ is a Cauchy sequence in measure in Ω. For every $\lambda > 0$, we obtain $\{|u_n - u_m| > \lambda\} \subset \{|u_n| > k\} \cup \{|u_m| > k\} \cup \{|T_k(u_n) - T_k(u_m)| > \lambda\}$, thus

$$meas\{|u_n - u_m| > \lambda\} \le meas\{|u_n| > k\} + meas\{|u_m| > k\} \\ + meas\{|T_k(u_n) - T_k(u_m)| > \lambda\}. \tag{23}$$

Using Hölder's inequality, Lemma 1 and (22), we have

$$k.meas\{|u_n| > k\} = \int_{\{|u_n|>k\}} |T_k(u_n)| dx \le \int_\Omega |T(u_n) dx$$
$$\le (meas(\Omega))^{\frac{1}{p^-}} \|T_k(u_n)\|_{L^{p^-}(\Omega)}$$
$$\le C(meas(\Omega))^{\frac{1}{p^-}} \|T_k(u_n)\|_{W_0^{1,\vec{p}}(\Omega, \vec{w})}$$
$$\le C(k + \|\psi^+\|_\infty + 1)^{\frac{1}{p^-}}.$$

Thus, $meas\{|u_n| > k\} \le C\left(\frac{1}{k^{-1+p^-}} + \frac{1 + \|\psi^+\|_\infty}{kp^-}\right)^{\frac{1}{p^-}} \to 0$ as $k \to +\infty$. Which means that, for each $\varepsilon > 0$, there exists k_0 such that for all $k > k_0$, we get

$$meas\{|u_n| > k\} \le \frac{\varepsilon}{3} \quad \text{and} \quad meas\{|u_m| > k\} \le \frac{\varepsilon}{3}. \tag{24}$$

As the sequence $(T_k(u_n))_n$ is bounded in $W_0^{1,\vec{p}}(\Omega, \vec{w})$, then there exists a subsequence $(T_k(u_n))_n$ such that $T(u_n)$ converges to v_k a.e. in Ω, weakly in $W_0^{1,\vec{p}}(\Omega, \vec{w})$ and strongly in $L^{p^-}(\Omega)$ as n goes to $+\infty$. Which implies that the sequence $(T_k(u_n))_n$ is a Cauchy sequence in measure in Ω, then for all $\lambda > 0$, there is n_0 such that

$$meas\{|T_k(u_n) - T_k(u_m)| > \lambda\} \le \frac{\varepsilon}{3}, \qquad \forall n, m \ge n_0. \tag{25}$$

Using (23), (24) and (25), then $\forall \lambda, \varepsilon > 0$, we have

$$meas\{|u_n - u_m| > \lambda\} \le \varepsilon \quad \text{for all } n, m \ge n_0.$$

Hence $(u_n)_n$ is a Cauchy sequence in measure in Ω, then there exists a subsequence denoted again by $(u_n)_n$ such that u_n converges to a measurable function u a.e. in Ω and

$$T_k(u_n) \rightharpoonup T(u) \quad \text{weakly in } W_0^{1,\vec{p}}(\Omega, \vec{w}) \quad \text{and a.e. in } \Omega \text{ for all } k > 0. \tag{26}$$

Now, we will prove that

$$\lim_{n\to\infty}\sum_{i=1}^{N}\int_{\Omega}[a_i(x,T_k(u_n),\nabla T_k(u_n))-a_i(x,T_k(u_n),\nabla T_k(u))](\partial_i T_k(u_n)-\partial_i T_k(u))dx=0.$$
(27)

Let us consider $v=u_n+T_1(u_n-T_m(u_n))^-$ as test function in approximate problem (13), we obtain

$$-\sum_{i=1}^{N}\int_{\Omega}a_i(x,u_n,\nabla u_n)\partial_i T_1(u_n-T_m(u_n))^-dx$$

$$-\sum_{i=1}^{N}\int_{\Omega}\phi_i^n(u_n)\partial_i T_1(u_n-T_m(u_n))^-dx\leq-\int_{\Omega}f_n T_1(u_n-T_m(u_n))^-dx.$$

Then

$$\sum_{i=1}^{N}\int_{\{-(m+1)\leq u_n\leq-m\}}a_i(x,u_n,\nabla u_n)\partial_i u_n dx$$

$$+\sum_{i=1}^{N}\int_{\{-(m+1)\leq u_n\leq-m\}}\phi_i(u_n)\partial_i u_n dx\leq-\int_{\Omega}f_n T_1(u_n-T_m(u_n))^-dx.$$

We pose $\Phi_i^n(s)=\int_0^s\phi_i^n(t)\chi_{\{-(m+1)\leq t\leq-m\}}dt$. By using the Green's formula, we obtain

$$\sum_{i=1}^{N}\int_{\{-(m+1)\leq u_n\leq-m\}}\phi_i(u_n)\partial_i u_n dx=\sum_{i=1}^{N}\int_{\Omega}\partial_i\Phi_i^n(u_n)dx=0.$$

Then, we have

$$\sum_{i=1}^{N}\int_{\{-(m+1)\leq u_n\leq-m\}}a_i(x,u_n,\nabla u_n)\partial_i u_n dx\leq-\int_{\Omega}f_n T_1(u_n-T_m(u_n))^-dx$$

According to Lebesgue's theorem, we have

$$\lim_{m\to+\infty}\limsup_{n\to+\infty}\int_{\Omega}f_n T_1(u_n-T_m(u_n))^-dx=0$$

Then, we get

$$\lim_{m\to+\infty}\limsup_{n\to+\infty}\sum_{i=1}^{N}\int_{\{-(m+1)\leq u_n\leq-m\}}a_i(x,u_n,\nabla u_n)\partial_i u_n dx=0.$$
(28)

Similarly, we choose $v=u_n-\eta T_1(u_n-T_m(u_n))^+$ as test function in approximate problem (13), we have

$$\lim_{m\to\infty}\limsup_{n\to\infty}\sum_{i=1}^{N}\int_{\{m\leq u_n\leq m+1\}}a_i(x,u_n,\nabla u_n)\partial_i u_n dx=0.$$
(29)

We define the following function for each $m > k$:

$$h_m(z) = \begin{cases} 1 & \text{if } |z| \leq m \\ 0 & \text{if } |z| \geq m+1 \\ m+1-|z| & \text{if } m \leq |z| \leq m+1, \end{cases}$$

By using in (13) the test function $\varphi = u_n - \eta(T_k(u_n) - T(u))^+ h_m(u_n)$, we obtain

$$\sum_{i=1}^{N} \int_{\Omega} a_i(x, u_n, \nabla u_n) \partial_i (T_k(u_n) - T(u))^+ h_m(u_n) dx$$

$$+ \sum_{i=1}^{N} \int_{\Omega} a_i(x, u_n \nabla u_n)(T_k(u_n) - T_k(u))^+ \partial_i u_n h'_m(u_n) dx$$

$$+ \sum_{i=1}^{N} \int_{\Omega} \phi_i^n(u_n) \partial_i (T_k(u_n) - T_k(u))^+ h_m(u_n) dx \tag{30}$$

$$+ \sum_{i=1}^{N} \int_{\Omega} \phi_i^n(u_n) \partial_i u_n (T_k(u_n) - T_k(u))^+ h'_m(u_n) dx$$

$$\leq \int_{\Omega} f_n(T_k(u_n) - T_k(u))^+ h_m(u_n) dx.$$

Using (28) and (29), we get the second integral in (30) converges to 0 when n and m tend to $+\infty$.

As $h_m(u_n) = 0$ if $|u_n| > m+1$. Then, we obtain

$$\sum_{i=1}^{N} \int_{\Omega} \phi_i^n(u_n) \partial_i (T_k(u_n) - T_k(u))^+ h_m(u_n) dx$$

$$= \sum_{i=1}^{N} \int_{\Omega} \phi_i(T_{m+1}(u_n)) h_m(u_n) \partial_i (T_k(u_n) - T_k(u))^+ dx.$$

By Lebesgue's theorem, we get $\phi_i^n(T_{m+1}(u_n)) h_m(u_n) \to \phi_i(T(u)) h_m(u)$ in $L^{p'_i}(\Omega, w_i^*)$ and $\partial_i T_k(u_n) \rightharpoonup \partial_i T(u)$ weakly in $L^{p_i}(\Omega, w_i)$ as n goes to $+\infty$, then the third integral in (30) converges to 0 when n and m tend to $+\infty$.

Combining (8), (28), (29) and Lebesgue's theorem, we get

$$\lim_{m \to +\infty} \lim_{n \to +\infty} \sum_{i=1}^{N} \int_{\{-(m+1) \leq u_n \leq -m\}} |\partial_i u_n|^{p_i} (T_k(u_n) - T_k(u))^+ w_i dx = 0,$$

and

$$\lim_{m \to +\infty} \lim_{n \to +\infty} \sum_{i=1}^{N} \int_{\{m \leq u_n \leq m+1\}} |\partial_i u_n|^{p_i} (T_k(u_n) - T_k(u))^+ w_i dx = 0.$$

We conclude that

$$\lim_{m \to +\infty} \lim_{n \to +\infty} \sum_{i=1}^{N} \int_{\Omega} a_i(x, u_n, \nabla u_n) \partial_i (T_k(u_n) - T_k(u))^+ h_m(u_n) dx \leq 0,$$

which implies that

$$\lim_{m\to+\infty}\lim_{n\to+\infty}\sum_{i=1}^{N}\int_{\{T_k(u_n)-T_k(u)\geq 0,|u_n|\leq k\}}a_i(x,u_n,\nabla u_n)\partial_i(T_k(u_n)-T_k(u))h_m(u_n)dx$$

$$-\lim_{m\to+\infty}\lim_{n\to+\infty}\sum_{i=1}^{N}\int_{\{T_k(u_n)-T_k(u)\geq 0,|u_n|>k\}}a_i(x,u_n,\nabla u_n)\partial_iT_k(u)h_m(u_n)dx\leq 0.$$

As $h_m(u_n)=0$ in $\{|u_n|>m+1\}$, then we obtain

$$\sum_{i=1}^{N}\int_{\{T_k(u_n)-T_k(u)\geq 0,|u_n|>k\}}a_i(x,u_n,\nabla u_n)\partial_iT_k(u)h_m(u_n)dx$$

$$=\sum_{i=1}^{N}\int_{\{T_k(u_n)-T_k(u)\geq 0,|u_n|>k\}}a_i(x,T_{m+1}(u_n),\nabla T_{m+1}(u_n))\partial_iT_k(u)h_m(u_n)dx.$$

Since $(a_i(x,T_{m+1}(u_n),\nabla T_{m+1}(u_n)))_{n\geq 0}$ is bounded in $L^{p'_i}(\Omega,w_i^*)$.
We have $a_i(x,T_{m+1}(u_n),\nabla T(u))$ converges to Y_m^i weakly in $L^{p'_i}(\Omega,w_i^*)$. Hence

$$\lim_{m\to+\infty}\lim_{n\to+\infty}\sum_{i=1}^{N}\int_{\{T_k(u_n)-T_k(u)\geq 0,|u_n|>k\}}a_i(x,T_{m+1}(u_n),\nabla T_{m+1}(u_n))\partial_iT_k(u)h_m(u_n)dx$$

$$=\lim_{m\to+\infty}\sum_{i=1}^{N}\int_{\{|u|>k\}}Y_m^i\partial_iT_k(u)h_m(u)dx=0,$$

as results

$$\lim_{m\to+\infty}\lim_{n\to+\infty}\sum_{i=1}^{N}\int_{\{T_k(u_n)-T_k(u)\geq 0\}}a_i(x,T_k(u_n),\nabla T_k(u_n))$$
$$\partial_i(T_k(u_n)-T_k(u))h_m(u_n)dx\leq 0. \quad (31)$$

Moreover, we have $a_i(x,T_k(u_n),\nabla T_k(u))h_m(u_n)\to a_i(x,T_k(u),\nabla T_k(u))h_m(u)$ in $L^{p'_i}(\Omega,w_i^*)$ and $\partial_i(T_k(u_n)-T_k(u))$ converges to 0 weakly in $L^{p_i}(\Omega,w_i)$, then

$$\lim_{m\to+\infty}\lim_{n\to+\infty}\sum_{i=1}^{N}\int_{\{T_k(u_n)-T_k(u)\geq 0\}}a_i(x,T_k(u_n),\nabla T_k(u))$$
$$\partial_i(T_k(u_n)-T_k(u))h_m(u_n)dx=0. \quad (32)$$

According to (10), (31) and (32), we deduce

$$\lim_{m\to+\infty}\lim_{n\to+\infty}\sum_{i=1}^{N}\int_{\{T_k(u_n)-T_k(u)\geq 0\}}\left[a_i(x,T_k(u_n),\nabla T_k(u_n))-a_i(x,T_k(u_n),\nabla T_k(u))\right]$$
$$\partial_i(T_k(u_n)-T_k(u))h_m(u_n)dx=0. \quad (33)$$

Similarly, we choose $\varphi=u_n+(T_k(u_n)-T_k(u))^-h_m(u_n)$ as test function in (13), we obtain

$$\lim_{m\to+\infty}\lim_{n\to+\infty}\sum_{i=1}^{N}\int_{\{T_k(u_n)-T_k(u)\leq 0\}}\left[a_i(x,T_k(u_n),\nabla T_k(u_n))-a_i(x,T_k(u_n),\nabla T_k(u))\right]$$
$$\partial_i(T_k(u_n)-T_k(u))h_m(u_n)dx=0. \quad (34)$$

Using (33) and (34), we have

$$\lim_{m \to +\infty} \lim_{n \to +\infty} \sum_{i=1}^{N} \int_{\Omega} \left[a_i(x, T_k(u_n), \nabla T_k(u_n)) - a_i(x, T_k(u_n), \nabla T_k(u)) \right]$$
$$\partial_i(T_k(u_n) - T_k(u)) h_m(u_n) dx = 0. \quad (35)$$

Now, we show

$$\lim_{m \to +\infty} \lim_{n \to +\infty} \sum_{i=1}^{N} \int_{\Omega} \left[a_i(x, T_k(u_n), \nabla T_k(u_n)) - a_i(x, T_k(u_n), \nabla T_k(u)) \right]$$
$$\partial_i(T_k(u_n) - T_k(u))(1 - h_m(u_n)) dx = 0.$$

Let $\varphi = u_n + T_k(u_n)^-(1 - h_m(u_n))$ as test function in approximate problem (1), we obtain

$$-\sum_{i=1}^{N} \int_{\Omega} a_i(x, u_n, \nabla u_n) \partial_i T_k(u_n)^-(1 - h_m(u_n)) dx$$

$$+\sum_{i=1}^{N} \int_{\Omega} a_i(x, u_n, \nabla u_n) \partial_i u_n T_k(u_n)^- h'_m(u_n) dx - \sum_{i=1}^{N} \int_{\Omega} \phi_i(u_n) \partial_i T_k(u_n)^-(1 - h_m(u_n)) dx$$

$$+\sum_{i=1}^{N} \int_{\Omega} \phi_i(u_n) \partial_i u_n T_k(u_n)^- h'_m(u_n) dx \leq -\int_{\Omega} f_n T_k(u_n)^-(1 - h_m(u_n)) dx. \quad (36)$$

Thanks to (28) and (29), we have

$$\lim_{m \to +\infty} \lim_{n \to +\infty} \sum_{i=1}^{N} \int_{\Omega} a_i(x, u_n, \nabla u_n) \partial_i u_n T_k(u_n)^- h'_m(u_n) dx = 0.$$

Thus, the second integral in (36) converges to 0 when n and m goes to $+\infty$. As $\partial_i T_k(u_n)^- \rightharpoonup \partial_i T_k(u)^-$ in $L^{p_i}(\Omega, w_i)$ and $\phi_i(T_k(u_n))(1 - h_m(u_n)) \to \phi_i(T_k(u))(1 - h_m(u))$ strongly in $L^{p'_i}(\Omega, w_i^*)$, we get

$$\lim_{m \to +\infty} \lim_{n \to +\infty} \sum_{i=1}^{N} \int_{\Omega} \phi_i(u_n) \partial_i T_k(u_n)^-(1 - h_m(u_n)) dx$$

$$= \lim_{m \to +\infty} \sum_{i=1}^{N} \int_{\Omega} \phi_i(T_k(u)) \partial_i T_k(u)^-(1 - h_m(u)) dx.$$

In view to Lebesgue's theorem, we get

$$\lim_{m \to +\infty} \sum_{i=1}^{N} \int_{\Omega} \phi_i(T_k(u)) \partial_i T_k(u)^-(1 - h_m(u)) dx = 0.$$

Hence, the third integral in (36) converges to 0 when m and n tends to $+\infty$.

We take $\Phi_i^n(t) = \int_0^t \phi_i(s) T_k(s)^- h'_m(s) ds$, in light of Green's Formula, we obtain

$$\sum_{i=1}^{N} \int_{\Omega} \phi_i^n(u_n) \partial_i u_n T_k(u_n)^- h'_m(u_n) dx = \sum_{i=1}^{N} \int_{\Omega} \partial_i \Phi_i^n(u_n) dx = 0.$$

Then the last integral of the left-hand side of (36) converges to 0 when n and m tend to $+\infty$. By using to Lebesgue dominated convergence theorem, we get the term of the right-hand side of (36) converges to 0 as m and n goes to $+\infty$. We Conclude

$$\lim_{m\to+\infty}\lim_{n\to+\infty}\sum_{i=1}^{N}\int_{\{u_n\leq0\}}a_i(x,u_n,\nabla u_n)\partial_i T_k(u_n)(1-h_m(u_n))dx=0. \tag{37}$$

Following this, for η small enough, we choose $\varphi=u_n-\eta T_k(u_n^+-\psi^+)(1-h_m(u_n))$ as test function in (13), we obtain

$$\sum_{i=1}^{N}\int_{\Omega}a_i(x,u_n,\nabla u_n)\partial_i T_k(u_n^+-\psi^+)(1-h_m(u_n))dx$$

$$-\sum_{i=1}^{N}\int_{\Omega}a_i(x,u_n,\nabla u_n)\partial_i u_n T_k(u_n^+-\psi^+)h_m'(u_n)dx$$

$$+\sum_{i=1}^{N}\int_{\Omega}\phi_i^n(u_n)\partial_i T_k(u_n^+-\psi^+)(1-h_m(u_n))dx$$

$$-\sum_{i=1}^{N}\int_{\Omega}\phi_i^n(u_n)\partial_i u_n T_k(u_n^+-\psi^+)h_m'(u_n)dx$$

$$\leq\int_{\Omega}f_n T_k(u_n^+-\psi^+)(1-h_m(u_n))dx. \tag{38}$$

From the Hölder inequality, (8), (28) and (29), we get

$$\lim_{m\to+\infty}\lim_{n\to+\infty}\sum_{i=1}^{N}\int_{\Omega}\phi_i^n(u_n)\partial_i u_n T_k(u_n^+-\psi^+)h_m'(u_n)dx=0.$$

By the Young inequality, we obtain

$$\sum_{i=1}^{N}\int_{\Omega}a_i(x,u_n,\nabla u_n)\partial_i T_k(u_n^+-\psi^+)(1-h_m(u_n))dx$$

$$\leq\sum_{i=1}^{N}\int_{\{-(m+1)\leq u_n\leq-m\}}a_i(x,u_n,\nabla u_n)\partial_i u_n T_k(u_n^+-\psi^+)dx$$

$$+\int_{\Omega}f_n T_k(u_n^+-\psi^+)(1-h_m(u_n))dx \tag{39}$$

$$+\sum_{i=1}^{N}\int_{\{u_n^+-\psi^+\leq k\}}\phi_i^n(u_n)\partial_i u_n^+(1-h_m(u_n))dx$$

$$+\sum_{i=1}^{N}\int_{\{u_n^+-\psi^+\leq k\}}\phi_i^n(u_n)\partial_i\psi^+(1-h_m(u_n))dx$$

Thank to (28), we get the first term on the right-hand converges to 0 when n and m tend to $+\infty$. By the Lebesgue dominated convergence theorem, we obtain the second part in the right-hand converges to 0 when m and n tend to $+\infty$.

As

$$\sum_{i=1}^{N} \int_{\{u_n^+ - \psi^+ \le k\}} \phi_i^n(u_n) \partial_i u_n^+ (1 - h_m(u_n)) dx$$

$$= \sum_{i=1}^{N} \int_{\Omega} \phi_i^n(T_{\{k+\|\psi^+\|_{L^\infty(O)}\}}(u_n)) \partial_i T_{\{k+\|\psi^+\|_{L^\infty(O)}\}}(u_n^+)(1 - h_m(u_n)) dx. \tag{40}$$

Since $\partial_i T_{\{k+\|\psi^+\|_{L^\infty(\Omega)}\}}(u_n^+) \rightharpoonup \partial_i T_{\{k+\|\psi^+\|_{L^\infty(\Omega)}\}}(u^+)$ weakly in $L^{p_i}(\Omega, w_i)$ and $\phi_i^n(T_{\{k+\|\psi^+\|_{L^\infty(\Omega)}\}}(u_n))(1 - h_m(u_n)) \rightarrow \phi_i(T_{\{k+\|\psi^+\|_{L^\infty(\Omega)}\}}(u))(1 - h_m(u))$ strongly in $L^{p_i'}(\Omega, w_i^*)$, we obtain

$$\sum_{i=1}^{N} \int_{\Omega} \phi_i^n(T_{\{k+\|\psi^+\|_{L^\infty(\Omega)}\}}(u_n)) \partial_i T_{\{k+\|\psi^+\|_{L^\infty(\Omega)}\}}(u_n^+)(1 - h_m(u_n)) dx$$

$$= \sum_{i=1}^{N} \int_{\Omega} \phi_i(T_{\{k+\|\psi^+\|_{L^\infty(\Omega)}\}}(u)) \partial_i T_{\{k+\|\psi^+\|_{L^\infty(\Omega)}\}}(u)(1 - h_m(u)) dx + \varepsilon(n).$$

Using the Lebesgue dominated convergencetheorem, we obtain

$$\lim_{m \to \infty} \sum_{i=1}^{N} \int_{\Omega} \phi_i(T_{\{k+\|\psi^+\|_{L^\infty(\Omega)}\}}(u)) \partial_i T_{\{k+\|\psi^+\|_{L^\infty(\Omega)}\}}(u)(1 - h_m(u)) dx = 0.$$

Hence, we get the third integral converges to 0 as m and n tend to $+\infty$. Similarly as (37), we have

$$\lim_{m \to +\infty} \lim_{n \to +\infty} \sum_{i=1}^{N} \int_{\{u_n > 0\}} a_i(x, u_n, \nabla u_n) \partial_i T_k(u_n)(1 - h_m(u_n)) dx = 0. \tag{41}$$

According to (37) and (41), we obtain

$$\lim_{m \to +\infty} \lim_{n \to +\infty} \sum_{i=1}^{N} \int_{\Omega} a_i(x, u_n, \nabla u_n) \partial_i T_k(u_n)(1 - h_m(u_n)) dx = 0. \tag{42}$$

Furthermore, we have

$$\sum_{i=1}^{N} \int_{\Omega} (a_i(x, T_k(u_n), \nabla T_k(u_n)) - a_i(x, T_k(u_n), \nabla T_k(u)))(\partial_i T_k(u_n) - \partial_i T_k(u)) dx$$

$$= \sum_{i=1}^{N} \int_{\Omega} (a_i(x, T_k(u_n), \nabla T_k(u_n)) - a_i(x, T_k(u_n), \nabla T_k(u)))(\partial_i T_k(u_n) - \partial_i T_k(u)) h(u_n) dx$$

$$+ \sum_{i=1}^{N} \int_{\Omega} (a_i(x, T_k(u_n), \nabla T_k(u_n))) \partial_i T_k(u_n)(1 - h_m(u_n)) dx$$

$$- \sum_{i=1}^{N} \int_{\Omega} (a_i(x, T_k(u_n), \nabla T_k(u_n))) \partial_i T_k(u)(1 - h_m(u_n)) dx$$

$$-\sum_{i=1}^{N}\int_{\Omega}(a_i(x,T_k(u_n),\nabla T_k(u)))(\partial_i T_k(u_n)-\partial_i T_k(u))(1-h_m(u_n))dx.$$

Combining (35) and (42), the first and the second integrals on the right-hand converge to 0 when m and n goes to ∞.

As $(a_i(x,T_k(u_n),\nabla T_k(u_n)))_n$ is bounded in $L^{p'_i}(\Omega,w_i^*)$ and $\partial_i T_k(u)(1-h_m(u_n)) \longrightarrow 0$ in $L^{p_i}(\Omega,w_i)$ when m and n goes to $+\infty$, hence the third term on the right-hand side converge to 0 as m and n goes to $+\infty$.

Where

$$a_i(x,T_k(u_n),\nabla T_k(u_n))(1-h_m(u_n)) \longrightarrow a_i(x,T_k(u),\nabla T_k(u))(1-h_m(u))$$

strongly in $L^{p'_i}(\Omega,w_i^*)$ and $\partial_i T_k(u_n) \rightharpoonup \partial_i T(u)$ weakly in $L^{p_i}(\Omega,w_i)$, we get the last integral on the right-hand side converge to 0 as m and n goes to $+\infty$. Then, we obtain (27).

Thanks to (26), (27) and Lemma 3, we have

$$T_k(u_n) \to T(u) \quad \text{strongly in} \quad W_0^{1,\overrightarrow{p}}(\Omega,\overrightarrow{w}) \quad \text{and a. e. in} \quad \Omega \quad \text{for all} k > 0.$$

$\square$

Step 4: Passing to the limit.

Let $\varphi \in K_\psi \cap L^\infty(\Omega)$, we choose $v = u_n - T_k(u_n - \varphi)$ as test function in approximate problem (13), we have

$$\sum_{i=1}^{N}\int_{\Omega}a_i(x,u_n,\nabla u_n)\partial_i T_k(u_n-\varphi)dx + \sum_{i=1}^{N}\int_{\Omega}\phi_i^n(u_n)\partial_i T_k(u_n-\varphi)dx$$
$$\leq \int_{\Omega}f_n T_k(u_n-\varphi)dx, \tag{43}$$

which implies that,

$$\sum_{i=1}^{N}\int_{\Omega}a_i(x,T_{k+\|\varphi\|_\infty}(u_n),\nabla T_{k+\|\varphi\|_\infty}(u_n))\partial_i T_k(u_n-\varphi)dx$$
$$+\sum_{i=1}^{N}\int_{\Omega}\phi_i(T_{k+\|\varphi\|_\infty}(u_n))\partial_i T_k(u_n-\varphi)dx \leq \int_{\Omega}f_n T_k(u_n-\varphi)dx.$$

As $T_k(u_n) \to T(u)$ strongly in $W_0^{1,\overrightarrow{p}}(\Omega,\overrightarrow{w})$ and a.e. in Ω for all $k > 0$, we obtain

$$a_i(x,T_{k+\|\varphi\|_\infty}(u_n),\nabla T_{k+\|\varphi\|_\infty}(u_n)) \rightharpoonup a_i(x,T_{k+\|\varphi\|_\infty}(u),\nabla T_{k+\|\varphi\|_\infty}(u))\text{weakly in}L^{p'_i}(\Omega,w_i^*)$$

$$\phi_i(T_{k+\|\varphi\|_\infty}(u_n)) \to \phi_i(T_{k+\|\varphi\|_\infty}(u)) \quad \text{strongly in } L^{p'_i}(\Omega,w_i^*)$$

and

$$\partial_i T_k(u_n-\varphi) \to \partial_i T_k(u-\varphi) \quad \text{strongly in} \quad L^{p_i}(\Omega,w_i).$$

Passing to the limit in (43) and this completes the proof of theorem 1.

References

1. Abbassi, A., Azroul, E., Barbara, A.: Degenerate $p(x)$-elliptic equation with second membre in L^1. Adv. Sci. Technol. Eng. Syst. J. **2**(5), 45–54 (2017)
2. Adams, R.: Sobolev Spaces. Academic Press, New York (1975)
3. Akdim, Y., Azroul, E., Benkirane, A.: Existence of solutions for quasilinear degenerate elliptic equations. Electron. J. Differ. Equ. (EJDE), **2001**, p. 19, Paper No. 71 (2001)
4. Akdim, Y., Allalou, C., Salmani, A.: Existence of solutions for some nonlinear elliptic anisotropic unilateral problems with lower order terms. Moroccan J. Pure Appl. Anal. **4**(2), 171–188 (2018)
5. Azroul, E., Benboubker, M.B., Hjiaj, H., Yazough, C.: Existence of solutions for a class of obstacle problems with L^1-data and without sign condition. Afrika Matematika **27**(5–6), 795–813 (2016)
6. Azroul, E., Benboubker, M.B., Ouaro, S.: The obstacle problem associated with nonlinear elliptic equations in generalized Sobolev spaces. Nonlinear Dyn. Syst. Theory **14**(3), 223–242 (2014)
7. Benilan, P., Boccardo, L., Gallouet, T., Gariepy, R., Pierre, M., Vazquez, J.L.: An L^1 theory of existence and uniqueness of nonlinear elliptic equations. Ann. Scuola Norm. Sup. Pisa **22**(2), 240–273 (1995)
8. Benkirane, A., Elmahi, A.: Strongly nonlinear elliptic unilateral problems having natural growth terms and L^1 data. Rendiconti di matematica, Serie VII **18**, 289–303 (1998)
9. Benkirane, A., Chrif, M., El Manouni, S.: Existence results for strongly nonlinear elliptic equations of infinite order. Z. Anal. Anwend. (J. Anal. Appl.) **26**, 303–312 (2007)
10. Boccardo, L., Gallouet, T., Marcellini, P.: Anisotropic equations in L^1. Differ. Integr. Equ. **1**, 209–212 (1996)
11. Brezis, H., Strauss, W.: Semilinear second-order elliptic equations in L^1. J. Math. Soc. Jpn. **25**(4), 565–590 (1973)
12. Chrif, M., El Manouni, S.: Anisotropic equations in weighted Sobolev spaces of higher order. Ricerche di matematica **58**(1), 1–14 (2009)
13. Lions, J.L.: Quelques méthodes de résolution des problèmes aux limites non linéaires. Dunod (1969)
14. Prignet, A.: Remarks on existence and uniqueness of solutions of elliptic problems with right-hand side measures. Rend. Mat. **15**, 321–337 (1995)
15. Salmani, A., Akdim, Y., Redwane, H.: Entropy solutions of anisotropic elliptic nonlinear obstacle problem with measure data. Ricerche di Matematica, **1–31** (2019)
16. Serrin, J.: Pathological solutions of elliptic differential equations. Ann. Scuola Norm. Sup. Pisa **18**, 385–387 (1964)
17. Yazouh, C., Azroul, E., Redwane, H.: Existence of solutions for some nonlinear elliptic unilateral problem with measure data. Electron. J. Qual. Theory Differ. Equ. **43**, 1–21 (2013)

Well-Posedness and Stability for the Viscous Primitive Equations of Geophysics in Critical Fourier-Besov-Morrey Spaces

A. Abbassi, C. Allalou, and Y. Oulha[✉]

Laboratory LMACS of Beni Mellal, University Sultan Moulay Slimane University
Morocco, Beni Mellal, Morocco
abbassi91@yahoo.fr, chakir.allalou@yahoo.fr, oulhayounes@gmail.com

Abstract. In this paper we study the Cauchy problem of the viscous primitive equations of geophysics in critical Fourier-Besov-Morrey spaces. By using the Fourier localization argument and the Littlewood-Paley theory, we prove that the Cauchy problem with Prankster number $P = 1$ is local well-posedness and global well-posedness when the initial data (u_0, θ_0) are small and we give a stability result for global solutions.

Keywords: Navier-Stokes equations · Global well-posedness ·
Analytic solutions · Coriolis force · Fourier-Besov-Morrey space

1 Introduction

In this paper, we study the initial value problem of the viscous primitive equations of geophysics in $\mathbb{R}^3$, which is a fundamental mathematical model in the field of fluid geophysics. The model reads as follows:

$$
\begin{cases}
\partial_t u + \nu \Delta u + \Omega e_3 \times u + (u.\nabla)u + \nabla p = g\theta e_3 & (t,x) \in \mathbb{R}^+ \times \mathbb{R}^3, \\[2mm]
\partial_t \theta + \mu \Delta \theta + (u.\nabla)\theta = -\mathcal{N}^2 u_3 & (t,x) \in \mathbb{R}^+ \times \mathbb{R}^3, \\[2mm]
\nabla.u = 0, \\[2mm]
u(0,x) = u_0(x) & x \in \mathbb{R}^3,
\end{cases}
\tag{1.1}
$$

where $u = u(t,x) = (u^1(t,x), u^2(t,x), u^3(t,x))$ and $p = p(t,x)$ denotes the unknown velocity field and the unknown pressure of the fluid at the point $(t,x) \in \mathbb{R}^+ \times \mathbb{R}^3$, respectively and θ is a scalar function representing the density fluctuation in the fluid (in the case of the ocean it depends on the temperature and the salinity, and in the case of the atmosphere it depends on the temperature), while $u_0 = u_0(x) = (u_0^1(x), u_0^2(x), u_0^3(x))$ denote the given initial velocity flied satisfying the compatibility condition $\nabla.u = 0$. ν, μ and g are positive

© Springer Nature Switzerland AG 2021
Z. Hammouch et al. (Eds.): SM2A 2019, LNNS 168, pp. 123–140, 2021.
https://doi.org/10.1007/978-3-030-62299-2_9

constants related to viscosity, diffusivity and gravity, respectively, $\Omega \in \mathbb{R}$ represents the speed of rotation around the vertical unit vector $e_3 = (0, 0, 1)$, which is called the Coriolis parameter, and "$\times$" represents the outer product, hence, $-\Omega e_3 \times u = (\Omega u_2, -\Omega u_1, 0)$. We recall that the Coriolis term has an another expression $-\Omega e_3 \times u = -\Omega J u$, where the skew-symmetric matrix J defined by

$$J = \begin{pmatrix} 0 & -1 & 0 \\ 1 & 0 & 0 \\ 0 & 0 & 0 \end{pmatrix}.$$

$\mathcal{N}$ is the stratification parameter, a nonnegative constant representing the Brunt-Visala wave frequency. The ratio $P := \frac{\nu}{\mu}$ is known as the Prandtl number and $B := \frac{\Omega}{\mathcal{N}}$ is essentially the "Burger" number of geophysics.

When $\theta \equiv 0$, $\mathcal{N} = 0$ and $\Omega = 0$, the problem (1.1) become the classical Navier-Stokes equation:

$$\begin{cases} u_t - \nu \Delta u + (u.\nabla)u + \nabla p = 0 & (t, x) \in \mathbb{R}^+ \times \mathbb{R}^3, \\ \nabla.u = 0, \\ u(0, x) = u_0(x) & x \in \mathbb{R}^3. \end{cases}$$

The existence of mild solutions and the regularity have been established locally in time and global for small initial data in various functional spaces, for example [5–8, 28, 29, 34, 36].

If only $\theta \equiv 0$, $\mathcal{N} = 0$ but $\Omega \neq 0$ the problem (1.1) corresponds to the usual Navier-Stokes equation with Coriolis force,

$$\begin{cases} u_t - \nu \Delta u + \Omega e_3 \times u + (u.\nabla)u + \nabla p = 0 & (t, x) \in \mathbb{R}^+ \times \mathbb{R}^3, \\ \nabla.u = 0, \\ u(0, x) = u_0(x) & x \in \mathbb{R}^3. \end{cases}$$

Hieber and Shibata [22] obtained the uniform global well-posedness for the Navier-Stokes equations with Coriolis force for small initial data in the Sobolev space $H^{\frac{1}{2}}(\mathbb{R}^3)$. Iwabuchi and Takada [26] proved the existence of global solutions for the Navier-Stokes equations with Coriolis force in Sobolev spaces $\dot{H}^s(\mathbb{R}^3)$ with $1/2 < s < 3/4$ if the speed of rotation Ω is large enough compared with the norm of initial data $\|u_0\|_{\dot{H}^s}$, they also obtained the global existence and the uniqueness of the mild solution for small initial data in the Fourier-Besov spaces $\dot{FB}_{1,2}^{-1}$ and proved the ill-posedness in the space $\dot{FB}_{1,q}^{-1}$, $2 < q \leq \infty$ for all $\Omega \in \mathbb{R}$ see [27]. El Baraka and Toumlilin [10] got global well posedness result with small initial data in $\mathcal{FN}_{p,\lambda,q}^{1-2\alpha+\frac{3}{p'}+\frac{\lambda}{p}}$ for $\alpha \neq 1$ and $\Omega = 0$, moreover, in [12] they generalize this result for $\alpha \neq 1$ and $\Omega \neq 0$ where they proved local well-posedness results and global well-posedness results with small initial data in Fourier-Besov-Morrey spaces.

When $\theta \neq 0$, $\mathcal{N} \neq 0$ and $\Omega \neq 0$, Babin, Maholov and Nicolaenko [3] proved the existance of global solution for problem (1.1) in $[H^s(\mathbb{T}^3)]^4$ with $s \geqslant 3/4$ for small initial data when the stratification parameter $\mathcal{N}$ is sufficiently large. Charve [18, 19] obtained the global well-posedness of problem (1.1) in $[\dot{H}^{\frac{1}{2}}(\mathbb{R}^3) \cap$

$\dot{H}^1(\mathbb{R}^3)]^4$ under the assumptions that both Ω and $\mathcal{N}$ are sufficiently large for arbitrary initial data, moreover we get the global well-posedness of (1.1) in less regular initial value spaces. [24] J.Sun and S.Cui proved that the Cauchy problem (1.1) with $P = 1$ is locally well-posed and globally well-posed when the initial data (u_0, θ_0) are small in Fourier-Bessov spaces $FB_{p,r}^{2-\frac{3}{p}}$ for $1 < p \leq \infty$, $1 \leq r < \infty$ and $FB_{1,r}^{-1}$ for $1 \leq r \leq 2$, they also proved that such problem is ill-posed in $FB_{1,r}^{-1}$ for $2 < r \leq \infty$.

We refer to [14,23,25,32] for rich literature about global-in-time well-posedness for fluid dynamics PDEs.

We first transform the Cauchy problem in to an equivalent Cauchy probleme. By setting $N := \mathcal{N}\sqrt{g}$, $v := (v^1, v^2, v^3, v^4) := (u^1, u^2, u^3, \frac{\sqrt{g}\theta}{\mathcal{N}})$, $v_0 := (v_0^1, v_0^2, v_0^3, v_0^4) := (u_0^1, u_0^2, u_0^3, \frac{\sqrt{g}\theta_0}{\mathcal{N}})$ and $\tilde{\nabla} := (\partial_1, \partial_2, \partial_3, 0)$, (1.1) can be rewritten into the following problem:

$$\begin{cases} v_t + \mathcal{A}v + \mathcal{B}v + \tilde{\nabla}p = -(v.\tilde{\nabla})v & (t, x) \in \mathbb{R}^+ \times \mathbb{R}^3, \\ \tilde{\nabla}v = 0, \\ v(0, x) = v_0(x) \ \ x \in \mathbb{R}^3. \end{cases} \tag{1.2}$$

Where

$$\mathcal{A} = \begin{pmatrix} -\nu \triangle & 0 & 0 & 0 \\ 0 & -\nu \triangle & 0 & 0 \\ 0 & 0 & -\nu \triangle & 0 \\ 0 & 0 & 0 & -\mu \triangle \end{pmatrix} \text{ and } \mathcal{B} = \begin{pmatrix} 0 & -\Omega & 0 & 0 \\ \Omega & 0 & 0 & 0 \\ 0 & 0 & 0 & -N \\ 0 & 0 & N & 0 \end{pmatrix}.$$

To solve the original problem (1.1), we may consider the following integral equation:

$$v(t) = T_{\Omega,N}(t)v_0 - \int_0^t T_{\Omega,N}(t-\tau)\tilde{\mathbb{P}}\tilde{\nabla} \cdot (v \otimes v)d\tau, \tag{1.3}$$

where, $\tilde{\mathbb{P}} = (\tilde{\mathbb{P}}_{ij})_{4\times 4}$ the Helmholtz projection onto the divergence-free vector fields defined by:

$$\tilde{\mathbb{P}}_{ij} = \begin{cases} \delta_{ij} + R_i R_j & 1 \leq i, j \leq 3 \\ \delta_{ij} & \text{otherwise,} \end{cases}$$

and $T_{\Omega,N}(.)$ denotes Stokes-Coriolis Stratification to the linear probleme of (1.2) via Fourier transform, which is given explicitly by

$$T_{\Omega,N}(t)f = \mathcal{F}^{-1}[\cos(\frac{|\xi|'}{|\xi|}t)M_1 + \sin(\frac{|\xi|'}{|\xi|}t)M_2 + M_3] * (e^{-\nu \triangle t}f).$$

Where

$$|\xi| := \sqrt{\xi_1^2 + \xi_2^2 + \xi_3^2} \text{ and } |\xi|' := |\xi|'_{\Omega,N} := \sqrt{N^2\xi_1^2 + N^2\xi_2^2 + \Omega^2\xi_3^2}$$

and

$$M_1 = \begin{pmatrix} \frac{\Omega^2 \xi_3^2}{|\xi|'^2} & 0 & -\frac{N^2 \xi_1 \xi_3}{|\xi|'^2} & \frac{\Omega N \xi_2 \xi_3}{|\xi|'^2} \\ 0 & \frac{\Omega^2 \xi_3^2}{|\xi|'^2} & -\frac{N^2 \xi_2 \xi_3}{|\xi|'^2} & -\frac{\Omega N \xi_1 \xi_3}{|\xi|'^2} \\ -\frac{\Omega^2 \xi_1 \xi_3}{|\xi|'^2} & -\frac{\Omega^2 \xi_2 \xi_3}{|\xi|'^2} & \frac{N^2(\xi_1^2+\xi_2^2)}{|\xi|'^2} & 0 \\ \frac{\Omega N \xi_2 \xi_3}{|\xi|'^2} & -\frac{\Omega N \xi_1 \xi_3}{|\xi|'^2} & 0 & \frac{N^2(\xi_1^2+\xi_2^2)}{|\xi|'^2} \end{pmatrix}.$$

$$M_2 = \begin{pmatrix} 0 & -\frac{\Omega \xi_3^2}{|\xi||\xi|'} & -\frac{\Omega \xi_2 \xi_3}{|\xi||\xi|'} & \frac{N \xi_1 \xi_3}{|\xi||\xi|'} \\ \frac{\Omega \xi_3^2}{|\xi||\xi|'} & 0 & -\frac{\Omega \xi_1 \xi_3}{|\xi||\xi|'} & \frac{N \xi_2 \xi_3}{|\xi||\xi|'} \\ -\frac{\Omega^2 \xi_2 \xi_3}{|\xi||\xi|'} & \frac{\Omega^2 \xi_1 \xi_3}{|\xi||\xi|'} & 0 & -\frac{N(\xi_1^2+\xi_3^2)}{|\xi||\xi|'} \\ -\frac{N \xi_1 \xi_3}{|\xi||\xi|'} & -\frac{N \xi_2 \xi_3}{|\xi||\xi|'} & \frac{N(\xi_1^2+\xi_3^2)}{|\xi||\xi|'} & 0 \end{pmatrix}.$$

$$M_3 = \begin{pmatrix} \frac{N^2 \xi_2^2}{|\xi|'^2} & -\frac{N^2 \xi_1 \xi_3}{|\xi|'^2} & 0 & -\frac{\Omega N \xi_1 \xi_2}{|\xi|'^2} \\ -\frac{N^2 \xi_1 \xi_2}{|\xi|'^2} & \frac{N^2 \xi_2^2}{|\xi|'^2} & 0 & \frac{\Omega N \xi_1 \xi_3}{|\xi|'^2} \\ 0 & 0 & 0 & 0 \\ -\frac{\Omega N \xi_2 \xi_3}{|\xi|'^2} & \frac{\Omega N \xi_1 \xi_3}{|\xi|'^2} & 0 & \frac{\Omega^2 \xi_3^2}{|\xi|'^2} \end{pmatrix}.$$

Note that, denoting by M_{jk}^l-th component of the matrix $M_l(\xi)$, it is obvious that non-vanishing M_{jk}^l satisfies

$$|M_{jk}^l| \le 2 \text{ for } \xi \in \mathbb{R}^3, j, k = 1, 2, 3, 4, l = 1, 2, 3.$$

Inspired by the work [10, 24], the aim of this paper is to prove the global existence and the stability of the global solution of the viscous primitive equations of geophysics in critical Fourier-Besov-Morrey spaces, using abstract lemma on the existence of fixed point solutions.

Lemma 1.1. *Let X be a Banach space with norm $\|.\|_X$ and $B : X \times X \longmapsto X$ be a bounded bilinear operator satisfying*

$$\|B(u,v)\|_X \le \eta \|u\|_X \|v\|_X$$

for all $u, v \in X$ and a constant $\eta > 0$. Then, if $0 < \varepsilon < \frac{1}{4\eta}$ and if $y \in X$ such that $\|y\|_X \le \varepsilon$, the equation $x := y + B(x,x)$ has a solution $\overline{x}$ in X such that $\|\overline{x}\|_X \le 2\varepsilon$. This solution is the only one in the ball $\overline{B}(0, 2\varepsilon)$. Moreover, the solution depends continuously on y in the sense: if $\|y'\|_X < \varepsilon$, $x' = y' + B(x', x')$, and $\|x'\|_X \le 2\varepsilon$, then

$$\|\overline{x} - x'\|_X \le \frac{1}{1 - 4\varepsilon\eta} \|y - y'\|_X.$$

2   Preliminaries and Main Results

To give the precise statements of our main results, we first recall the definitions of the Morrey space $M_p^\lambda(\mathbb{R}^n)$, Besov space $B_{p,q}^s(\mathbb{R}^n)$ and Fourier-Besov-Morrey space $\mathcal{F}\dot{\mathcal{N}}_{p,\lambda,q}^s(\mathbb{R}^n)$ were introduced by Ferreira and Lima [16] in order to analyze a class of active scalar equations. As usual we denote by the space of Schwartz functions on $\mathbb{R}^3$, and by the space of tempred distributions on $\mathbb{R}^3$. Choose two nonnegative smooth radial functions χ, φ satisfying

$$\operatorname{supp}\varphi \subset \{\xi \in \mathbb{R}^n : \frac{3}{4} \le |\xi| \le \frac{8}{3}\}, \quad \sum_{j\in\mathbb{Z}} \varphi(2^{-j}\xi) = 1, \quad \xi \in \mathbb{R}^n\setminus\{0\},$$

$$\operatorname{supp}\chi \subset \{\xi \in \mathbb{R}^n : |\xi| \le \frac{4}{3}\}, \quad \chi(\xi) + \sum_{j\ge 0} \varphi(2^{-j}\xi) = 1, \quad \xi \in \mathbb{R}^n.$$

We denote $\varphi_j(\xi) = \varphi(2^{-j}\xi)$ and $\mathcal{P}$ the set of all polynomials. The space of tempered distributions is denoted by S'. The homogeneous dyadic blocks $\dot{\Delta}_j$ and $\dot{S}_j$ are defined for all $j \in \mathbb{Z}$ by

$$\dot{\Delta}_j u = \varphi(2^{-j}D)u = 2^{jn}\int h(2^j y)u(x-y)\,dy,$$

$$\dot{S}_j u = \sum_{k\le j-1} \dot{\Delta}_k u = \chi(2^{-j}D)u = 2^{jn}\int \tilde{h}(2^j y)u(x-y)\,dy,$$

where $h = \mathcal{F}^{-1}\varphi$ and $\tilde{h} = \mathcal{F}^{-1}\chi$.
We defined the function spaces $M_p^\lambda(\mathbb{R}^n)$.

Definition 2.1. [28,34]. For $1 \le p < \infty$, $0 \le \lambda < n$, the Morrey spaces $M_p^\lambda = M_p^\lambda(\mathbb{R}^n)$ is defined by

$$M_p^\lambda(\mathbb{R}^n) = \{f \in L_{loc}^p(\mathbb{R}^n); \|f\|_{M_p^\lambda} < \infty\},$$

where

$$\|f\|_{M_p^\lambda} = \sup_{x_0\in\mathbb{R}^n} \sup_{r>0} r^{-\frac{\lambda}{p}}\|f\|_{L^p(B(x_0,r))},$$

with $B(x_0,r)$ the ball in $\mathbb{R}^n$ with center x_0 and radius r.
The space M_p^λ endowed with the norm $\|f\|_{M_p^\lambda}$ is a Banach space.
If $1 \le p_1, p_2, p_3 < \infty$ and $0 \le \lambda_1, \lambda_2, \lambda_3 < n$ with $\frac{1}{p_3} = \frac{1}{p_1}+\frac{1}{p_2}$ and $\frac{\lambda_3}{p_3} = \frac{\lambda_1}{p_1}+\frac{\lambda_2}{p_2}$, then we have the Hölder inequality

$$\|fg\|_{M_{p_3}^{\lambda_3}} \le \|f\|_{M_{p_1}^{\lambda_1}}\|g\|_{M_{p_2}^{\lambda_2}}.$$

Also, for $1 \le p < \infty$ and $0 \le \lambda < n$,

$$\|\varphi * g\|_{M_p^\lambda} \le \|\varphi\|_{L^1}\|g\|_{M_p^\lambda}, \tag{2.1}$$

for all $\varphi \in L^1$ and $g \in M_p^\lambda$.

Bernstein type lemma in Fourier variables in Morrey spaces.

Lemma 2.2. *[16]. Let $1 \leq q \leq p < \infty$, $0 \leq \lambda_1, \lambda_2 < n$, $\frac{n-\lambda_1}{p} \leq \frac{n-\lambda_2}{q}$ and let γ be a multi-index. If $supp(\widehat{f}) \subset \{|\xi| \leq A2^j\}$, then there is a constant $C > 0$ independent of f and j such that*

$$\|(i\xi)^\gamma \widehat{f}\|_{M_q^{\lambda_2}} \leq C2^{j|\gamma|+j(\frac{n-\lambda_2}{q}-\frac{n-\lambda_1}{p})}\|\widehat{f}\|_{M_p^{\lambda_1}}. \tag{2.2}$$

Then, we define the function spaces $\mathcal{F}\dot{\mathcal{N}}^s_{p,\lambda,q}(\mathbb{R}^n)$.

Definition 2.3. (Homogeneous Besov-Morrey spaces) Let $s \in \mathbb{R}$, $1 \leq p < \infty$, $1 \leq q \leq \infty$ and $0 \leq \lambda < n$, the space $\dot{\mathcal{N}}^s_{p,\lambda,q}(\mathbb{R}^n)$ is defined by

$$\dot{\mathcal{N}}^s_{p,\lambda,q}(\mathbb{R}^n) = \left\{ u \in \mathcal{Z}'(\mathbb{R}^n); \ \|u\|_{\dot{\mathcal{N}}^s_{p,\lambda,q}(\mathbb{R}^n)} < \infty \right\}.$$

Here

$$\|u\|_{\dot{\mathcal{N}}^s_{p,\lambda,q}(\mathbb{R}^n)} = \begin{cases} \left\{ \sum_{j\in\mathbb{Z}} 2^{jqs}\|\dot{\Delta}_j u\|^q_{M_p^\lambda} \right\}^{1/q} & for \ q < \infty, \\ \sup_{j\in\mathbb{Z}} 2^{js}\|\dot{\Delta}_j u\|_{M_p^\lambda} & for \ q = \infty. \end{cases}$$

The space $\mathcal{Z}'(\mathbb{R}^n)$ denotes the topological dual of the space $\mathcal{Z}(\mathbb{R}^n) = \{f \in \mathcal{S}(\mathbb{R}^n); \partial^\alpha \widehat{f}(0) = 0 \text{ for every multi-index } \alpha\}$ and can be identified to the quotient space $\mathcal{S}'(\mathbb{R}^n)/\mathcal{P}$, where $\mathcal{P}$ represents the set of all polynomials on $\mathbb{R}^n$. We refer to [37, chap. 8] and [15] for more details.

Definition 2.4. (Homogeneous Fourier-Besov-Morrey spaces)
Let $s \in \mathbb{R}$, $0 \leq \lambda < n$, $1 \leq p < \infty$ and $1 \leq q \leq \infty$. The space $\mathcal{F}\dot{\mathcal{N}}^s_{p,\lambda,q}(\mathbb{R}^n)$ denotes the set of all $u \in \mathcal{Z}'(\mathbb{R}^n)$ such that

$$\|u\|_{\mathcal{F}\dot{\mathcal{N}}^s_{p,\lambda,q}(\mathbb{R}^n)} = \left\{ \sum_{j\in\mathbb{Z}} 2^{jqs}\|\widehat{\dot{\Delta}_j u}\|^q_{M_p^\lambda} \right\}^{1/q} < \infty, \tag{2.3}$$

with appropriate modifications made when $q = \infty$.
Note that the space $\mathcal{F}\dot{\mathcal{N}}^s_{p,\lambda,q}(\mathbb{R}^n)$ equipped with the norm (2.3) is a Banach space. Since $M_p^0 = L^p$, we have $\mathcal{F}\dot{\mathcal{N}}^s_{p,0,q} = F\dot{B}^s_{p,q}$, $\mathcal{F}\dot{\mathcal{N}}^s_{1,0,q} = F\dot{B}^s_{1,q} = \dot{\mathcal{B}}^s_q$ and $\mathcal{F}\dot{\mathcal{N}}^{-1}_{1,0,1} = \chi^{-1}$ where $\dot{\mathcal{B}}^s_q$ is the Fourier-Herz space and χ^{-1} is the Lei-Lin space [4, 11].

Now, we give the definition of the mixed space-time spaces.

Definition 2.5. Let $s \in \mathbb{R}$, $1 \leq p < \infty$, $1 \leq q, \rho \leq \infty$, $0 \leq \lambda < n$, and $I = [0, T)$, $T \in (0, \infty]$. The space-time norm is defined on $u(t, x)$ by

$$\|u(t,x)\|_{\mathcal{L}^\rho(I,\mathcal{F}\dot{\mathcal{N}}^s_{p,\lambda,q})} = \left\{ \sum_{j\in\mathbb{Z}} 2^{jqs}\|\widehat{\dot{\Delta}_j u}\|^q_{L^\rho(I,M_p^\lambda)} \right\}^{1/q},$$

and denote by $\mathcal{L}^\rho(I, \mathcal{F}\dot{\mathcal{N}}^s_{p,\lambda,q})$ the set of distributions in $S'(\mathbb{R} \times \mathbb{R}^n)/\mathcal{P}$ with finite $\|\cdot\|_{\mathcal{L}^\rho(I,\mathcal{F}\dot{\mathcal{N}}^s_{p,\lambda,q})}$ norm.

Theorem 2.6. *Let Prandtl number $P = 1$, i.e., $\mu = \nu$, $\Omega \in \mathbb{R}$, $0 \leq \lambda < 3$, $1 \leq q \leq 2$.*
For $\max\{1, \frac{3-\lambda}{2}\} \leq p < \infty$, there exists a positive time T such that for $v_0 = (u_0, \theta_0) \in \mathcal{F}\dot{\mathcal{N}}_{p,\lambda,q}^{-1+\frac{3}{p'}+\frac{\lambda}{p}}$ and $\nabla.u_0 = 0$, the problem (1.1) admits a unique local solution $(u, \theta) \in \mathcal{L}^4\left([0,T), \mathcal{F}\dot{\mathcal{N}}_{p,\lambda,q}^{-\frac{1}{2}+\frac{3}{p'}+\frac{\lambda}{p}}\right)$.
Furthermore $1 \leq p < \infty$ there exists a constant $C_0(p, q)$ such that for any $v_0 = (u_0, \theta_0) \in \mathcal{F}\dot{\mathcal{N}}_{p,\lambda,q}^{-1+\frac{3}{p'}+\frac{\lambda}{p}}$ satisfying $\nabla.u_0 = 0$ and $\|(u_0, \frac{\sqrt{g}\theta_0}{N})\|_{\mathcal{F}\dot{\mathcal{N}}_{p,\lambda,q}^{-1+\frac{3}{p'}+\frac{\lambda}{p}}} < C_0\mu$, the problem (1.1) admits a unique global solution

$$(u, \theta) \in \mathcal{L}^\infty\left([0,\infty); \mathcal{F}\dot{\mathcal{N}}_{p,\lambda,q}^{-1+\frac{3}{p'}+\frac{\lambda}{p}}\right) \cap \mathcal{L}^1\left([0,\infty), \mathcal{F}\dot{\mathcal{N}}_{p,\lambda,q}^{1+\frac{3}{p'}+\frac{\lambda}{p}}\right),$$

and it satisfies

$$\left\|(u, \frac{\sqrt{g}\theta}{N})\right\|_{\mathcal{L}^\infty([0,\infty);\mathcal{F}\dot{\mathcal{N}}_{p,\lambda,q}^{-1+\frac{3}{p'}+\frac{\lambda}{p}})} + \mu\left\|(u, \frac{\sqrt{g}\theta}{N})\right\|_{\mathcal{L}^1([0,\infty),\mathcal{F}\dot{\mathcal{N}}_{p,\lambda,q}^{1+\frac{3}{p'}+\frac{\lambda}{p}})}$$
$$\leq 2C\left\|(u_0, \frac{\sqrt{g}\theta_0}{N})\right\|_{\mathcal{F}\dot{\mathcal{N}}_{p,\lambda,q}^{-1+\frac{3}{p'}+\frac{\lambda}{p}}},$$

where C is a positive constant.

Theorem 2.7. *Let T^* denote the maximal time of existence of a solution $v = (u, \theta)$ in*
$$\mathcal{L}^\infty\left([0,T^*); \mathcal{F}\dot{\mathcal{N}}_{p,\lambda,q}^{-1+\frac{3}{p'}+\frac{\lambda}{p}}\right) \cap \mathcal{L}^1\left([0,T^*), \mathcal{F}\dot{\mathcal{N}}_{p,\lambda,q}^{1+\frac{3}{p'}+\frac{\lambda}{p}}\right). \text{ If } T^* < \infty, \text{ then}$$

$$\|v\|_{\mathcal{L}^1([0,T^*),\mathcal{F}\dot{\mathcal{N}}_{p,\lambda,q}^{1+\frac{3}{p'}+\frac{\lambda}{p}})} = \infty.$$

Besides; if $v' = (u', \theta') \in C(\mathbb{R}^+, \mathcal{F}\dot{\mathcal{N}}_{p,\lambda,q}^{-1+\frac{3}{p'}+\frac{\lambda}{p}})$ is a global solution of (1.1), and for all $v'_0 \in \mathcal{F}\dot{\mathcal{N}}_{p,\lambda,q}^{-1+\frac{3}{p'}+\frac{\lambda}{p}}$ satisfying

$$\|v'_0 - v_0\|_{\mathcal{F}\dot{\mathcal{N}}_{p,\lambda,q}^{-1+\frac{3}{p'}+\frac{\lambda}{p}}} < C_0\frac{\mu}{8}\exp\left\{-\int_0^\infty \frac{1}{C_0}(\|\mathcal{B}\| + \|v\|_{\mathcal{F}\dot{\mathcal{N}}_{p,\lambda,q}^{1+\frac{3}{p'}+\frac{\lambda}{p}}})\right\} \quad (2.4)$$

for some constant C_0 sufficiently small and $\|\mathcal{B}\|$ is matrix norm, then the viscous primitive equations starting from v_0 has a global solution v fulfilling the inequality

$$\|v'(t) - v(t)\|_{\mathcal{F}\dot{\mathcal{N}}_{p,\lambda,q}^{-1+\frac{3}{p'}+\frac{\lambda}{p}}} + \frac{\mu}{2}\|v'(s) - v(s)\|_{\mathcal{L}^1([0,t),\mathcal{F}\dot{\mathcal{N}}_{p,\lambda,q}^{1+\frac{3}{p'}+\frac{\lambda}{p}})}$$
$$< C\|v'_0 - v_0\|_{\mathcal{F}\dot{\mathcal{N}}_{p,\lambda,q}^{-1+\frac{3}{p'}+\frac{\lambda}{p}}}\exp\left\{\int_0^\infty C(\|\mathcal{B}\| + \|v\|_{\mathcal{F}\dot{\mathcal{N}}_{p,\lambda,q}^{1+\frac{3}{p'}+\frac{\lambda}{p}}})\right\}$$

where C is a positive constant.

3   Well-Posedness

In this section we present the proof of Theorem 2.6. To this end, we establish some basic estimates.

Lemma 3.1. *Let* $T > 0$, $s \in \mathbb{R}$, $0 \leq \lambda < 3$, $1 \leq p < \infty$, $1 \leq q, \rho, r \leq \infty$ *and* $f \in \mathcal{L}^r([0,T), \mathcal{F}\dot{\mathcal{N}}^s_{p,\lambda,q})$. *There exists a constant* $C > 0$ *such that*

$$\Big\| \int_0^t T_{\Omega,N}(t-\tau) f(\tau) d\tau \Big\|_{\mathcal{L}^\rho([0,T), \mathcal{F}\dot{\mathcal{N}}^s_{p,\lambda,q})} \leq C \|f\|_{\mathcal{L}^r\left([0,T), \mathcal{F}\dot{\mathcal{N}}^{s-2-\frac{2}{\rho}+\frac{2}{r}}_{p,\lambda,q}\right)}.$$

Proof: Set $1 + \frac{1}{\rho} = \frac{1}{\tilde{\rho}} + \frac{1}{r}$. The definition of the space-time norm of $\mathcal{L}^\rho([0,T), \mathcal{F}\dot{\mathcal{N}}^s_{p,\lambda,q})$ and Young's inequality give

$$\Big\| \int_0^t T_{\Omega,N}(t-\tau) f(\tau) d\tau \Big\|_{\mathcal{L}^\rho([0,T), \mathcal{F}\dot{\mathcal{N}}^s_{p,\lambda,q})}$$

$$= \Big\{ \sum_{j \in \mathbb{Z}} 2^{jqs} \Big(\int_0^T \big\| \varphi_j \int_0^t \mathcal{F}(T_{\Omega,N}(t-\tau)f)(\tau) d\tau \big\|^\rho_{M^\lambda_p} dt \Big)^{\frac{q}{\rho}} \Big\}^{1/q}$$

$$\leq C \Big\{ \sum_{j \in \mathbb{Z}} 2^{jqs} \Big(\int_0^T \big\| \varphi_j \int_0^t e^{-\mu|\xi|^2(t-\tau)} \hat{f}(\tau) d\tau \big\|^\rho_{M^\lambda_p} dt \Big)^{\frac{q}{\rho}} \Big\}^{1/q}$$

$$\leq C \Big\{ \sum_{j \in \mathbb{Z}} 2^{jqs} \Big(\int_0^T \big\| \varphi_j \int_0^t e^{-\mu 2^{2j}(t-\tau)} \hat{f}(\tau) d\tau \big\|^\rho_{M^\lambda_p} dt \Big)^{\frac{q}{\rho}} \Big\}^{1/q}$$

$$\leq C \Big\{ \sum_{j \in \mathbb{Z}} 2^{jqs} \Big(\int_0^T e^{-t\mu\tilde{\rho}2^{2j}} dt \Big)^{\frac{q}{\tilde{\rho}}} \big\| \varphi_j \hat{f}(\tau) \big\|^q_{L^r([0,T), M^\lambda_p)} \Big\}^{1/q}$$

$$\leq C \Big\{ \sum_{j \in \mathbb{Z}} 2^{jq(s-2-\frac{2}{\tilde{\rho}}+\frac{2}{r})} \big\| \varphi_j \hat{f}(\tau) \big\|^q_{L^r([0,T), M^\lambda_p)} \Big\}^{1/q}$$

$$\leq C \|f\|_{\mathcal{L}^r([0,T), \mathcal{F}\dot{\mathcal{N}}^{s-2-\frac{2}{\rho}+\frac{2}{r}}_{p,\lambda,q})}.$$

Lemma 3.2. *Let* $T > 0$, $0 \leq \lambda < 3$, $1 \leq p < \infty$, $1 \leq q \leq \infty$, $s \in \mathbb{R}$ *and* $u_0 \in \mathcal{F}\mathcal{N}^{-1+\frac{3}{p'}+\frac{\lambda}{p}}_{p,\lambda,q}(\mathbb{R}^3)$. *Then there exists a constant* $C > 0$ *such that*

$$\|T_{\Omega,N}(.)v_0\|_{\mathcal{L}^\rho\left([0,T), \mathcal{F}\dot{\mathcal{N}}^{s+\frac{2}{\rho}}_{p,\lambda,q}\right)} \leq C \|v_0\|_{\mathcal{F}\dot{\mathcal{N}}^s_{p,\lambda,q}}, \tag{3.1}$$

Proof: Since Supp $\psi_j \subset \{\xi \in \mathbb{R}^3 : 2^{j-1} \leq |\xi| \leq 2^{j+1}\}$, one has

$$\|\Delta_j \widehat{T_{\Omega,N}}(.)v_0\|_{M^\lambda_p} \leq C e^{-\mu 2^{2j} t} \|\widehat{\psi_j} \widehat{v_0}\|_{M^\lambda_p}.$$

for all $t \geq 0$, which yields that

$$\|\Delta_j \widehat{T_{\Omega,N}}(.)v_0\|_{L^\rho([0,T), M^\lambda_p)} \leq C \Big(\frac{1 - e^{-\mu 2^{2j} \rho T}}{\mu 2^{2j} \rho} \Big)^{\frac{1}{\rho}} \|\widehat{\psi_j} \widehat{v_0}\|_{M^\lambda_p}.$$

Thus, we have

$$\|T_{\Omega,N}(.)v_0\|_{\mathcal{L}^p\left([0,T),\mathcal{F}\dot{\mathcal{N}}_{p,\lambda,q}^{s+\frac{2}{p}}\right)} \leq C\|v_0\|_{\mathcal{F}\dot{\mathcal{N}}_{p,\lambda,q}^s}.$$

Proposition 3.3. *[10] Let* $1 \leq p < \infty$, $1 \leq q \leq 2$, $\frac{1}{2} \leq \alpha \leq 1 + \frac{3}{2p'} + \frac{\lambda}{2p}$ *and* $0 \leq \lambda < 3$. *Set*

$$X = \mathcal{L}^\infty\left([0,\infty),\mathcal{F}\dot{\mathcal{N}}_{p,\lambda,q}^{1-2\alpha+\frac{3}{p'}+\frac{\lambda}{p}}\right) \cap \mathcal{L}^1\left([0,\infty),\mathcal{F}\dot{\mathcal{N}}_{p,\lambda,q}^{1-2\alpha+\frac{3}{p'}+\frac{2\alpha}{\rho}+\frac{\lambda}{p}}\right),$$

with the norm

$$\|u\|_X = \|u\|_{\mathcal{L}^\infty\left([0,\infty),\mathcal{F}\dot{\mathcal{N}}_{p,\lambda,q}^{1-2\alpha+\frac{3}{p'}+\frac{\lambda}{p}}\right)} + \mu\|u\|_{\mathcal{L}^1\left([0,\infty),\mathcal{F}\dot{\mathcal{N}}_{p,\lambda,q}^{1-2\alpha+\frac{3}{p'}+\frac{2\alpha}{\rho}+\frac{\lambda}{p}}\right)}.$$

There exists a constant $C = C(p,q) > 0$ *depending on* α, p, q *such that*

$$\|\nabla.(u \otimes v)\|_{\mathcal{L}^1\left([0,\infty),\mathcal{F}\dot{\mathcal{N}}_{p,\lambda,q}^{1-4\alpha+\frac{3}{p'}+\frac{2\alpha}{\rho}+\frac{\lambda}{p}}\right)} \leq C\mu^{-1}\|u\|_X\|v\|_X. \qquad (3.2)$$

Proposition 3.4. *[10] Let* $0 \leq \lambda < 3$, $\max\{1, \frac{3-\lambda}{2}\} \leq p < \infty$, $1 \leq q \leq 2$, $I = [0,T)$, $0 < T \leq \infty$ *and* $\frac{2}{3} < \alpha \leq \frac{2}{3} + \frac{1}{p'} + \frac{\lambda}{3p}$. *Set*

$$Y = \mathcal{L}^4(I, \mathcal{F}\dot{\mathcal{N}}_{p,\lambda,q}^{1-\frac{3}{2}\alpha+\frac{3}{p'}+\frac{\lambda}{p}}),$$

there exists a constant $C = C(p,q) > 0$ *depending on* p, q *such that*

$$\|uv\|_{\mathcal{L}^2(I,\mathcal{F}\dot{\mathcal{N}}_{p,\lambda,q}^{2-3\alpha+\frac{3}{p'}+\frac{\lambda}{p}})} \leq C\|u\|_Y\|v\|_Y. \qquad (3.3)$$

Proof of Theorem 2.6. For the local existence, we set

$$Y = \mathcal{L}^4(I, \mathcal{F}\dot{\mathcal{N}}_{p,\lambda,q}^{-\frac{1}{2}+\frac{3}{p'}+\frac{\lambda}{p}}), \ I = [0,T).$$

Here, as usual, we begin with the mild integral equation

$$v(t) = T_{\Omega,N}(t)v_0 - \int_0^t T_{\Omega,N}(t-\tau)\tilde{\mathbb{P}}\tilde{\nabla} \cdot (v \otimes v)d\tau, \qquad (3.4)$$

and we consider the bilinear operator B given by

$$B(v,v') = \int_0^t T_{\Omega,N}(t-\tau)\tilde{\mathbb{P}}\tilde{\nabla} \cdot (v \otimes v')d\tau.$$

According to Lemma 3.1 and Proposition 3.4 with $\alpha = 1$, we obtain

$$\|B(v,v')\|_{\mathcal{L}^4(I,\dot{\mathcal{FN}}_{p,\lambda,q}^{-\frac{1}{2}+\frac{3}{p'}+\frac{\lambda}{p}})}$$

$$= \|\int_0^t T_{\Omega,N}(t-\tau)\tilde{\mathbb{P}}\tilde{\nabla}\cdot(v\otimes v')d\tau\|_{\mathcal{L}^4(I,\dot{\mathcal{FN}}_{p,\lambda,q}^{-\frac{1}{2}+\frac{3}{p'}+\frac{\lambda}{p}})}$$

$$\leq C\|\tilde{\nabla}.(v\otimes v')\|_{\mathcal{L}^2(I,\dot{\mathcal{FN}}_{p,\lambda,q}^{-2+\frac{3}{p'}+\frac{\lambda}{p}})}$$

$$\leq C\|vv'\|_{\mathcal{L}^2(I,\dot{\mathcal{FN}}_{p,\lambda,q}^{-1+\frac{3}{p'}+\frac{\lambda}{p}})}$$

$$\leq C\|v\|_Y\|v'\|_Y .$$

Lemma 3.2 yields

$$\|T_{\Omega,N}(t)v_0\|_Y \leq C\|v_0\|_{\dot{\mathcal{FN}}_{p,\lambda,q}^{-1+\frac{3}{p'}+\frac{\lambda}{p}}} . \tag{3.5}$$

Now, we shall decompose the initial data u_0 into two terms

$$v_0 = \mathcal{F}^{-1}(\chi_{B(0,\delta)}\hat{v_0}) + \mathcal{F}^{-1}(\chi_{B^C(0,\delta)}\hat{v_0}) := v_{0,1} + v_{0,2},$$

where $\delta = \delta(v_0) > 0$ is a real number. Since $v_{0,2}$ converge to 0 in $\dot{\mathcal{FN}}_{p,\lambda,q}^{-1+\frac{3}{p'}+\frac{\lambda}{p}}$ as $\delta \to +\infty$, by (3.5) there exists δ large enough such that

$$\|T_{\Omega,N}(t)v_{0,2}\|_Y \leq \frac{1}{8C} .$$

For the first term $v_{0,1}$,

$$\|T_{\Omega,N}(t)v_{0,1}\|_Y \leq \|2^{j(-\frac{1}{2}+\frac{3}{p'}+\frac{\lambda}{p})}\|\varphi_j e^{-\mu t|\xi|^2}\chi_{B(0,\delta)}\hat{v_0}\|_{L^4(I,M_p^\lambda)}\|_{\ell^q}$$

$$\leq \|2^{j(-\frac{1}{2}+\frac{3}{p'}+\frac{\lambda}{p})}\|\sup_{\xi\in B(0,\delta)} e^{-\mu t|\xi|^2}|\xi|^{\frac{1}{2}}\|_{L^4([0,T))}\|\varphi_j|\xi|^{-\frac{1}{2}}\hat{v_0}\|_{M_p^\lambda}\|_{\ell^q}$$

$$\leq C\delta^{\frac{1}{2}}T^{\frac{1}{4}}\|v_0\|_{\dot{\mathcal{FN}}_{p,\lambda,q}^{-1+\frac{3}{p'}+\frac{\lambda}{p}}} .$$

Thus for arbitrary v_0 in $\dot{\mathcal{FN}}_{p,\lambda,q}^{-1+\frac{3}{p'}+\frac{\lambda}{p}}$, (3.4) has a unique local solution in Y on $[0,T)$ where

$$T \leq \left(\frac{1}{8C^2\delta^{\frac{1}{2}}\|u_0\|_{\dot{\mathcal{FN}}_{p,\lambda,q}^{-1+\frac{3}{p'}+\frac{\lambda}{p}}}}\right)^4 .$$

For the global existence, we will again use Lemma 1.1 to ensure the existence of global mild solution with small initial data in the Banach space X given by

$$X = \mathcal{L}^\infty([0,\infty),\dot{\mathcal{FN}}_{p,\lambda,q}^{-1+\frac{3}{p'}+\frac{\lambda}{p}}) \cap \mathcal{L}^1([0,\infty),\dot{\mathcal{FN}}_{p,\lambda,q}^{1+\frac{3}{p'}+\frac{\lambda}{p}}) .$$

According to Lemma 3.1 and Proposition 3.3, we obtain

$$\|B(v,v')\|_{\mathcal{L}^1([0,\infty),\mathcal{F}\dot{\mathcal{N}}_{p,\lambda,q}^{1+\frac{3}{p'}+\frac{\lambda}{p}})}$$

$$= \left\| \int_0^t T_{\Omega,N}(t-\tau)\tilde{\mathbb{P}}\tilde{\nabla}\cdot(v\otimes v')d\tau \right\|_{\mathcal{L}^1([0,\infty),\mathcal{F}\dot{\mathcal{N}}_{p,\lambda,q}^{1+\frac{3}{p'}+\frac{\lambda}{p}})}$$

$$\leq C\|\tilde{\nabla}.(v\otimes v')\|_{\mathcal{L}^1([0,\infty),\mathcal{F}\dot{\mathcal{N}}_{p,\lambda,q}^{1-2\alpha+\frac{3}{p'}+\frac{\lambda}{p}})}$$

$$\leq C\mu^{-1}\|v'\|_X\|v\|_X .$$

Similarly,

$$\|B(v,v')\|_{\mathcal{L}^\infty([0,\infty),\mathcal{F}\dot{\mathcal{N}}_{p,\lambda,q}^{1-2\alpha+\frac{3}{p'}+\frac{\lambda}{p}})}$$

$$= \left\| \int_0^t T_{\Omega,N}(t-\tau)\tilde{\mathbb{P}}\tilde{\nabla}\cdot(v\otimes v')d\tau \right\|_{\mathcal{L}^\infty([0,\infty),\mathcal{F}\dot{\mathcal{N}}_{p,\lambda,q}^{-1+\frac{3}{p'}+\frac{\lambda}{p}})}$$

$$\leq C\|\tilde{\nabla}.(v\otimes v')\|_{\mathcal{L}^1([0,\infty),\mathcal{F}\dot{\mathcal{N}}_{p,\lambda,q}^{-1+\frac{3}{p'}+\frac{\lambda}{p}})}$$

$$\leq C\mu^{-1}\|v\|_X\|v'\|_X .$$

Finally,

$$\|B(v,v')\|_X \leq C\mu^{-1}\|v\|_X\|v'\|_X .$$

Lemma 3.2 yields

$$\|T_{\Omega,N}(t)v_0\|_X \leq C\|v_0\|_{\mathcal{F}\dot{\mathcal{N}}_{p,\lambda,q}^{1-2\alpha+\frac{3}{p'}+\frac{\lambda}{p}}} .$$

If $\|v_0\|_{\mathcal{F}\dot{\mathcal{N}}_{p,\lambda,q}^{-1+\frac{3}{p'}+\frac{\lambda}{p}}} < C_0\mu$ with $C_0 = \frac{1}{4C^2}$, then (1.1) has a unique global solution $u \in X$ satisfying

$$\|v\|_{\mathcal{L}^\infty([0,\infty);\mathcal{F}\dot{\mathcal{N}}_{p,\lambda,q}^{-1+\frac{3}{p'}+\frac{\lambda}{p}})} + \mu\|v\|_{\mathcal{L}^1([0,\infty),\mathcal{F}\dot{\mathcal{N}}_{p,\lambda,q}^{1+\frac{3}{p'}+\frac{\lambda}{p}})} \leq 2C\|v_0\|_{\mathcal{F}\dot{\mathcal{N}}_{p,\lambda,q}^{-1+\frac{3}{p'}+\frac{\lambda}{p}}} .$$

4   Stability of Global Solutions

In this section we prove Theorem 2.7. Let T^* be the maximal existence time of a solution u of (1.1) in
$\mathcal{L}^\infty\left([0,T^*);\mathcal{F}\dot{\mathcal{N}}_{p,\lambda,q}^{-1+\frac{3}{p'}+\frac{\lambda}{p}}\right)\cap\mathcal{L}^1\left([0,T^*),\mathcal{F}\dot{\mathcal{N}}_{p,\lambda,q}^{1+\frac{3}{p'}+\frac{\lambda}{p}}\right)$. In order to prove a blow-up criterion of the solution given by Theorem 2.6, assume that $T^* < \infty$ and $\|v\|_{\mathcal{L}^1([0,T^*),\mathcal{F}\dot{\mathcal{N}}_{p,\lambda,q}^{1+\frac{3}{p'}+\frac{\lambda}{p}})} < \infty$, then we can find $0 < T_0 < T^*$ satisfying

$$\|v\|_{\mathcal{L}^1([T_0,T^*),\mathcal{F}\dot{\mathcal{N}}_{p,\lambda,q}^{1+\frac{3}{p'}+\frac{\lambda}{p}})} < \frac{1}{2} .$$

For $t \in [T_0, T^*)$, we explicitly consider the integral equation

$$v(t) = T_{\Omega,N}(t)v(T_0) - \int_{T_0}^{t} T_{\Omega,N}(t-\tau)\tilde{\mathbb{P}}\tilde{\nabla} \cdot (v \otimes v)d\tau,$$

we obtain

$$|\widehat{v}(t,\xi)| \le e^{-\mu|\xi|^2 t}|\widehat{v}(T_0,\xi)| + \int_{T_0}^{t} e^{-\mu(t-s)|\xi|^2}|\tilde{\mathbb{P}}\tilde{\nabla} \cdot (v \otimes v)(s,\xi)| \, ds.$$

The same reasoning as in the proof of Proposition 3.3 gives

$$\|v\|_{\mathcal{L}^\infty([T_0,t),\mathcal{F}\dot{\mathcal{N}}_{p,\lambda,q}^{-1+\frac{3}{p'}+\frac{\lambda}{p}})} \lesssim \|v(T_0)\|_{\mathcal{F}\dot{\mathcal{N}}_{p,\lambda,q}^{-1+\frac{3}{p'}+\frac{\lambda}{p}}}$$
$$+ \|v\|_{\mathcal{L}^\infty([T_0,t),\mathcal{F}\dot{\mathcal{N}}_{p,\lambda,q}^{-1+\frac{3}{p'}+\frac{\lambda}{p}})} \|v\|_{\mathcal{L}^1([T_0,t),\mathcal{F}\dot{\mathcal{N}}_{p,\lambda,q}^{1+\frac{3}{p'}+\frac{\lambda}{p}})}.$$

It follows that

$$\|v\|_{\mathcal{L}^\infty([T_0,t),\mathcal{F}\dot{\mathcal{N}}_{p,\lambda,q}^{-1+\frac{3}{p'}+\frac{\lambda}{p}})} \lesssim \|v(T_0)\|_{\mathcal{F}\dot{\mathcal{N}}_{p,\lambda,q}^{-1+\frac{3}{p'}+\frac{\lambda}{p}}} + \frac{1}{2}\|v\|_{\mathcal{L}^\infty([T_0,t),\mathcal{F}\dot{\mathcal{N}}_{p,\lambda,q}^{-1+\frac{3}{p'}+\frac{\lambda}{p}})}.$$

We can deduce that

$$\sup_{T_0 \le s \le t} \|v\|_{\mathcal{F}\dot{\mathcal{N}}_{p,\lambda,q}^{-1+\frac{3}{p'}+\frac{\lambda}{p}}} \lesssim 2\|v(T_0)\|_{\mathcal{F}\dot{\mathcal{N}}_{p,\lambda,q}^{-1+\frac{3}{p'}+\frac{\lambda}{p}}}, \forall t \in [T_0, T^*).$$

Setting

$$M = \max\left(2\|v(T_0)\|_{\mathcal{F}\dot{\mathcal{N}}_{p,\lambda,q}^{-1+\frac{3}{p'}+\frac{\lambda}{p}}}, \max_{t \in [0,T_0]} \|v\|_{\mathcal{F}\dot{\mathcal{N}}_{p,\lambda,q}^{-1+\frac{3}{p'}+\frac{\lambda}{p}}}\right),$$

we have

$$\|v(t)\|_{\mathcal{F}\dot{\mathcal{N}}_{p,\lambda,q}^{-1+\frac{3}{p'}+\frac{\lambda}{p}}} \lesssim M, \ \forall t \in [0, T^*).$$

On the other side

$$v(t) = e^{-t\mu(-\Delta)}u_0 - \Omega\int_0^t e^{-\mu(t-\tau)(-\Delta)}\tilde{\mathbb{P}}Bv(\tau)d\tau - \int_0^t e^{-\mu(t-\tau)(-\Delta)}\tilde{\mathbb{P}}\tilde{\nabla} \cdot (v \otimes v)(\tau)d\tau.$$

Then,

$$v(t') - v(t) = (e^{-\mu t'(-\Delta)}v_0 - e^{-\mu t(-\Delta)}v_0)$$

$$- \Big(\int_0^{t'} e^{-\mu(t'-\tau)(-\Delta)} \tilde{\mathbb{P}}\tilde{\nabla} \cdot (v \otimes v)(\tau)d\tau - \int_0^t e^{-\mu(t-\tau)(-\Delta)} \tilde{\mathbb{P}}\tilde{\nabla} \cdot (v \otimes v)(\tau)d\tau \Big)$$

$$- \Omega \Big(\int_0^{t'} e^{-\mu(t'-\tau)(-\Delta)} \tilde{\mathbb{P}}\mathcal{B}v(\tau)d\tau - \int_0^t e^{-\mu(t-\tau)(-\Delta)} \tilde{\mathbb{P}}\mathcal{B}v(\tau)d\tau \Big)$$

$$= [e^{-\mu t'(-\Delta)}v_0 - e^{-\mu t(-\Delta)}v_0] - \Big[\int_t^{t'} e^{-\mu(t'-\tau)(-\Delta)} \tilde{\mathbb{P}}\tilde{\nabla} \cdot (v \otimes v)(\tau)d\tau \Big]$$

$$- \Big[\int_0^t e^{-\mu(t-\tau)(-\Delta)} (e^{-\mu(t'-t)(-\Delta)} - 1)\tilde{\mathbb{P}}\tilde{\nabla} \cdot (v \otimes v)(\tau)d\tau \Big]$$

$$- \Omega \Big[\int_t^{t'} e^{-\mu(t'-\tau)(-\Delta)} \tilde{\mathbb{P}}\mathcal{B}v(\tau)d\tau \Big]$$

$$- \Omega \Big[\int_0^t e^{-\mu(t-\tau)(-\Delta)} (e^{-\mu(t'-t)(-\Delta)} - 1)\tilde{\mathbb{P}}\mathcal{B}v(\tau)d\tau \Big]$$

$$:= J_1 + J_2 + J_3 + J_4 + J_5 .$$

We will estimate J_1, J_2, J_3, J_4 and J_5;

$$\|J_1\|_{\mathcal{F}\dot{\mathcal{N}}_{p,\lambda,q}^{-1+\frac{3}{p'}+\frac{\lambda}{p}}} = \Big\| 2^{j(-1+\frac{3}{p'}+\frac{\lambda}{p})} \|\varphi_j(e^{-\mu t'|\xi|^2} - e^{-\mu t|\xi|^2})\hat{u}_0\|_{M_p^\lambda} \Big\|_{\ell^q}$$

$$\leq \Big\| 2^{j(-1+\frac{3}{p'}+\frac{\lambda}{p})} \|\varphi_j(e^{-\mu(t'-t)|\xi|^2} - 1)\hat{u}_0\|_{M_p^\lambda} \Big\|_{\ell^q} ,$$

$$\|J_2\|_{\mathcal{F}\dot{\mathcal{N}}_{p,\lambda,q}^{-1+\frac{3}{p'}+\frac{\lambda}{p}}} \leq \Big\| 2^{j(-1+\frac{3}{p'}+\frac{\lambda}{p})} \int_t^{t'} \|\varphi_j e^{-\mu(t'-\tau)|\xi|^2} \mathcal{F}(\tilde{\nabla} \cdot v \otimes v)(\tau)\|_{M_p^\lambda} d\tau \Big\|_{\ell^q}$$

$$\leq \Big\| 2^{j(\frac{3}{p'}+\frac{\lambda}{p})} \int_t^{t'} \|\varphi_j \mathcal{F}(v \otimes v)(\tau)\|_{M_p^\lambda} d\tau \Big\|_{\ell^q} ,$$

$$\|J_3\|_{\mathcal{F}\dot{\mathcal{N}}_{p,\lambda,q}^{-1+\frac{3}{p'}+\frac{\lambda}{p}}} \leq$$

$$\Big\| 2^{j(-1+\frac{3}{p'}+\frac{\lambda}{p})} \times \int_0^t \|\varphi_j e^{-\mu(t'-\tau)|\xi|^2}(1 - e^{-\mu(t'-t)|\xi|^2}) \mathcal{F}(\tilde{\nabla} \cdot v \otimes v)(\tau)\|_{M_p^\lambda} d\tau \Big\|_{\ell^q}$$

$$\leq \Big\| 2^{j(\frac{3}{p'}+\frac{\lambda}{p})} \int_0^t \|\varphi_j(e^{-\mu(t'-t)|\xi|^2} - 1)\mathcal{F}(v \otimes v)(\tau)\|_{M_p^\lambda} d\tau \Big\|_{\ell^q} ,$$

$$\|J_4\|_{\mathcal{F}\dot{\mathcal{N}}_{p,\lambda,q}^{-1+\frac{3}{p'}+\frac{\lambda}{p}}} \lesssim \Big\| 2^{j(-1+\frac{3}{p'}+\frac{\lambda}{p})} \int_t^{t'} \|\varphi_j e^{-\mu(t'-\tau)|\xi|^2} \mathcal{F}(\mathcal{B}v)(\tau)\|_{M_p^\lambda} d\tau \Big\|_{\ell^q}$$

$$\lesssim \Big\| 2^{j(-1+\frac{3}{p'}+\frac{\lambda}{p})} \int_t^{t'} \|\varphi_j \mathcal{F}(\mathcal{B}v)(\tau)\|_{M_p^\lambda} d\tau \Big\|_{\ell^q} ,$$

and

$$\|J_5\|_{\mathcal{F}\dot{\mathcal{N}}_{p,\lambda,q}^{-1+\frac{3}{p'}+\frac{\lambda}{p}}} \lesssim$$

$$\Big\| 2^{j(-1+\frac{3}{p'}+\frac{\lambda}{p})} \times \int_0^t \|\varphi_j e^{-\mu(t'-\tau)|\xi|^2}(1 - e^{-\mu(t'-t)|\xi|^{2\alpha}}) \mathcal{F}(\mathcal{B}v)(\tau)\|_{M_p^\lambda} d\tau \Big\|_{\ell^q}$$

$$\lesssim \Big\| 2^{j(-1+\frac{3}{p'}+\frac{\lambda}{p})} \int_0^t \|\varphi_j(e^{-\mu(t'-t)|\xi|^{2\alpha}} - 1)\mathcal{F}(\mathcal{B}v)(\tau)\|_{M_p^\lambda} d\tau \Big\|_{\ell^q} .$$

The dominated convergence theorem gives

$$\limsup_{t,t' \nearrow T^*, t \leq t'} \|v(t) - v(t')\|_{\mathcal{F}\dot{\mathcal{N}}_{p,\lambda,q}^{-1+\frac{3}{p'}+\frac{\lambda}{p}}} = 0.$$

This means that $v(t)$ satisfies the Cauchy criterion at T^*. As $\mathcal{F}\dot{\mathcal{N}}_{p,\lambda,q}^{-1+\frac{3}{p'}+\frac{\lambda}{p}}$ is a Banach space, then there exists an element v^* in $\mathcal{F}\dot{\mathcal{N}}_{p,\lambda,q}^{-1+\frac{3}{p'}+\frac{\lambda}{p}}$ such that $v(t) \to v^*$ in $\mathcal{F}\dot{\mathcal{N}}_{p,\lambda,q}^{-1+\frac{3}{p'}+\frac{\lambda}{p}}$ as $t \to T^*$. Set $v(T^*) = v^*$ and consider the probleme (1.2) starting by v^*. By the well-posedness we obtain a solution existing on a larger time interval than $[0, T^*)$, which is a contradiction. Now, let $v \in \mathcal{C}\big([0,T^*); \mathcal{F}\dot{\mathcal{N}}_{p,\lambda,q}^{-1+\frac{3}{p'}+\frac{\lambda}{p}}\big) \cap \mathcal{L}^1\big([0,T^*), \mathcal{F}\dot{\mathcal{N}}_{p,\lambda,q}^{1+\frac{3}{p'}+\frac{\lambda}{p}}\big)$ be the maximal solution of (1.1) corresponding to the initial condition v_0'. We want to prove $T^* = \infty$. Put $w = v' - v$ and $w_0 = v_0' - v_0$. We have

$$w_t + \mathcal{A}w + \mathcal{B}w + w \cdot \tilde{\nabla}w + v \cdot \tilde{\nabla}w + w \cdot \tilde{\nabla}v = -\tilde{\nabla}p.$$

We first apply $\tilde{\mathbb{P}}$ to the above equation, then we have

$$w_t + \mathcal{A}w = -\tilde{\mathbb{P}}\mathcal{B}w - \tilde{\mathbb{P}}\tilde{\nabla}.(w \otimes w) - \tilde{\mathbb{P}}\tilde{\nabla}.(v \otimes w) - \tilde{\mathbb{P}}\tilde{\nabla}.(w \otimes v).$$

Due to Duhamel's formula, we write

$$|\widehat{w}(t,\xi)| \leq e^{-\mu|\xi|^2 t}|\widehat{w}(0,\xi)| + \int_0^t e^{-\mu(t-s)|\xi|^2}|\mathcal{F}(\tilde{\mathbb{P}}\tilde{\nabla}.(w \otimes w))(s,\xi)|\,\mathrm{d}s$$

$$+ \int_0^t e^{-\mu(t-s)|\xi|^2}|\mathcal{F}(\tilde{\mathbb{P}}\tilde{\nabla}.(v \otimes w))(s,\xi)|\,\mathrm{d}s$$

$$+ \int_0^t e^{-\mu(t-s)|\xi|^2}|\mathcal{F}(\tilde{\mathbb{P}}\tilde{\nabla}.(w \otimes v))(s,\xi)|\,\mathrm{d}s$$

$$+ \int_0^t e^{-\mu(t-s)|\xi|^2}|\mathcal{F}(\tilde{\mathbb{P}}\mathcal{B}w)(s,\xi)|\,\mathrm{d}s.$$

Then, for $t \in [0, T^*)$ we get

$$\mu\|w\|_{\mathcal{L}^1([0,t),\mathcal{F}\dot{\mathcal{N}}_{p,\lambda,q}^{1+\frac{3}{p'}+\frac{\lambda}{p}})} \leq C\Big\{\|w_0\|_{\mathcal{F}\dot{\mathcal{N}}_{p,\lambda,q}^{-1+\frac{3}{p'}+\frac{\lambda}{p}}} + \|\tilde{\nabla}.(w \otimes w)\|_{\mathcal{L}^1([0,t),\mathcal{F}\dot{\mathcal{N}}_{p,\lambda,q}^{-1+\frac{3}{p'}+\frac{\lambda}{p}})}$$

$$+ \|\tilde{\nabla}.(v \otimes w)\|_{\mathcal{L}^1([0,t),\mathcal{F}\dot{\mathcal{N}}_{p,\lambda,q}^{-1+\frac{3}{p'}+\frac{\lambda}{p}})}$$

$$+ \|\tilde{\nabla}.(w \otimes v)\|_{\mathcal{L}^1([0,t),\mathcal{F}\dot{\mathcal{N}}_{p,\lambda,q}^{-1+\frac{3}{p'}+\frac{\lambda}{p}})}$$

$$+ \|\mathcal{B}w\|_{\mathcal{L}^1([0,t),\mathcal{F}\dot{\mathcal{N}}_{p,\lambda,q}^{-1+\frac{3}{p'}+\frac{\lambda}{p}})}\Big\}.$$

Similarly,

$$\|w\|_{\mathcal{L}^\infty([0,t),\mathcal{F}\dot{\mathcal{N}}_{p,\lambda,q}^{-1+\frac{3}{p'}+\frac{\lambda}{p}})} \le \|w_0\|_{\mathcal{F}\dot{\mathcal{N}}_{p,\lambda,q}^{-1+\frac{3}{p'}+\frac{\lambda}{p}}} + \|\tilde{\nabla}.(w \otimes w)\|_{\mathcal{L}^1([0,t),\mathcal{F}\dot{\mathcal{N}}_{p,\lambda,q}^{-1+\frac{3}{p'}+\frac{\lambda}{p}})}$$
$$+ \|\tilde{\nabla}.(v \otimes w)\|_{\mathcal{L}^1([0,t),\mathcal{F}\dot{\mathcal{N}}_{p,\lambda,q}^{-1+\frac{3}{p'}+\frac{\lambda}{p}})}$$
$$+ \|\tilde{\nabla}.(w \otimes v)\|_{\mathcal{L}^1([0,t),\mathcal{F}\dot{\mathcal{N}}_{p,\lambda,q}^{-1+\frac{3}{p'}+\frac{\lambda}{p}})}$$
$$+ \|\mathcal{B}w\|_{\mathcal{L}^1([0,t),\mathcal{F}\dot{\mathcal{N}}_{p,\lambda,q}^{-1+\frac{3}{p'}+\frac{\lambda}{p}})}.$$

Consequently, for $t \in [0, T^*)$ we get

$$\|w(t)\|_{\mathcal{F}\dot{\mathcal{N}}_{p,\lambda,q}^{-1+\frac{3}{p'}+\frac{\lambda}{p}}} + \mu\|w\|_{\mathcal{L}^1([0,t),\mathcal{F}\dot{\mathcal{N}}_{p,\lambda,q}^{1+\frac{3}{p'}+\frac{\lambda}{p}})}$$
$$\le C\Big\{ \|w_0\|_{\mathcal{F}\dot{\mathcal{N}}_{p,\lambda,q}^{-1+\frac{3}{p'}+\frac{\lambda}{p}}} + \|\tilde{\nabla}.(w \otimes w)\|_{\mathcal{L}^1([0,t),\mathcal{F}\dot{\mathcal{N}}_{p,\lambda,q}^{-1+\frac{3}{p'}+\frac{\lambda}{p}})}$$
$$+ \|\tilde{\nabla}.(v \otimes w)\|_{\mathcal{L}^1([0,t),\mathcal{F}\dot{\mathcal{N}}_{p,\lambda,q}^{-1+\frac{3}{p'}+\frac{\lambda}{p}})}$$
$$+ \|\tilde{\nabla}.(w \otimes v)\|_{\mathcal{L}^1([0,t),\mathcal{F}\dot{\mathcal{N}}_{p,\lambda,q}^{-1+\frac{3}{p'}+\frac{\lambda}{p}})}$$
$$+ \|\mathcal{B}w\|_{\mathcal{L}^1([0,t),\mathcal{F}\dot{\mathcal{N}}_{p,\lambda,q}^{-1+\frac{3}{p'}+\frac{\lambda}{p}})}\Big\}$$
$$\lesssim \|w_0\|_{\mathcal{F}\dot{\mathcal{N}}_{p,\lambda,q}^{-1+\frac{3}{p'}+\frac{\lambda}{p}}} + L_1 + L_2 + L_3.$$

Where

$$L_1 = \|\tilde{\nabla}.(w \otimes w)\|_{\mathcal{L}^1([0,t),\mathcal{F}\dot{\mathcal{N}}_{p,\lambda,q}^{-1+\frac{3}{p'}+\frac{\lambda}{p}})},$$
$$L_2 = \|\tilde{\nabla}.(v \otimes w)\|_{\mathcal{L}^1([0,t),\mathcal{F}\dot{\mathcal{N}}_{p,\lambda,q}^{-1+\frac{3}{p'}+\frac{\lambda}{p}})} + \|\tilde{\nabla}.(w \otimes v)\|_{\mathcal{L}^1([0,t),\mathcal{F}\dot{\mathcal{N}}_{p,\lambda,q}^{-1+\frac{3}{p'}+\frac{\lambda}{p}})}$$

and $L_3 = \|\mathcal{B}w\|_{\mathcal{L}^1([0,t),\mathcal{F}\dot{\mathcal{N}}_{p,\lambda,q}^{-1+\frac{3}{p'}+\frac{\lambda}{p}})}$. The same calculus in the proof of Proposition 3.3 gives

$$L_1 \lesssim \|w\|_{\mathcal{L}^\infty([0,t),\mathcal{F}\dot{\mathcal{N}}_{p,\lambda,q}^{-1+\frac{3}{p'}+\frac{\lambda}{p}})} \|w\|_{\mathcal{L}^1([0,t),\mathcal{F}\dot{\mathcal{N}}_{p,\lambda,q}^{1+\frac{3}{p'}+\frac{\lambda}{p}})},$$
$$L_2 \lesssim \int_0^t \|w\|_{\mathcal{F}\dot{\mathcal{N}}_{p,\lambda,q}^{-1+\frac{3}{p'}+\frac{\lambda}{p}}} \|v\|_{\mathcal{F}\dot{\mathcal{N}}_{p,\lambda,q}^{1+\frac{3}{p'}+\frac{\lambda}{p}}},$$
$$L_3 \lesssim \|\mathcal{B}\|\|w\|_{\mathcal{L}^1([0,t),\mathcal{F}\dot{\mathcal{N}}_{p,\lambda,q}^{-1+\frac{3}{p'}+\frac{\lambda}{p}})}.$$

Then

$$\|w(t)\|_{\mathcal{F}\dot{\mathcal{N}}_{p,\lambda,q}^{-1+\frac{3}{p'}+\frac{\lambda}{p}}} + \mu\|w\|_{\mathcal{L}^1([0,t),\mathcal{F}\dot{\mathcal{N}}_{p,\lambda,q}^{1+\frac{3}{p'}+\frac{\lambda}{p}})}$$

$$\leq C\Big\{ \|w_0\|_{\mathcal{F}\dot{\mathcal{N}}_{p,\lambda,q}^{-1+\frac{3}{p'}+\frac{\lambda}{p}}} + \|w\|_{\mathcal{L}^\infty([0,t),\mathcal{F}\dot{\mathcal{N}}_{p,\lambda,q}^{-1+\frac{3}{p'}+\frac{\lambda}{p}})} \|w\|_{\mathcal{L}^1([0,t),\mathcal{F}\dot{\mathcal{N}}_{p,\lambda,q}^{1+\frac{3}{p'}+\frac{\lambda}{p}})}$$

$$+ \int_0^t \|w\|_{\mathcal{F}\dot{\mathcal{N}}_{p,\lambda,q}^{-1+\frac{3}{p'}+\frac{\lambda}{p}}} \|v\|_{\mathcal{F}\dot{\mathcal{N}}_{p,\lambda,q}^{1+\frac{3}{p'}+\frac{\lambda}{p}}} + \|\mathcal{B}\|\|w\|_{\mathcal{L}^1([0,t),\mathcal{F}\dot{\mathcal{N}}_{p,\lambda,q}^{-1+\frac{3}{p'}+\frac{\lambda}{p}})} \Big\}.$$

Put

$$T = \sup\{t \in [0,T^*), \|w\|_{\mathcal{L}^\infty([0,t],\mathcal{F}\dot{\mathcal{N}}_{p,\lambda,q}^{-1+\frac{3}{p'}+\frac{\lambda}{p}})} < \frac{\mu}{4C}\}. \qquad (4.1)$$

For $t \in [0,T)$, we have

$$\|w(t)\|_{\mathcal{F}\dot{\mathcal{N}}_{p,\lambda,q}^{-1+\frac{3}{p'}+\frac{\lambda}{p}}} + \frac{\mu}{2}\|w\|_{\mathcal{L}^1([0,t),\mathcal{F}\dot{\mathcal{N}}_{p,\lambda,q}^{-1+\frac{3}{p'}+\frac{\lambda}{p}})}$$

$$\leq C\Big\{ \|w_0\|_{\mathcal{F}\dot{\mathcal{N}}_{p,\lambda,q}^{-1+\frac{3}{p'}+\frac{\lambda}{p}}} + \int_0^t \|w\|_{\mathcal{F}\dot{\mathcal{N}}_{p,\lambda,q}^{-1+\frac{3}{p'}+\frac{\lambda}{p}}} (\|\mathcal{B}\| + \|u\|_{\mathcal{F}\dot{\mathcal{N}}_{p,\lambda,q}^{1+\frac{3}{p'}+\frac{\lambda}{p}}}) \Big\}.$$

Gronwall's Lemma yields

$$\|w(t)\|_{\mathcal{F}\dot{\mathcal{N}}_{p,\lambda,q}^{-1+\frac{3}{p'}+\frac{\lambda}{p}}} + \frac{\mu}{2}\int_0^t \|w\|_{\mathcal{F}\dot{\mathcal{N}}_{p,\lambda,q}^{1+\frac{3}{p'}+\frac{\lambda}{p}}}$$

$$\leq C\|w_0\|_{\mathcal{F}\dot{\mathcal{N}}_{p,\lambda,q}^{-1+\frac{3}{p'}+\frac{\lambda}{p}}} \exp\Big\{ \int_0^t C(\|\mathcal{B}\| + \|u\|_{\mathcal{F}\dot{\mathcal{N}}_{p,\lambda,q}^{1+\frac{3}{p'}+\frac{\lambda}{p}}}) \Big\}$$

$$\leq C\|w_0\|_{\mathcal{F}\dot{\mathcal{N}}_{p,\lambda,q}^{-1+\frac{3}{p'}+\frac{\lambda}{p}}} \exp\Big\{ \int_0^\infty C(\|\mathcal{B}\| + \|v\|_{\mathcal{F}\dot{\mathcal{N}}_{p,\lambda,q}^{1+\frac{3}{p'}+\frac{\lambda}{p}}}) \Big\}.$$

Thus if we take C_0 sufficiently small in (2.4), we have

$$\|w(t)\|_{\mathcal{F}\dot{\mathcal{N}}_{p,\lambda,q}^{-1+\frac{3}{p'}+\frac{\lambda}{p}}} + \frac{\mu}{2}\|w\|_{\mathcal{L}^1([0,t),\mathcal{F}\dot{\mathcal{N}}_{p,\lambda,q}^{-1+\frac{3}{p'}+\frac{\lambda}{p}})} < \frac{\mu}{8C},$$

which contradicts the Definition (4.1).

Then $T = T^*$ and $\|w\|_{\mathcal{L}^1([0,T^*),\mathcal{F}\dot{\mathcal{N}}_{p,\lambda,q}^{1+\frac{3}{p'}+\frac{\lambda}{p}})} < \infty$, therefore $T^* = \infty$. This completes the proof of Theorem 2.7.

References

1. Babin, A., Mahalov, A., Nicolaenko, B., Long-time averaged Euler and Navier-Stokes equations for rotating fluids. In: Advance Series Nonlinear Dynamic of Nonlinear Waves in Fluids, World Scientific Publications, River Edge, NJ, Hannover, vol. 7 1995, pp. 145–157 (1994)

2. Babin, A., Mahalov, A., Nicolaenko, B.: Regularity and integrability of 3D Euler and Navier-Stokes equations for rotating fluids. Asymptot. Anal. **15**, 103–150 (1997)
3. Babin, A., Mahalov, A., Nicolaenko, B.: On the regularity of three-dimensional rotating Euler-Boussinesq equations. Math. Models Methods Appl. Sci. **9**, 1089–1121 (1999)
4. Cannone, M., Wu, G.: Global well-posedness for Navier-Stokes equations in critical Fourier-Herz spaces. Nonlinear Anal. **75**, 3754 (2012)
5. Germain, P., Pavlovic, N., Staffilani, G.: Regularity of solutions to the Navier-Stokes equations evolving from small data in BMO^{-1}. Int. Math. Res. Not. IMRN **2007**, rnm087 (2007)
6. Giga, Y., Miyakawa, T.: Navier-Stokes flow in $\mathbb{R}^3$ with measures as initial vorticity and Morry spaces. Commun. Partial Differ. Equ. **14**, 577–618 (1989)
7. Kato, T.: Strong L^p-solutions of the Navier–Stokes in $\mathbb{R}^3$ with applications to weak solutions. Math. Z. **187**, 471–480 (1984)
8. Koch, H., Tataru, D.: Well-posedness for the Navier-Stokes equations. Adv. Math. **157**, 22–35 (2001)
9. Benameur, J.: Long time decay to the Lei-Lin solution of 3D Navier-Stokes equations. J. Math. Anal. Appl. **422**(1), 424–434 (2015)
10. El Baraka, A., Toumlilin, M.: Global well-posedness for fractional Navier-Stokes equations in critical Fourier-Besov-Morrey spaces. Moroccan J. Pure and Appl. Anal. **3**(1), 1–14 (2017)
11. El Baraka, A., Toumlilin, M.: Global well-posedness and decay results for 3D generalized magneto-hydrodynamic equations in critical Fourier-Besov-Morrey spaces. Electron. J. Differ. Equ. **2017**(65), 1–20 (2017)
12. El Baraka, A., Toumlilin, M.: Uniform well-posedness and stability for fractional Navier-Stokes equations with Coriolis force in critical Fourier-Besov-Morrey spaces. Open J. Math. Anal. **3**(1), 70–89 (2019)
13. Toumlilin, M.: Global well-posedness and analyticity for generalized porous medium equation in critical Fourier-Besov-Morrey spaces open. J. Math. Anal. **3**(2), 71–80 (2019)
14. Kozono, H., Yamazaki, M.: Semilinear heat equations and the Navier-Stokes equation with distributions in new function spaces as initial data. Comm. Partial Differ. Equ. **19**(5–6), 959–1014 (1994)
15. Ferreira, L.C.F., Angulo-Castillo, V.: Global well-posedness and asymptotic behavior for Navier-Stokes-Coriolis equations in homogeneous Besov spaces. Asympt. Anal. **112**, 37–58 (2019)
16. Ferreira, L.C.F., Lima, L.S.M.: Self-similar solutions for active scalar equations in Fourier-Besov-Morrey spaces. Monatsh. Math. **175**(4), 491–509 (2014)
17. Aurazo-Alvarez, L.L.: On long-time solutions for Boussinesq-type models in Besov and Fourier-Besov-Morrey spaces, Ph.D. Thesis, University of Campinas, p. 107 (Advisor: Lucas C. F.Ferreira) (2020)
18. Charve, F.: Global well-posedness and asymptotics for a geophysical fluid system. Comm. Partial Differ. Equ. **29**, 1919–1940 (2004)
19. Charve, F.: Global well-posedness for the primitive equations with less regular initial data. Ann. Fac. Sci. Toulouse Math. **17**, 221–238 (2008)
20. Charve, F.: Asymptotics and lower bound for the lifespan of solutions to the Primitive Equations. arXiv:1411.6859
21. Ibrahim, S., Yoneda, T.: Long-time solvability of the Navier-Stokes-Boussinesq equations with almostperiodic initial large data. J. Math. Sci. Univ. Tokyo **20**, 1–25 (2013)

22. Hieber, M., Shibata, Y.: The Fujita-Kato approach to the Navier-Stokes equations in the rotational framework. Math. Z. **265**, 481–491 (2010)
23. Cannone, M.: A generalization of a theorem by Kato on Navier-Stokes equations. Rev. Mat. Iberoamericana **13**(3), 515–541 (1997)
24. Sun, J., Cui, S.: Sharp well-posedness and ill-posedness of the three-dimensional primitive equations of geophysics in Fourier-Besov spaces. Nonlinear Anal. Real World Appl. **48**, 445–465 (2019)
25. Iwabuchi, T.: Global well-posedness for Keller-Segel system in Besov type spaces. J. Math. Anal. Appl. **379**, 930–948 (2011)
26. Iwabuchi, T., Takada, R.: Global solutions for the Navier-Stokes equations in the rotational framework. Mathematische Annalen **357**(2), 727–741 (2013)
27. Iwabuchi, T., Takada, R.: Global well-posedness and ill-posedness for the Navier-Stokes equations with the Coriolis force in function spaces of Besov type. J. Funct. Anal. **267**(5), 1321–1337 (2014)
28. Kato, T.: Strong solutions of the Navier-Stokes equations in Morrey spaces. Bol. Soc. Brasil Mat. **22**(2), 127–155 (1992)
29. Lei, Z., Lin, F.: Global mild solutions of Navier-Stokes equations. Comm. Pure Appl. Math. **64**(9), 1297–1304 (2011)
30. Lemarié-Rieusset, P.G.: The Navier-Stokes Problem in the 21st Century. CRC Press, Boca Raton (2016)
31. Majda, A.: Introduction to PDEs and Waves for the Atmosphere and Ocean, Courant Lecture Notes in Mathematics (2003)
32. Konieczny, P., Yoneda, T.: On dispersive effect of the Coriolis force for the stationary Navier-Stokes equations. J. Differ. Equ. **250**(10), 3859–3873 (2011)
33. Sickel, W.: Smoothness spaces related to Morrey spaces -a survey I. Eurasian Math. J. **3**(3), 110–149 (2012)
34. Taylor, M.E.: Analysis on Morrey spaces and applications to Navier-Stokes and other evolution equations. Commun. Partial Differ. Equ. **17**, 1407–1456 (1992)
35. Liu, Q., Zhao, J.: Global well-posedness for the generalized magneto-hydrodynamic equations in the critical Fourier-Herz spaces. J. Math. Anal. Appl. **420**, 1301–1315 (2014)
36. Xiao, W., Chen, J., Fan, D., Zhou, X.: Global well-posedness and long time decay of fractional Navier-Stokes equations in Fourier Besov spaces. In: Abstract and Applied Analysis, vol. 2014. Hindawi Publishing Corporation
37. Yuan, W., Sickel, W., Yang, D.: Morrey and Campanato Meet Besov, Lizorkin and Triebel. Lecture Notes in Mathematics, vol. 2005. Springer, Berlin, Germany (2010)
38. Zhao, H., Wang, Y.: A remark on the Navier-Stokes equations with the Coriolis force. Math. Methods Appl. Sci. **40**(18), 7323–7332 (2017)

Regional Controllability of a Class
of Time-Fractional Systems

Asmae Tajani$^{(\boxtimes)}$, Fatima-Zahrae El Alaoui, and Ali Boutoulout

TSI Team, Faculty of Sciences, Moulay Ismail University, 11201 Meknes, Morocco
tajaniasmae1993@gmail.com, fzelalaoui2011@yahoo.fr, boutouloutali@yahoo.fr

Abstract. The main purpose of this paper is to develop the concept of
regional controllability for an important class of Caputo time-fractional
semi-linear systems using the analytical approach, where the dynamic
of the considered system is generates by an analytical semigroup. This
approach use the fixed point techniques and semigroup theory. Finally,
we present some numerical simulations to approve our theoretical results.

1 Introduction

Fractional Calculus has received a considerable amount of interest in the last
years, its main purpose is the investigation of the notions of derivation and
integration of real or complex order. Many problems in physics, chemistry, engineering and control theory are represented by fractional equations (see [4,14]
and [9]), which are being used in modeling the anomalous behavior of problems
occurring in the real world . Fractional operators (integration and differentiation) have an important advantage, which is the nonlocal property, where the
current state, of a fractional system, depends on historical and past states. Many
researchers worked on the existence of solutions for initial and boundary value
fractional differential equations (see [13,18]), Zhou and Jiao discussed the existence of mild solution for fractional evolution and neutral evolution equations
in Banach spaces based on a probability density function and semigroup theory
(see [20] and [19]), seeing this big interest on fractional order systems, it is natural to study and analyze these kinds of systems as an extension or a general
case of classical dynamical systems (ie. systems with integer order derivatives).

The analysis of dynamical systems consists of many branches and various
concepts, Controllability being one amongst others. The concept of controllability consists of steering a system into a desired state (exactly or approximately)
at time T from an arbitrary initial state. The concept in hand has a very vast
literature for various type of systems (linear, Semi-linear, nonlinear...), for more
informations (see [3,6,8,11,17,21]). In many practical applications their exists
states which are not reachable, also sometimes we only need to control the system
on a particular region, in this cases the regional controllability concept should
be considered (see [12,15,16] and references therein).

Regional controllability's purpose is to steer a system into a desired state only
in a subregion of the whole evolution domain, this notion is a general case of

© Springer Nature Switzerland AG 2021
Z. Hammouch et al. (Eds.): SM2A 2019, LNNS 168, pp. 141–155, 2021.
https://doi.org/10.1007/978-3-030-62299-2_10

'global' Controllability. This notion is developed by several researchers to cover various types of systems. In particular, recently Ge,Chen and Kou discussed the regional controllability for time-fractional sub-diffusion systems with Caputo and Riemann-Liouville fractional derivatives (see [7]).

The concept in hand, namely regional controllability for nonlinear fractional systems, is in an initial stage and needs some more research, thus the motivation for this work, is to develop this theory for semi-linear time-fractional systems with Caputo derivative by using the analytical approach, which is based on the fixed point techniques and semigroup theory .

This paper is presented as follows, in Sect. 2, we introduce some preliminaries, definitions and results which will be used throughout this work. In Sect. 3, by using some properties of analytical semigroup and under suitable assumptions we show that the considered system is regionally controllable by a control that will be given later. In Sect. 4 , we provide an algorithm, which is based on the steps of the used approach. The Sect. 5 is devoted to present successful numerical results illustrating the theoretical ones. Finally a conclusion shall be giving.

2   Preliminaries and Considered System

In this section, we recall some basic definitions and properties used throughout this paper.

Definition 1 [10]. The (left) Caputo fractional derivative of a function y at a point t of order $\alpha \in]0,1]$ is defined as follows :

$$^C\mathrm{D}_{0+}^\alpha y(t) = \frac{1}{\Gamma(1-\alpha)} \int_0^t (t-s)^{-\alpha} \frac{d}{ds}(y(s))ds, \qquad 0 \le t < T. \qquad (1)$$

We have the following two propositions.

Proposition 1 [1]. Let X and Y be two Banach spaces. Let's consider $f \in L_{loc}^1(0,T;X)$ and $\mathscr{T} : [0,T] \to \mathscr{L}(X,Y)$ be a strongly continuous function. Then the convolution

$$(\mathscr{T} * f)(t) := \int_0^t \mathscr{T}(t-s)f(s)ds,$$

exists in the bochner sense and defines a continuous function from $[0,T]$ into Y

Proposition 2 [1](Young's Inequality). Let's consider $p,q,s \ge 1$ such that $\frac{1}{q} + \frac{1}{p} = 1 + \frac{1}{s}$.

If $\mathscr{T} \in L^p(0,T;\mathscr{L}(X,Y))$ and $f \in L^q(0,T;X)$, then

$$\mathscr{T} * f \in L^s(0,T;Y) \qquad and \qquad \|\mathscr{T} * f\|_{L^s(0,T;Y)} \le \|f\|_{L^q(0,T;X)} \cdot \|\mathscr{T}\|_{L^p(0,T;\mathscr{L}(X,Y))}.$$

Let w be a measurable function defined from $[0, T]$ to $\mathbb{R}^+$. The Weighted Lebesgue space ([2]) associate to w is defined by:

$$L_w^p[0, T] := \left\{ f \in L_{loc}^p[0, T] \ \Big| \ \int_0^T w(t)|f(t)|^p dt < +\infty \right\}, \qquad p \geq 1$$

which is a Banach space endowed with the norm :

$$\|f\|_{L_w^p[0,T]} = \left[\int_0^T w(t)|f(t)|^p dt \right]^{\frac{1}{p}}.$$

For $0 < \alpha \leq 1$, let $w(t) = t^{\alpha-1}$, we denote $L_w^p[0, T] := L_{\alpha-1}^p(0, T)$ and we have the following inclusion $L_{\alpha-1}^p(0, T) \subset L^p[0, T]$.

Let's consider $n \in \mathbb{N}^*$, Ω an open bounded subset of $\mathbb{R}^n$ with smooth enough boundary $\partial\Omega$ and let $\alpha \in]0, 1]$. For a time $T > 0$, set $Q = \Omega \times]0, T]$ and $\Sigma = \partial\Omega \times]0, T]$. Let's consider the following fractional semi-linear evolution equation:

$$\begin{cases} {}^C D_{0+}^\alpha y_u(x, t) + A y_u(x, t) = N y_u(x, t) + Bu(t) \text{ in } Q, \\ \\ y_u(\xi, t) = 0 \qquad\qquad\qquad\qquad\qquad \text{on } \Sigma, \qquad (2) \\ \\ y_u(x, 0) = y_0(x) \qquad\qquad\qquad\qquad\quad\ \text{in } \Omega, \end{cases}$$

where

- $-A$ is the infinitesimal generator of a C_0 semi-group $\{\mathscr{S}(t)\}_{t\geq 0}$ on the Hilbert space $X = L^2(\Omega)$.
- N a nonlinear operator.
- B is the control operator from $\mathscr{U}$ into X which is linear.
- u is given in $U = L^2(0, T, \mathscr{U})$ and $y_0 \in X$.

Without loss of generality, we denote $y_u(., t) := y_u(t)$.

Definition 2 [5, 19]. A mild solution of the system (2) is any function y_u in $C(0, T; X)$ satisfying the following integral equation :

$$y_u(t) = S_\alpha(t)y_0 + \int_0^t (t-\tau)^{\alpha-1} K_\alpha(t-\tau) Ny(\tau)d\tau + \int_0^t (t-\tau)^{\alpha-1} K_\alpha(t-\tau) Bu(\tau)d\tau,$$
$$(3)$$

where

$$S_\alpha(t) = \int_0^\infty \phi_\alpha(\theta)\mathscr{S}(t^\alpha\theta)d\theta,$$

$$K_\alpha(t) = \alpha \int_0^\infty \theta\phi_\alpha(\theta)\mathscr{S}(t^\alpha\theta)d\theta,$$

$$\phi_\alpha(\theta) = \frac{1}{\alpha} \theta^{-1-\frac{1}{\alpha}} W_\alpha(\theta^{-\frac{1}{\alpha}}) \qquad \text{for all } \theta \quad \text{positive,}$$

and

$$W_\alpha(\theta) = \frac{1}{\pi} \sum_{n=1}^{\infty} (-1)^{n-1} \theta^{-n\alpha-1} \frac{\Gamma(n\alpha+1)}{n!} sin(n\pi\alpha).$$

we have the following proposition.

Proposition 3 *[20]. For all $\beta \geq -1$, we have*

$$\int_0^\infty \theta^\beta \phi_\alpha(\theta) d\theta = \frac{\Gamma(1+\beta)}{\Gamma(1+\alpha\beta)},$$

then we have the following remark.

Remark 1. If $\beta = 0$, we can see that ϕ_α is a probability density.

Let $\omega \subset \Omega$ be a subregion with positive Lebesgue measure. The restriction operator in ω is defined as follows:

$$\chi_\omega : L^2(\Omega) \longrightarrow L^2(\omega)$$

$$y \longmapsto y_{|\omega}$$

and we denote its adjoint by χ_ω^*.
The mild solution defined by (3) can be written :

$$y_u(t) = S_\alpha(t)y_0 + L_\alpha(t)Ny_u(.) + L_\alpha(t)Bu(.), \tag{4}$$

where

$$L_\alpha(t)y(.) = \int_0^t (t-\tau)^{\alpha-1} K_\alpha(t-\tau)y(\tau)d\tau.$$

We also define the restriction of the controllability operator in ω by:

$$H_\omega^\alpha : U \longrightarrow L^2(\omega)$$

$$u \longmapsto \chi_\omega L_\alpha(T)Bu.$$

Definition 3. The system (2) is said to be exactly (respectively, approximately) ω-controllable if for all $y_d \in L^2(\omega)$ (respectively, for all $\epsilon > 0$ and for all $y_d \in L^2(\omega)$), we can find a control $u \in U$ such that $\chi_\omega y_u(T) = y_d$ (respectively, $||\chi_\omega y_u(T) - y_d||_{L^2(\omega)} \leq \epsilon$).

Problem: For any state y_d in $L^2(\omega)$, is it possible to find a control u^* that steer the system (2) in a finite time T to y_d only in the subregion ω ?

We consider the following linear system associate to the nonlinear system (2):

$$\begin{cases} {}^{C}D_{0+}^{\alpha}y(x,t) + Ay(x,t) = Bu(t) \text{ in } Q, \\ \\ y(\xi,t) = 0 \qquad\qquad\qquad \text{on } \Sigma, \\ \\ y(x,0) = y_0(x) \qquad\qquad \text{in } \Omega, \end{cases} \qquad (5)$$

which we suppose, for the rest of this work, to be approximately ω-controllable. Then we give the following proposition.

Proposition 4.   *If the following hypotheses hold*

- $[y_d - \chi_\omega S_\alpha(T)y_0 - \chi_\omega L_\alpha(T)Ny_{u^*}(.)] \in Im(H_\omega^\alpha),$
- $\quad Im(H_\omega^\alpha) \quad$ *a closed subset.*

Then the system (2) is exactly ω- controllable by the control

$$u^*(.) = H_\omega^{\alpha^\dagger}[y_d - \chi_\omega S_\alpha(T)y_0 - \chi_\omega L_\alpha(T)Ny_{u^*}(.)].$$

Where

$$H_\omega^{\alpha^\dagger} := H_\omega^{\alpha^*}\left(H_\omega^\alpha H_\omega^{\alpha^*}\right)^{-1} \qquad \text{is the Pseudo-inverse operator of } H_\omega^\alpha.$$

Proof. Using the expression (3), the solution of system (2) controlled by u^* is giving by the following formula

$$y_{u^*}(t) = S_\alpha(t)y_0 + L_\alpha(t)Ny_{u^*}(.) + L_\alpha(t)BH_\omega^{\alpha^\dagger}[y_d - \chi_\omega S_\alpha(T)y_0 - \chi_\omega L_\alpha(T)Ny_{u^*}(.)],$$

hence

$$\chi_\omega y_{u^*}(T) = \chi_\omega S_\alpha(T)y_0 + \chi_\omega L_\alpha(T)Ny_{u^*}(.) + H_\omega^\alpha H_\omega^{\alpha^\dagger}[y_d - \chi_\omega S_\alpha(T)y_0 - \chi_\omega L_\alpha(T)Ny_{u^*}(.)],$$

since

$$[y_d - \chi_\omega S_\alpha(T)y_0 - \chi_\omega L_\alpha(T)Ny_{u^*}(.)] \in Im(H_\omega^\alpha),$$

and $\quad H_\omega^\alpha H_\omega^{\alpha^\dagger}$ is the orthogonal projection on $Im(H_\omega^\alpha)$. Then

$$\chi_\omega y_{u^*}(T) = \chi_\omega S_\alpha(T)y_0 + \chi_\omega L_\alpha(T)Ny_{u^*}(.) + y_d - \chi_\omega S_\alpha(T)y_0 - \chi_\omega L_\alpha(T)Ny_{u^*}(.) = y_d.$$

In the next section, we will study the regional controllability of the system (2) in $Im(H_\omega^\alpha)$ endowed with the norm

$$||y_d||_{Im(H_\omega^\alpha)} = ||H_\omega^{\alpha^\dagger}y_d||_U$$

Remark 2. $||.||_{Im(H_\omega^\alpha)}$ defines a semi-norm on $Im(H_\omega^\alpha)$ but it becomes a norm if the linear system (5) is approximately ω-controllable .

Proof. It is sufficient to show that

$$||y_d||_{\mathrm{Im}(H_\omega^\alpha)} = 0 \Longrightarrow y_d = 0$$

We have

$$||y_d||_{\mathrm{Im}(H_\omega^\alpha)} = 0 \Longrightarrow ||H_\omega^{\alpha^\dagger} y_d||_U = 0$$

$$\Longrightarrow H_\omega^{\alpha^\dagger} y_d = 0$$

$$\Longrightarrow (H_\omega^{\alpha*} H_\omega^\alpha) H_\omega^{\alpha^\dagger} y_d = 0$$

$$\Longrightarrow H_\omega^{\alpha*} y_d = 0.$$

Since the linear system (5) is approximately ω-controllable, then $\ker(H_\omega^{\alpha*}) = \{0\}$ by [7], therefore $y_d = 0$.

3 Analytical Approach

We consider the system (2) with $y_0 = 0$, moreover, let $-A$ the infinitesimal generator of an analytic semigroup of bounded linear operator $(T(t))_{t\geq 0}$ on X.

Let 0 be an element of the resolvent set of $-A$, then it is possible to define the fractional power A^ν for any ν belongs to the interval $]0,1]$. $X^\nu := D(A^\nu)$ is a Banach space, which is dense in X, endowed with the graph norm: $||.||_{X^\nu} = ||A^\nu(.)||_X$.

Remark 3. For the sake of simplification, we choose the order of fractional power of A to be the same as the order of fractional derivative.

We have the following proposition.

Proposition 5. *[12] For all $\alpha \in]0,1]$, the following properties are satisfied*

(i) $\exists \ C_\alpha > 0$ *such that* $||A^\alpha T(t)||_{\mathscr{L}(X,X)} \leq C_\alpha t^{-\alpha}$ $0 < t \leq T$.

(ii) $\forall \ t \in [0,T]$, *we have*

$$||\mathrm{K}_\alpha(t)||_{\mathscr{L}(X,X^\alpha)} \leq \frac{\alpha C_\alpha}{t^{\alpha^2}} \times \frac{\Gamma(2-\alpha)}{\Gamma(1+\alpha(1-\alpha))} := f_\alpha(t).$$

Corollary 1. *Let's consider $H(t) = t^{\alpha-1} \mathrm{K}_\alpha(t)$ and $q \geq 1$. If $f_\alpha \in \mathrm{L}_{\alpha-1}^q(0,T)$, then*

$$H \in L^q(0,T;\mathscr{L}(X,X^\alpha)) \quad and \quad ||H(.)||_{L^q(0,T;\mathscr{L}(X,X^\alpha))} \leq ||f_\alpha(.)||_{\mathrm{L}_{\alpha-1}^q(0,T)}.$$

Hypotheses :
We assume that the following conditions hold.

(i)   For all $p, s \geq 1$, there exists $q \geq 1$ such that

$$\frac{1}{q} = 1 + \frac{1}{p} - \frac{1}{s} \quad \text{and} \quad f_\alpha \in \mathrm{L}_{\alpha-1}^q(0, T). \tag{6}$$

(ii)   Let $N : L^p(0, T; X^\alpha) \longrightarrow L^s(0, T; X)$ be the nonlinear operator satisfying

$$\begin{cases} N(0) = 0, \\ \\ ||Nx - Ny||_{L^s(0,T;X)} \leq k(||x||, ||y||)||x - y||_{L^p(0,T;X^\alpha)}, \end{cases} \tag{7}$$

where $k : \mathbb{R}^+ \times \mathbb{R}^+ \longrightarrow \mathbb{R}^+$ is such that $\lim\limits_{(\theta_1,\theta_2)\to(0,0)} k(\theta_1, \theta_2) = 0$.

We define the operator

$$\Psi(y_d, u) = \mathrm{H}_\omega^{\alpha^\dagger}(y_d - \chi_\omega L_\alpha(T) N y_u).$$

Then the regional controllability problem becomes a fixed point problem of the function $\Psi(y_d, .)$, where y_d is an element of $\mathrm{Im}(H_\omega^\alpha)$

Theorem 1. *If the hypotheses (i) and (ii) hold and*

(iii)

$$||L_\alpha(.)Bu||_{L^p(0,T;X^\alpha)} \leq \beta||u||_U, \qquad \beta > 0, \tag{8}$$

(iv)

$$||\chi_\omega K_\alpha(.)||_{\mathcal{L}(X,Im(H_\omega^\alpha))} = g_\alpha \in L_{\alpha-1}^r(0, T), \quad \frac{1}{r} + \frac{1}{s} = 1, \tag{9}$$

are satisfied, then the following assertions hold.

1. *There exists $a > 0$, $\rho = \rho(a) > 0$ and $m = m(a) > 0$ such that for any state y_d in $\mathcal{B}(0, \rho) \subset Im(H_\omega^\alpha)$ there exists u^* in $\mathcal{B}(0, m)$ that steers the system (2) to y_d in ω. Where $\mathcal{B}(0, k)$ is a ball with center 0 and radius k.*
2. *The mapping*

$$F : \mathcal{B}(0, \rho) \longrightarrow U$$

$$y_d \longmapsto u^*,$$

is a lipschitz mapping.

Proof. 1- Based on hypothesis (ii), we have

$$\lim_{(\theta_1,\theta_2)\to(0,0)} k(\theta_1, \theta_2) = 0,$$

then $\exists\, a > 0$, $\exists\, \nu > 0$ such that

$$k(\theta_1, \theta_2) < \nu < \frac{1}{\beta||g_\alpha||_{L_{\alpha-1}^r(0,T)} + ||f_\alpha||_{L_{\alpha-1}^q(0,T)}} \qquad \forall \theta_1, \theta_2 \leq a,$$

which gives

$$\sup_{\theta_i \le a} k(\theta_1, \theta_2) \le \nu < \frac{1}{\beta \|g_\alpha\|_{L^r_{\alpha-1}(0,T)} + \|f_\alpha\|_{L^q_{\alpha-1}(0,T)}}.$$

Let's consider $A_1 = \beta \|g_\alpha\|_{L^r_{\alpha-1}(0,T)} \sup_{\theta_i \le a} k(\theta_1, \theta_2)$ and $A_2 = \|f_\alpha\|_{L^q_{\alpha-1}(0,T)}$ $\sup_{\theta_i \le a} k(\theta_1, \theta_2)$.

We have $A_1 < 1$ and $A_2 < 1$.

If we set

$$m = \frac{a}{\beta} \left(1 - \|f_\alpha\|_{L^q_{\alpha-1}(0,T)} \sup_{\theta \le a} k(\theta, 0)\right),$$

then m is positive.

In fact,

$$\|f_\alpha\|_{L^q_{\alpha-1}(0,T)} \sup_{\theta \le a} k(\theta, 0) \le A_2 < 1.$$

Moreover, the following function

$$f : \mathscr{B}(0, m) \longrightarrow \mathscr{B}(0, a)$$

$$u \longmapsto y_u$$

is a Lipschitz mapping with constant $\dfrac{\beta}{1 - A_2}$.

For that, by the equation (4) and corollary (1), for all $u, v \in B(0, m)$, we have

$$\|y_u - y_v\|_{L^p(0,T;X^\alpha)} = \|L_\alpha(.)N(y_u - y_v) + L_\alpha(.)B(u - v)\|_{L^p(0,T;X^\alpha)}$$

$$\le \|(H * N(y_u - y_v))(.)\|_{L^p(0,T;X^\alpha)} + \|L_\alpha(.)B(u - v)\|_{L^p(0,T;X^\alpha)},$$

using hypotheses (ii) and (iii) we get

$$\|y_u - y_v\|_{L^p(0,T;X^\alpha)} \le A_2 \|y_u - y_v\|_{L^p(0,T;X^\alpha)} + \beta \|u - v\|_U,$$

hence f is a lipschitz mapping with constant $\dfrac{\beta}{1 - A_2}$.

Next we show that $\Psi(y_d, .)$ has a unique fixed point in $\mathscr{B}(0, m)$.

Let's consider $y_d \in \mathrm{Im}\,(H^\alpha_\omega)$ and $u, v \in \mathscr{B}(0, m)$, we have

$$\|\Psi(y_d, u) - \Psi(y_d, v)\|_U = \|\chi_\omega L_\alpha(T)(Ny_u - Ny_v)\|_{\mathrm{Im}\,H^\alpha_\omega}$$

$$\le \|g_\alpha\|_{L^r_{\alpha-1}(0,T)} \|Ny_u - Ny_v\|_{L^s(0,T;X)}$$

$$\le \|g_\alpha\|_{L^r_{\alpha-1}(0,T)} \sup_{(\theta_i \le a)} k(\theta_1, \theta_2) \|y_u - y_v\|_{L^p(0,T;X^\alpha)},$$

since f is lipschitz, then

$$\|\Psi(y_d, u) - \Psi(y_d, v)\|_U \le \frac{\beta A_1}{1 - A_2} \|u - v\|_U. \tag{10}$$

If we denote $A_3 := \dfrac{\beta A_1}{1 - A_2}$, we have $A_3 < 1$, thus $\Psi(y_d, .)$ is a strict contraction mapping.

For $u \in \mathscr{B}(0, m)$, we have $y_u \in \mathscr{B}(0, a)$ and

$$\|\Psi(y_d, u)\|_U = \|y_d - \chi_\omega L_\alpha(T) N y_u\|_{\mathrm{Im}\,(H_\omega^\alpha)}$$

$$\leq \|y_d\|_{\mathrm{Im}\,(H_\omega^\alpha)} + \|\chi_\omega L_\alpha(T) N y_u\|_{\mathrm{Im}\,(H_\omega^\alpha)}$$

$$\leq \|y_d\|_{\mathrm{Im}\,(H_\omega^\alpha)} + \|g_\alpha\|_{L_{\alpha-1}^r(0,T)} a \sup_{(\theta \leq a)} k(\theta, 0),$$

therefore, if

$$\|y_d\|_{\mathrm{Im}\,(H_\omega^\alpha)} \leq m - \|g_\alpha\|_{L_{\alpha-1}^r(0,T)} a \sup_{(\theta \leq a)} k(\theta, 0),$$

then $\Psi(y_d, u) \in \mathscr{B}(0, m)$.

We set $\rho = \dfrac{a}{\beta}(1 - (\|f_\alpha\|_{L_{\alpha-1}^q(0,T)} + \beta\|g_\alpha\|_{L_{\alpha-1}^r(0,T)}) \sup_{\theta \leq a} k(\theta, 0))$,

hence, if $y_d \in \mathscr{B}(0, \rho) \subset \mathrm{Im}\,(H_\omega^\alpha)$, we deduce from the Picard fixed point theorem that $\Psi(y_d, .)$ admits a unique fixed point $u^* \in \mathscr{B}(0, m)$.

We remark that u^* obtained is solution of the exact regional controllability problem.

2- Let z_d and y_d in $\mathscr{B}(0, \rho)$, we have

$$F(z_d) - F(y_d) = \Psi(z_d, F(z_d)) - \Psi(z_d, F(y_d)) + \Psi(z_d, F(y_d)) - \Psi(y_d, F(y_d)),$$

since

$$\|\Psi(z_d, F(z_d)) - \Psi(z_d, F(y_d))\|_U \leq A_3 \|F(z_d) - F(y_d)\|_U,$$

$$\|\Psi(z_d, F(y_d)) - \Psi(y_d, F(y_d))\|_U = \|z_d - y_d\|_{\mathrm{Im}\,(H_\omega^\alpha)}.$$

Then

$$\|F(z_d) - F(y_d)\|_U \leq \dfrac{1}{1 - A_3}\|z_d - y_d\|_{\mathrm{Im}\,(H_\omega^\alpha)},$$

therefore, F satisfies the Lipschitz condition.

Moreover, the Picard fixed point theorem gives also the existence of a sequence which converges to the control u^*.

We give the following proposition.

Proposition 6. *The sequence*

$$\begin{cases} u_0 & = 0 \\ \\ u_{n+1} = H_\omega^{\alpha^\dagger}(y_d - \chi_\omega L_\alpha(T) N y_{u_n}), \end{cases} \tag{11}$$

converges to u^* *in* $\mathscr{B}(0, m) \subset U$.

Proof. Let's consider $n, k \in \mathbb{N}^*$ we have

$$||u_{n+k} - u_n||_U \leq \sum_{l=n}^{n+k-1} ||u_{l+1} - u_l||_U.$$

By the Inequality (10) we obtain

$$||u_{l+1} - u_l||_U = ||\Psi(y_d, u_l) - \Psi(y_d, u_{l-1})||_U \leq A_3||u_l - u_{l-1}||_U \leq A_3^l||u_1||_U,$$

which yields

$$||u_{n+k} - u_n||_U \leq \sum_{l=n}^{n+k-1} A_3^l||u_1||_U \leq \frac{1 - A_3^k}{1 - A_3} A_3^n ||u_1||_U,$$

hence since $A_3^n \to 0$, we conclude that $\lim\limits_{n \to +\infty} ||u_{n+k} - u_n||_U = 0$.

The sequence $(u_n)_n$ is a Cauchy sequence on $\mathscr{B}(0, m)$, then $(u_n)_n$ converges to u in $\mathscr{B}(0, m)$.

Passing to the limit in (11), we have $u = \Psi(y_d, u)$, since $\Psi(y_d, .)$ has a unique fixed point in $B(0, m)$, then $u = u^*$.

4   Algorithm

In this section, we present an algorithm which has as objective, finding a control that steering the considered system to the desired state only in ω, this leads to some numerical simulations which will be presented in the next section.

Algorithm 1

Initialization:
Fractional order of derivative α.
The region ω.
Actuator (D, f).
$r_1 = y_d$.
Error estimate ε.
Calculation of $u_1 = H_\omega^{\dagger \alpha} r_1$ and obtain $y_{u_1}(T)$.
repeat
$\quad r_n = r_{n-1} + (y_d - \chi_\omega y_{u_{n-1}}(T)), \quad n \geq 2.$
$\quad$Calculation of $u_n = H_\omega^{\alpha \dagger} r_n$.
$\quad$Solve the semi-linear system (2) controlled by u_n.
until
$$||\chi_\omega y_{u_n}(T) - y_d||_{Im(H_\omega^\alpha)} < \varepsilon.$$

.

---

5   Numerical Results

In this section, we present two numerical simulations illustrating our theoretical result where the first one is done by using zonal actuator and the second example is giving by using a pointwise actuator.

5.1   Case of Zonal Actuator

Let's consider the following sub-diffusion one-dimensional system with order $\alpha = 0.7$:

$$\begin{cases} {}^{C}D_{0^{+}}^{0.7}z(x,t) - \dfrac{\partial^{2z}(x,t)}{\partial x^{2}} = \chi_{D}u(t) + \displaystyle\sum_{j=1}^{\infty}(<z,\varphi_{j}>)^{2}\varphi_{j}(x) & \text{in } [0,1]\times\,]0,3] \\ z(x,t) = 0 & \text{on}\{0,1\}\times\,]0,3] \\ z(x,0) = 0 & \text{in } [0,1], \end{cases}$$

$$(12)$$

where $\varphi_{j}(x) = \sqrt{2}\sin(j\pi x)$.

The control operator in the system (12) is given by a zonal actuator (D,f) where $D = [0.2\,,0.3]$ and $f = 1$.

We consider the region $\omega\ =\,]0.4\,,0.68]$ and the desired state $z_{d}(x)\ =\ 5.3\,x^{2}\,(x-1)^{2}\,(x-0.4)$.

Using the previous algorithm, we obtain the following results:

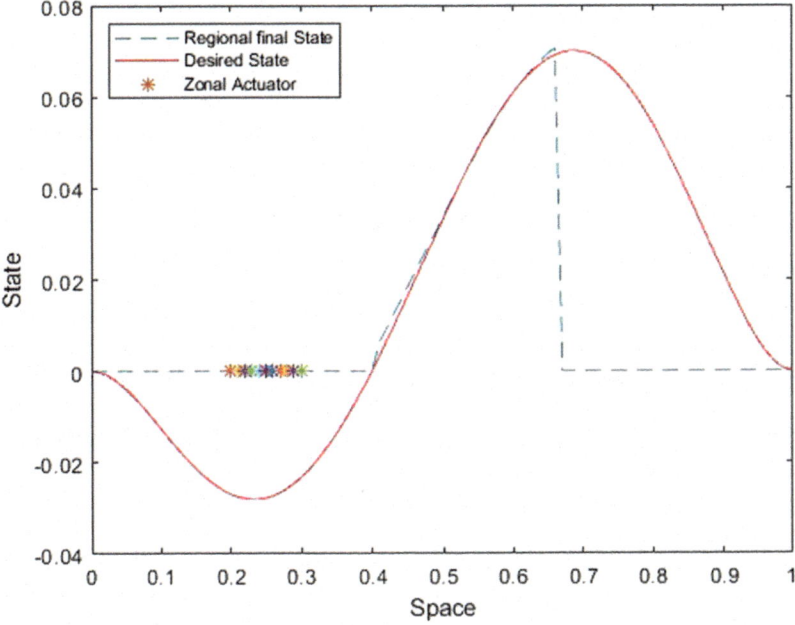

Fig. 1. Desired state and estimate final one in $\omega = [0.4\,,0.68]$

In the subregion $]0.4, 0.68]$, we can see that the regional final state and the desired state z_d are very close with error $|| \chi_\omega z_u(x,t) - z_d ||_{L^2(\omega)} = 7.05 \times 10^{-4}$. Here we have the evolution of the control function.

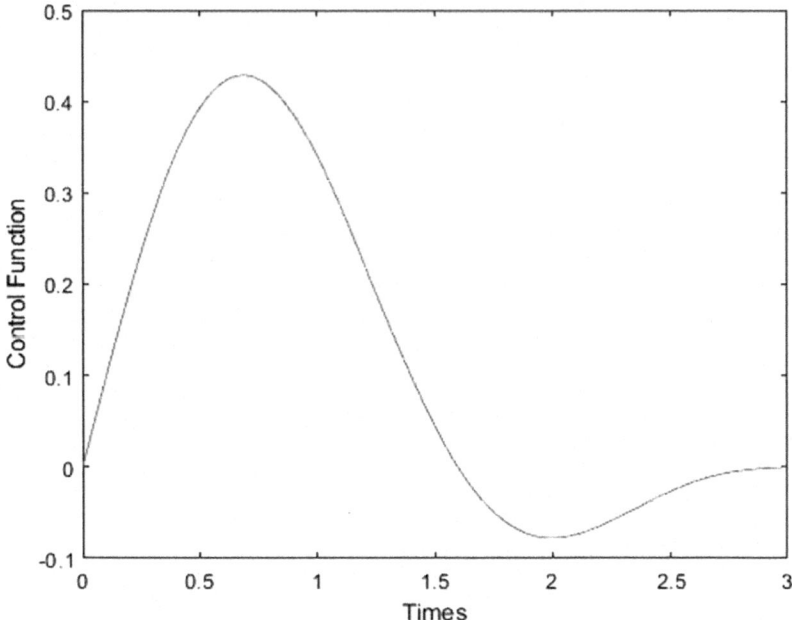

Fig. 2. Control function.

with a transfer cost $|| u^* ||^2_{L^2(0,T)} = 0.13$.

5.2 Case of Pointwise Actuator

We will treat the same kind of system with $\alpha = 0.8$ and a pointwise actuator located in $b = 0.4$, which amounts to consider the following system:

$$\begin{cases} {}^C D_{0+}^{0.8} z(x,t) - \dfrac{\partial^2}{\partial x^2} z(x,t) = \delta_b u(t) + \displaystyle\sum_{j=1}^{\infty} (< z, \varphi_j >)^2 \varphi_j(x) & \text{in } [0,1] \times]0,2] \\ z(\xi, t) = 0 & \text{on } \{0,1\} \times]0,2] \\ z(x,0) = 0 & \text{in } [0,1], \end{cases}$$

where $\varphi_j(x) = \sqrt{2} \sin(j\pi x)$.

The subregion under consideration is $\omega =]0.45 , 0.7]$. Let's consider the following desired state:

$$z_d(x) = x(x-1)(3.5x - 0.2)(0.6 - x)(x - 0.1).$$

Using the proposed algorithm, we obtain the following figure:

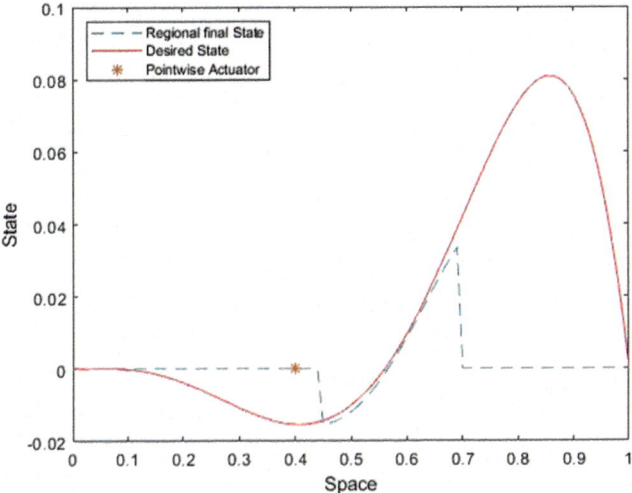

Fig. 3. Desired and estimate final states in ω.

We can see that the final state is very close to the desired state in the sub-region $]0.45, 0.7]$ with an error of

$$\| \chi_\omega z_u(t) - z_d \|_{L^2(\omega)} = 6.04 \times 10^{-4}.$$

The following figure shows the evolution of the control function. with a transfer cost $\| u^* \|^2_{L^2(0,T)} = 0.03$.

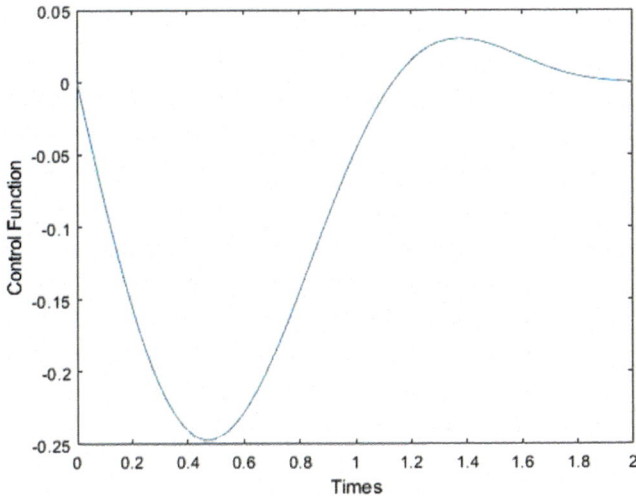

Fig. 4. Control function.

6 Conclusion

In this work, we have studied the regional controllability of Caputo time-fractional sub-diffusion system with analytical approach, which is a technical one, based on fixed point techniques and semigroup theory, we also presented an algorithm based on our theoretical results, which leads to successful numerical results. As a future work, we are working on the concept of boundary and gradient regional controllability for the same type of systems.

References

1. Arendt, W., Batty, C.J.K., Hieber, M., Neubrander, F.: Vector-valued Laplace Transforms and Cauchy Problems, 2nd edn. Birkhäuser, Basel (2011)
2. Cholewa, J.W., Dlotko, T.: Cauchy problems in weighted lebesgue spaces. Czechoslov. Math. J. **54**, 991–1013 (2004)
3. Do, V.: A note on approximate controllability of semilinear systems. Syst. Control. Lett. **12**, 365–371 (1989)
4. Dzieliński, A., Sierociuk, D., Sarwas, G.: Some applications of fractional order calculus. Bull. Pol. Acad. Sci-Tech. **58**, 583–592 (2010)
5. El-Borai, M.M.: Some probability densities and fundamental solutions of fractional evolution equations. Chaos Soliton. Fract. **14**, 433–440 (2002)
6. El Jai, A., Pritchard, A.J.: Sensors and actuators in distributed systems analysis. Int. J. Control **46**, 1139–1153 (2007)
7. Fudong, G., YangQuan, C., Chunhai, K.: Regional Analysis of Time-Fractional Diffusion Processes. Springer International Publishing, New York (2018)
8. Glowinski, R., Lions, J.L.: Exact and approximate controllability for distributed parameter systems. Acta Numer. **4**, 159–328 (1995)
9. Hilfer, R.: Applications of Fractional Calculus in Physics. Default Book Series (2000)
10. Kilbas, A.A., Srivastava, H.M., Trujillo, J.J.: Theory and Applications of Fractional Differential Equations. Elsevier, Amsterdam (2006)
11. Lions, J.L.: On the controllability of distributed systems. Proc. Natl. Acad. Sci. U. S. A. **94**, 4828–4835 (1997)
12. Mahmudov, N.I., Zorlu, S.: On the approximate controllability of fractional evolution equations with compact analytic semigroup. J. Comput. Appl. Math. **259**, 194–204 (2014)
13. N'Guérékata, G.M.: A Cauchy problem for some fractional abstract differential equations with nonlocal conditions. Nonlinear Anal. **5**, 1873–1876 (2009)
14. Qureshi, S., Yusuf, A., Shaikh, A.A., Inc, M., Baleanu, D.: Fractional modeling of blood ethanol concentration system with real data application. Chaos **29**, 1–8 (2019)
15. Zerrik, E., Kamal, A.: Output controllability for semi-linear distributed systems. J. Dyn. Control Syst. **13**, 289–306 (2007)
16. Zerrik, E., El Jai, A., Boutoulout, A.: Actuators and regional boundary controllability of parabolic system. Int. J. Sys. Sci. **31**, 73–82 (2000)
17. Zhou, H.X.: Controllability properties of linear and semilinear abstract control systems. SIAM J. Control. Optim. **22**, 405–422 (1984)
18. Zhou, Y., Jiao, F.: Existence and uniqueness for p-type fractional abstract differential equations. Nonlinear Anal. **71**, 2724–2733 (2009)

19. Zhou, Y., Jiao, F.: Existence of mild solutions for fractional neutral evolution equations. Comput. Math. Appl. **59**, 1063–1077 (2010)
20. Zhou, Y., Zhang, L., Shen, X.H.: Existence of mild solutions for fractional evolution equations. J. Integral. Equ. Appl. **25**, 557–586 (2013)
21. Zuazua, E.: Exact controllability for semilinear wave equations in one space dimension. Ann. Inst. Henri Poincaré **10**, 109–129 (1993)

Quadratic Optimal Control for Bilinear Systems

Soufiane Yahyaoui$^{(\boxtimes)}$ and Mohamed Ouzahra

Laboratory M2PA, University Sidi Mohamed Ben Abdellah ENS FES, Fes, Morocco
soufiane.yahyaoui@usmba.ac.ma, m.ouzahra@yahoo.fr

Abstract. In this work, we will investigate the quadratic optimal control for bilinear systems. We will first study the existence of a solution for the considered optimal control. Then, we will focus on a special class of bilinear systems for which the quadratic optimal control can be expressed in a feedback law form. The approach relies on the conditions of optimality and linear semi-group theory.

Keywords: Quadratic cost · Optimal control · Bilinear systems

1 Introduction

The subject of this paper is to study the quadratic optimal control for infinite dimensional bilinear systems. The importance of bilinear systems lies in the fact that they represent a theoretical model for real processes (natural or industrial) and that they represent a first generalization of linear systems. Optimal control theory consists in seeking the best control strategy among others from the set of admissible controls, that is the one that enables us to reach a precise objective (reach a desired state, minimize a cost or energy, etc.). Quadratic cost functions have a wide use in differential geometry, statistics, special relativity, solid mechanics, etc. Optimal control problems have been the subject of several works. In [7], Kalman has studied the problem of quadratic optimal control for linear finite-dimensional systems. He characterized the control in term of the system's state and the solution of the Riccati equation. These results have been generalized by Banks and Yu [4] to a class of infinite dimensional semi-linear systems. The case of distributed bilinear systems has been considered by Alami [2]. Another approach, based on the Pontryagin's maximum principle, has been developed by Pontryagin [10] for finite-dimensional systems. These results have been generalized by different authors (e.g. [1,5,8,11,12]) for a class of semi-linear systems with a variety of cost functions. The main limit of the approach based on the adjoint equation is that such an equation involves the unknown optimal control. This together with the nonlinear dependence of the state w.r.t to control becomes somehow embarrassing when one looks for the explicit expression of optimal control.

In this work, we consider the problem of optimal control that minimizes a given quadratic cost. In Sect. 2, we will first state our quadratic optimal problem

© Springer Nature Switzerland AG 2021
Z. Hammouch et al. (Eds.): SM2A 2019, LNNS 168, pp. 156–163, 2021.
https://doi.org/10.1007/978-3-030-62299-2_11

for bilinear systems and then study the existence of an optimal control. In Sect. 3, we will focus on a special class of bilinear system for which the quadratic optimal control can be expressed in a feedback law form.

2   Quadratic Optimal Control

2.1   Problem Statement

Let us consider the following bilinear system

$$\begin{cases} \dot{y}(t) = Ay(t) + u(t)By(t) \\ y(0) = y_0 \in X \end{cases} \tag{1}$$

where

- $A : D(A) \subset X \mapsto X$ is the infinitesimal generator of a linear C_0- semigroup $S(t)$ of isometries on a real Hilbert space X whose inner product and the corresponding norm are denoted respectively by $\langle .,. \rangle$ and $\|.\|$,
- $B : X \mapsto X$ is a linear bounded operator,
- $u(\cdot)$ is a scalar valued control that belongs to the control space $L^2(0, T)$ and $y(\cdot)$ is the corresponding mild solution with initial state $y_0 \in X$.

The quadratic cost function J to be minimized is defined by

$$J(u) = 2\|y(T)\| + \int_0^T \|y(t)\|^2 dt + \int_0^T u(t)^2 dt, \tag{2}$$

for any admissible control u, i.e. for which the corresponding solution y exists and $J(u)$ makes sense.

It is well known that for any $u \in L^2(0, T)$ the system (1) admits a unique mild solution (see [3]) and we have $J(u) < +\infty$.

In the sequel we take $U_{ad} := L^2(0, T)$ (the set of admissible control).

The optimal control problem may be stated as follows

$$\begin{cases} min J(u) \\ u \in U_{ad} \end{cases} \tag{3}$$

Let us introduce the following time varying cost function

$$J(u)(t) = 2\|y(t)\| + \int_0^t \|y(s)\|^2 ds + \int_0^t u(s)^2 ds, \quad t \in [0, T]. \tag{4}$$

2.2   Existence of the Optimal Control

In this subsection, we prove the existence of a square integrable control u that minimizes the cost (2).

Theorem 1. *There exists an optimal control solution of the problem (3).*

Proof: Since the set $\{J(u)/u \in U_{ad}\} \subset \mathbb{R}^+$ is not empty and bounded from below, it admits a lower bound J^*. Let $(u_n)_{n\in\mathbb{N}}$ be a minimizing sequence such that $J(u_n) \to J^*$.

By the coercivity of the mapping $R : u \mapsto \int_0^T \|u(t)\|^2 dt$, we deduce that the sequence (u_n) is bounded, so it admits a subsequence still denoted by (u_n) as well, which weakly converges to $u^* \in U_{ad}$.

Let y_n and y^* be the solutions of (1) respectively corresponding to u_n and u^*. From Theorem 3.6 of [3] we have

$$\lim_{n\to+\infty} \|y_n(t) - y^*(t)\| = 0, \ \forall t \in [0, T]. \tag{5}$$

Since the norm $\|.\|$ is lower semi-continuous, it follows from (5) that for all $t \in [0, T]$

$$\|y^*(t)\|^2 \le \lim_{n\to+\infty} inf \|y_n(t)\|^2.$$

Applying Fatou's lemma we obtain

$$\int_0^T \|y^*(t)\|^2 dt \le \lim_{n\to+\infty} inf \int_0^T \|y_n(t)\|^2 dt. \tag{6}$$

Taking into account that R is convex and lower semi-continuous with respect to the weak topology, we get (see Corollary III.8 of [6])

$$R(u^*) \le \lim_{n\to+\infty} inf \, R(u_n). \tag{7}$$

Combining the formulas (5), (6) and (7) we deduce that

$$J(u^*) = 2\|y^*(T)\| + \int_0^T \|y(t)\|^2 dt + \int_0^T u^*(t)^2 dt$$

$$\le 2 \lim_{n\to+\infty} inf \, \|y_n(T)\| + \lim_{n\to+\infty} inf \int_0^T \|y_n(t)\|^2 dt + \lim_{n\to+\infty} inf \int_0^T u_n(t)^2 dt$$

$$\le \lim_{n\to+\infty} inf \, J(u_n)$$

$$\le J^*.$$

So we conclude that $J(u^*) = J^*$ and hence u^* is a solution of the problem (3).

3   Feedback Optimal Control

In the previous section, we have established an existence result of the quadratic optimal control for the problem (3). However, this result does not provide any information about the expression of the optimal control. This section consists of expressing the quadratic optimal control as an explicit feedback for a class of bilinear systems with $B = I$. Thus the system (1) becomes

$$\begin{cases} \dot{y}(t) = Ay(t) + u(t)y(t) \\ y(0) = y_0 \in X \end{cases} \tag{8}$$

Theorem 2. *The feedback control defined by*

$$u^*(t) = -\|y^*(t)\|, \tag{9}$$

is an optimal control for the problem (3).

Proof: Observing that the mapping $y \to -\|y\|y$ is locally Lipschitz, we deduce that the system (8), controlled by (9), has a unique mild solution defined on a maximal sub-interval $[0, T]$, which is given by the following variation of constants formula

$$y^*(t) = S(t)y_0 - \int_0^t S(t-s)\|y^*(s)\|y^*(s)ds.$$

Moreover, the control (9) results in decreasing norm state which implied that the optimal solution is global (see [9] p.185).

Let $v \in U_{ad}$ and let y_v be the respective solution to system (8). Here, we will show that $J(u^*) \le J(v)$. To this end, two cases will be discussed.

Case 1: $y^*(t) \neq 0, \quad \forall t \in [0, T]$.
Case 1.1: $y_v(t) \neq 0, \forall t \in [0, T]$.
Let $A_n = nA(nI - A)^{-1}$ be the Yoshida approximation of the operator A, and let y_{v_n} be the respective solution to (8) with A_n instead of A. Since the operator A_n is bounded, it follows that $y_{v_n} \in H^1(0, T)$.

Multiplying the system (8) by the state y_{v_n}, and using that A generates an isometric semi-group we deduce that

$$\|y_{v_n}(t)\|\frac{d}{dt}\|y_{v_n}(t)\| = \frac{1}{2}\frac{d}{dt}\|y_{v_n}(t)\|^2 = v(t)\|y_{v_n}(t)\|^2, \quad \forall t \in (0, T).$$

Taking into account that $\|y_v(t)\| \neq 0$, for all $t \in [0, T]$ and that $y_{v_n} \longrightarrow y_v$ (strongly) as $n \to +\infty$, we deduce that

$$\exists N \in \mathbb{N}, \forall n \ge N, \quad \|y_{v_n}(t)\| \neq 0, \forall t \in [0, T]$$

so that

$$\frac{d}{dt}\|y_{v_n}(t)\| = v(t)\|y_{v_n}(t)\|, \quad \forall t \in (0, T). \tag{10}$$

Integrating (10) over $[0, T]$, we get

$$2(\|y_{v_n}(T)\| - \|y_0\|) = \int_0^T 2v(t)\|y_{v_n}(t)\| dt$$
$$= \int_0^T \left((v(t) + \|y_{v_n}(t)\|)^2 - (v^2(t) + \|y_{v_n}(t)\|^2) \right) dt. \quad (11)$$

Using the fact that $y_{v_n} \longrightarrow y_v$ (strongly) as $n \to +\infty$, we deduce by taking the limit in the relation (11) that

$$J(v) - 2\|y_0\| = \int_0^T (v(t) + \|y_v(t)\|)^2 dt \geq 0. \quad (12)$$

In particular for $v = u^*$ we get

$$J(u^*) - 2\|y_0\| = \int_0^T (u^*(t) + \|y^*(t)\|)^2 dt. \quad (13)$$

This, together with the expression (9), gives

$$J(u^*) - 2\|y_0\| = 0. \quad (14)$$

Combining (12) and (14) we conclude that

$$J(v) \geq J(u^*).$$

Case 1.2: $\exists t_1 \in (0, T) \ / y_v(t_1) = 0.$
In this case we have $y_v(t) = 0, \quad \forall t \in [t_1, T].$
Let us define t' by

$$t' = \inf\{t \in [0, T] \ / y_v(t) = 0\}.$$

By the continuity of the state y_v we deduce that $y_v(t') = 0$ and that $y_v(t) \neq 0, \quad \forall t \in [0, t').$
Following the same method as in the previous case, we conclude that

$$J(v)(t') \geq J(u^*)(t'). \quad (15)$$

Moreover, it comes from the expression of $J(v)$ that

$$J(v)(T) = \int_0^T (v^2(t) + \|y_v(t)\|^2) dt = J(v)(t') + \int_{t'}^T v^2(t) dt \geq J(v)(t'). \quad (16)$$

Since $y^*(t) \neq 0$ for all $t \in [0, t']$ we deduce, according to Case 1.1, that

$$J(u^*)(T) = J(u^*)(t') = 2\|y_0\|. \quad (17)$$

Combining (15), (16) and (17) we conclude that

$$J(v)(T) \geq J(v)(t') \geq J(u^*)(t') = J(u^*)(T).$$

Case 2: $\exists t_1 \in [0, T] \quad /y^*(t_1) = 0.$
Here we have $y^*(t) = 0, \quad \forall t \in [t_1, T]$.

Then, taking $t' = \inf\{t \in [0, T] \quad /y^*(t) = 0\}$, we deduce that $y^*(t') = 0$ and $y^*(t) \neq 0, \quad \forall t \in [0, t').$

According to Case 1 we have

$$J(v)(t') \geq J(u^*)(t'), \forall v \in U_{ad}. \tag{18}$$

Using the fact that $y^*(t) = 0$, for all $t \in [t', T]$, we derive from the expression of $J(u^*)$ that

$$J(u^*)(T) = \int_0^T (u^{*2}(t) + \|y^*(t)\|^2)dt = \int_0^{t'} (u^{*2}(t) + \|y^*(t)\|^2)dt = J(u^*)(t'). \tag{19}$$

In the sequel, we will show that $J(v)(T) \geq J(v)(t')$, which amounts to showing that

$$\int_{t'}^T (v^2(t) + \|y_v(t)\|^2)dt + 2\|y_v(T)\| - 2\|y_v(t')\| \geq 0.$$

Case 2.1: $y_v(t) \neq 0, \quad \forall t \in [t', T].$
According to Case 1.1 we have

$$\int_{t'}^T (v^2(t) + \|y_v(t)\|^2)dt + 2\|y_v(T)\| - 2\|y_v(t')\| = \int_{t'}^T (v(t) + \|y_v(t)\|)^2 dt \geq 0.$$

It follows that

$$J(v)(T) \geq J(v)(t'). \tag{20}$$

Combining (18), (19) and (20) we obtain that

$$J(v)(T) \geq J(u^*)(T).$$

Case 2.2 : $\exists t_1 \in [t', T] \quad /y_v(t_1) = 0.$
In this case we have $y_v(t) = 0, \quad \forall t \in [t_1, T]$.

Then, letting $t'' = \inf\{t \in [t', T] \quad /y_v(t) = 0\}$, we deduce that $y_v(t'') = 0$ and that $y_v(t) \neq 0, \quad \forall t \in [t', t'')$. Here again we get from Case 1.1

$$J(v)(t'') \geq J(v)(t').$$

Since $y_v(t) = 0$ for all $t \in [t'', T]$, we can see that

$$J(v)(T) = \int_0^T (v^2(t) + \|y_v(t)\|^2)dt = J(v)(t'') + \int_{t''}^T v^2(t)dt \geq J(v)(t'').$$

We conclude that

$$J(v)(T) \geq J(v)(t'') \geq J(v)(t'). \tag{21}$$

Combining (18), (19) and (21) we get that

$$J(v)(T) \geq J(u^*)(T).$$

Then, we conclude that u^* is an optimal control for the problem (3).

4 Examples

4.1 Wave Equation

Let us consider the following wave equation

$$\begin{cases} \frac{\partial^2}{\partial t^2} z(t,x) = \Delta z(t,x) + u(t)a(x)z(t,x), & t \in [0,T] \text{ and } x \in \Omega = (0,1) \\ z(t,0) = z(t,1) = 0, & t \in [0,T] \\ z(0,x) = z_0(x), & x \in \Omega \end{cases}$$

where $a(.) \in L^\infty(\Omega)$ and $u \in L^2(0,T)$. This system has the form of the system (1) if we take $X = H_0^1(\Omega) \times L^2(\Omega)$ with $\langle (y_1,z_1),(y_2,z_2)\rangle_X = \langle y_1,y_2\rangle_{H_0^1(\Omega)} + \langle z_1,z_2\rangle_{L^2(\Omega)}$ and

$$A = \begin{pmatrix} 0 & I \\ \Delta & 0 \end{pmatrix} \text{ with } D(A) = H_0^1(\Omega) \cap H^2(\Omega) \times H_0^1(\Omega) \text{ and } B = \begin{pmatrix} 0 & 0 \\ a & 0 \end{pmatrix}.$$

Here B is a linear bounded operator on X and A is the infinitesimal generator of a linear C_0- semi-group $S(t)$ of isometries and $y(t) = (z(t),\dot{z}(t))$.

The quadratic cost function is given by

$$J(u) = 2(\|z(T)\|_{H_0^1(\Omega)}^2 + \|\dot{z}(T)\|_{L^2(\Omega)}^2)^{\frac{1}{2}} + \int_0^T (\|z(t)\|_{H_0^1(\Omega)}^2 + \|\dot{z}(t)\|_{L^2(\Omega)}^2)dt + \int_0^T u(t)^2 dt.$$
$$(22)$$

According to Theorem 1, there exists an optimal control u^* that minimizes the quadratic cost (22).

4.2 The Transport Equation

Let us consider the following transport problem

$$\begin{cases} \frac{\partial}{\partial t} y(t,x) = -\frac{\partial}{\partial x} y(t,x) + u(t)y(t,x) & t \in [0,T] \text{ and } x \in \Omega = (0,+\infty) \\ y(t,0) = 0, \ t \in [0,T] \\ y(0,x) = y_0(x), \ x \in \Omega \end{cases}$$

where $u \in L^2(0,T)$ is the control and $y(t) = y(t,.) \in L^2(\Omega)$ is the state. The operator $A = -\frac{\partial}{\partial x}$ with domain $D(A) = H_0^1(\Omega)$ generates a C_0 semi-group $S(t)$ of isometries on $X = L^2(\Omega)$.

According to Theorem 2, the feedback control $u(t) = -\|y(t)\|$ minimizes the following functional cost

$$J(u) = (\int_0^{+\infty} y(T,x)^2 dx)^{\frac{1}{2}} + \int_0^T (\int_0^{+\infty} y(t,x)^2 dx)dt + \int_0^T u(t)^2 dt.$$

5   Conclusion

In this paper, we investigated the quadratic optimal control problem for a class of bilinear infinite dimensional systems . We formulated optimality conditions in the general case, then we showed that the optimal control can be expressed as a feedback law for a class of bilinear systems. The established results are applied to wave and transport equations. As a natural continuation of the present work is to extend the obtained results to a larger class of bilinear systems, and to study the problems of controllability and stability of bilinear systems using a quadratic optimal control.

References

1. Addou, A., Benbrik, A.: Existence and uniqueness of optimal control for a distributed parameter bilinear system. J. Dyn. Control Syst. **8**(2), 141–152 (2002)
2. El Alami, N.: Analyse et commande optimale des systèmes bilinéaires distribués.Application aux procédés énergétiques, Thèse, Université de Perpignan (1986)
3. Ball, J.M., Marsden, J.E., Slemrod, M.: Controllability for distributed bilinear systems. SIAM J. Control Optim. **20**(4), 575–597 (1982)
4. Banks, S., Yew, M.: On a class of suboptimal controls for infinite-dimensional bilinear systems. Syst. Control Lett. **5**, 327–333 (1985)
5. Bradly, M.E., Lenhart, S.: Bilinear optimal control of a Kirchhoff plate. Syst. Control Lett. **22**, 27–38 (1994)
6. Brezis, H.: Analyse fonctionnelle : Théorie et Applications. Masson, Paris (1987)
7. Kalman, R.E.: Contributions to the theory of optimal control. Bol. Soc. Mat. Mexicano **5**(2), 102–119 (1960)
8. Li, X., Yong, J.: Optimal Control Theory for Infinite-Dimensional Systems. Springer Science & Business Media, Berlin (2012)
9. Pazy, A.: Semi-groups of Linear Operators and Applications to Partial Differential Equations. Springer-Verlag, New York (1983)
10. Pontryagin, L.S., Boltyanski, V.G., Gamkrelidze, R.V., Mischenko, E.F.: Mathematical Theory of Optimal Processes. Wiley, New York (1962)
11. Zerrik, E., El Boukhari, N.: Regional optimal control for a class of semilinear systems with distributed controls. Int. J. Control, 896-907 (2018)
12. Zerrik, E., El Boukhari, N.: Constrained optimal control for a class of semilinear infinite dimensional systems. J. Dyn. Control Syst. **24**(1), 65–81 (2018)

Regional Observability of Linear Fractional Systems Involving Riemann-Liouville Fractional Derivative

Khalid Zguaid$^{(\boxtimes)}$, Fatima Zahrae El Alaoui, and Ali Boutoulout

Faculty of Sciences, Moulay Ismail University, 11201 Mekens, Morocco
zguaid.khalid@gmail.com, fzelalaoui2011@yahoo.fr, boutouloutali@yahoo.fr

Abstract. In this paper, we study the concept of regional observability, more precisely the regional reconstruction of the initial state of a linear fractional system on a subregion ω of the evolution domain Ω. We use the Hilbert uniqueness method in order to reconstruct the initial state of the given system, which consists of transforming the reconstruction problem into a solvability one. After presenting an algorithm that allows us to reconstruct the regional initial state, we give, at the end, two successful numerical results, in order to backup our theoretical work, each with a different type of sensor and with a reasonable value of error.

1 Introduction

Let Ω be a bounded domain in $\mathbb{R}^n$, $n \geq 1$, with smooth enough boundary $\partial\Omega$, $[0,T]$ a time interval, $\alpha \in [0,1]$ and A a second order, linear, differential operator. We consider the following fractional system :

$$
\begin{cases}
^{RL}\mathscr{D}_{0+}^{\alpha} y(x,t) = Ay(x,t) & in\ \Omega \times [0,T], \\
y(\xi,t) = 0 & on\ \partial\Omega \times [0,T], \\
\lim_{t \mapsto 0^+} \mathscr{I}_{0+}^{1-\alpha} y(x,t) = y_0(x) & in\ \Omega,
\end{cases}
\tag{1}
$$

where $^{RL}\mathscr{D}_{0+}^{\alpha}$ is the left sided Riemann-Liouville fractional derivative of order α. This kind of systems are called fractional diffusion equations or processes, which mean some kind of a diffusion phenomena governed by evolution equations involving fractional derivatives with respect to time and whose solution is given by means of a probability density function [21]. These systems were and are still being widely investigated because, as the theory of continuous time random walks (CTRW) states, they provide a better characterization of anomalous diffusion processes, they also give a better performance compared with conventional diffusion systems [13].

Not only for this kind of systems, Fractional Calculus is a valuable and useful tool especially in modeling real world phenomena in the fields of physics, engineering, aerospace, visco-elasticity, electricity, chemistry, control theory and so forth. For more details see [3, 18, 23, 25].

For instance, in [19], a time-fractional diffusion system for signal smoothing is used, it was mentioned that the fractional model has another adjustable time-fractional

© Springer Nature Switzerland AG 2021
Z. Hammouch et al. (Eds.): SM2A 2019, LNNS 168, pp. 164–178, 2021.
https://doi.org/10.1007/978-3-030-62299-2_12

derivative order to control the diffusion process. In the same work, already simulated signals were used in order to compare between the classical diffusion equation and the fractional one. It is claimed that the fractional diffusion filtering, which was applied to nuclear magnetic resonance (NMR) spectrum smoothing, has more advantageous results than those of the classical diffusion filtering, it was also stated that its performance is higher than that of the classical smoothing methods.

Fractional Calculus (FC) is a wide discipline of mathematics which has been around since 300 years ago, the first traces of this subject goes back to Leibniz and L'hospital in their discussion about the meaning of $\dfrac{d^n}{dx^n}$ if $n = \dfrac{1}{2}$. The first attempt to give a logical definition is due to Liouville in 1832, and since then a lot of researchers came up with new and different definitions of fractional derivatives, we mention : Riemann-Liouville, Caputo, Riez, Caputo-Fabrizo, Atangana-Baleanu and many other ones. We refer the reader seeking more information about fractional calculus and its properties to see the following books and the references therein [15, 17, 23].

One thing that some find hard to grasp is the initial conditions in a Riemann-Liouville type time-fractional system, which are given as a limit of an integral, we refer any one wondering about this issue to the work of Heymans and Podlubny [14] where they demonstrated that, in fact, one can give a significant physical meaning to such types of initial conditions, and it is very much possible to attribute some values to those kinds of conditions by using appropriate measurements and observations. They also gave a series of concrete examples where these initial conditions make complete sense.

A very important discipline of mathematics that we are dealing with in this paper, is control theory, this field of study plays a serious role in linking between mathematics and technology and it includes several notions such as controllability, stability, observability, stabilization and many more.

In this paper, we deal with observability, precisely the regional observability of a time fractional diffusion system written in terms of Riemann-Liouville time-fractional derivative. The concept of observability was introduced for the first time by the Hungarian-American engineer Rudolf Kalman [16]. This notion has as goal the possibility of finding and reconstructing the initial state of the considered system in a finite time using only the outputs (measurements). This concept has been thoroughly investigated and it also possesses a large literature for various types of system (linear, semilinear...). For more information see [9, 24, 26, 27] and the references therein.

We shall point out the fact that in case of distributed or diffusion systems not all states are observable, hence the necessity of introducing a more weaker notion to cut back the losses for non observable systems, we are speaking about regional observability which also consists of finding and reconstructing the initial state of a system but only in a desired subregion of the evolution domain $\omega \subset \Omega$. Regional observability had seen light for the first time in the nineties with professors El Jai and Afifi for discrete systems see [1], and El Jai and Zerrik for continuous systems see [2, 12]. Afterwards this concept started to be developed by Badraoui, Boutoulout, El Alaoui, Bourray, Zouiten and Torres to cover various types of systems and cases see [4–8, 10, 28, 30].

Lately, regional observability for time-fractional diffusion systems was being studied, see [13], which does not only cover the regional observability but regional analysis in general (regional controllability, regional stability, regional detectability...) for time fractional diffusion processes.

The main goal of this work is to reconstruct the initial state of the considered system using an extension of the Hilbert uniqueness method (HUM), which was firstly introduced by Lions in [20]. This approach relies on the concept of duality for integer order distributed parameter systems, this duality comes from Green's formula. This property fails to work for non-integer order systems, yet we can derive a similar property of duality where the adjoint or dual system is given in terms of Caputo fractional derivative. This relation is obtained with the help of fractional green's formula see [22].

This paper will be organized as follows : After this introduction, we give, in Sect. 2, some preliminary results to be used along this work also as a quick follow up of the considered system. In Sect. 3, we show the formulation and steps of the HUM approach and in Sect. 4, we propose an algorithm that reconstructs the initial state of our system in a desired subregion. As for Sect. 5, we present two successful numerical simulations to back up our work, the first is given with a pointwise sensor and second with a zonal sensor.

2   Considered System and Preliminaries

In this section, we shall introduce some basic definitions needed to present our main result. We give a quick reminder of some necessary notions and properties of fractional calculus followed by other tool of control theory.

We recall the following definitions.

Definition 1 [17]. We call the left sided fractional integral of order $\alpha \in [0, 1]$ of a function $y(x,t)$ at $t \in [0, T]$ for all $x \in \Omega$, the following integral formula :

$$\left(\mathscr{I}_{0^+}^{\alpha} y(x,.) \right)(t) = \frac{1}{\Gamma(\alpha)} \int_0^t (t-s)^{\alpha-1} y(x,s)ds,$$

where $\Gamma(\alpha) := \int_0^{+\infty} t^{\alpha-1} e^{-t} dt$ is the Euler's gamma function.

Definition 2 [17]. We define the left sided Riemann-Liouville fractional derivative of order $\alpha \in [0, 1]$ of a function $y(x,t)$ in $t \in [0, T]$ for all $x \in \Omega$, by :

$$\left({}^{RL}\mathscr{D}_{0^+}^{\alpha} y(x,.) \right)(t) := \frac{d}{dt} \left(\mathscr{I}_{0^+}^{1-\alpha} y(x,.) \right)(t) = \frac{1}{\Gamma(1-\alpha)} \frac{d}{dt} \int_0^t (t-s)^{-\alpha} y(x,s)ds.$$

Definition 3 [17]. The right sided Caputo fractional derivative of order $\alpha \in [0, 1]$ of a function $y(x,t)$ in $t \in [0, 1]$ for all $x \in \Omega$ is given as follows :

$$\left({}^{C}\mathscr{D}_{T^-}^{\alpha} y(x,.) \right)(t) = \frac{-1}{\Gamma(1-\alpha)} \int_t^T (s-t)^{-\alpha} \frac{\partial}{\partial s} y(x,s)ds.$$

Let's denote, if there is no confusion,

$$\left(\mathscr{I}_{0+}^{\alpha} y(x,.)\right)(t) := \mathscr{I}_{0+}^{\alpha} y(x,t), \quad \left(^{RL}\mathscr{D}_{0+}^{\alpha} y(x,.)\right)(t) := {}^{RL}\mathscr{D}_{0+}^{\alpha} y(x,t) \text{ and}$$

$$\left(^{C}\mathscr{D}_{T-}^{\alpha} y(x,.)\right)(t) := {}^{C}\mathscr{D}_{T-}^{\alpha} y(x,t).$$

Let $A : D(A) \subseteq L^2(\Omega) \to L^2(\Omega)$ be a second order, linear, differential operator, which generates a C_0-semigroup $\{S(t)\}_{t \geq 0}$ on $L^2(\Omega)$ and $C : D(C) \subseteq L^2(\Omega) \longrightarrow \mathcal{O}$ a linear, possibly unbounded, operator called the observation operator, where $\mathcal{O}$ is the observation space.

We consider the system (1) augmented with the output equation,

$$z(t) = Cy(.,t), \quad t \in [0,T]. \tag{2}$$

We give now the definition of the mild solution for the above system.

Definition 4 [29]. We say that a function $y \in C(0,T;L^2(\Omega))$ is a mild solution of (1) if the following formula is satisfied:

$$y(x,t) = t^{\alpha-1} R_{\alpha}(t) y_0(x), \quad \forall (x,t) \in \Omega \times [0,T], \tag{3}$$

where

$$R_{\alpha}(t) = \alpha \int_0^{+\infty} \theta \xi_{\alpha}(\theta) S(t^{\alpha}\theta) d\theta, \quad t \in [0,T],$$

$$\xi_{\alpha}(\theta) = \frac{1}{\alpha} \theta^{-1-\frac{1}{\alpha}} \varpi_{\alpha}\left(\theta^{-\frac{1}{\alpha}}\right), \quad \theta \in]0,+\infty[.$$

and

$$\varpi_{\alpha}(\theta) = \frac{1}{\pi} \sum_{n=1}^{+\infty} (-1)^{n-1} \theta^{-n\alpha-1} \frac{n\alpha+1}{n!} \sin(n\pi\alpha), \quad \theta \in]0,+\infty[.$$

Note that ξ_{α} is probability density, that is,

$$\xi_{\alpha}(\theta) \geq 0, \ \forall \theta \in]0,+\infty[\quad and \quad \int_0^{+\infty} \xi_{\alpha}(\theta) d\theta = 1,$$

and that the output function can be written as follows,

$$z(t) = t^{\alpha-1} C R_{\alpha}(t) y_0(.) = K_{\alpha}(t) y_0.$$

The operator $K_{\alpha}(.) : L^2(\Omega) \longrightarrow L^2([0,T],\mathcal{O})$ appears to be a linear operator, it is called the observability operator and it plays an important role in the characterization of observability.

Remark 1. The operator $K_{\alpha}(.)$ is bounded if C is bounded.

In order to study the concept of regional observability we need to use the adjoint operator of $K_{\alpha}(.)$ which is not always defined, precisely in the case when the operator C is not bounded, so for $K_{\alpha}^*(.)$ to be well defined we need to assume that C is an admissible observation operator, then we have the following definition.

Definition 5 [31]. We say that the operator C is an admissible observation operator for R_α if,

$$\exists M > 0, \text{ such that } \int_0^T \|CR_\alpha(t)x\|_\mathscr{O}^2 \, dt \leq M\|x\|_{L^2(\Omega)}^2 \quad \forall x \in D(A).$$

Remark 2. If C is bounded then it is also an admissible observation operator.

For the rest of this work we suppose that C is an admissible observation operator, in this case the adjoint operator $K_\alpha(.)$ is given as follows,

$$K_\alpha^*(.) : L^2(0,T;\mathscr{O}) \longrightarrow L^2(\Omega)$$

$$q \longmapsto \int_0^T t^{\alpha-1} R_\alpha^*(t) C^* q(t) dt.$$

Let $\omega \subset \Omega$ be a subregion with positive Lebesgue measure, we define the restriction operator in ω by,

$$\chi_\omega : L^2(\Omega) \longrightarrow L^2(\omega),$$

$$y \longmapsto y_{|\omega},$$

and χ_ω^* denotes its adjoint, and we have the following definition.

Definition 6 [13]. The system (1) together with the output (2) is said to be approximately regionally observable in ω (or approximately ω-observable) if

$$\mathscr{I}m\left(\chi_\omega K_\alpha^*(.)\right) = L^2(\omega),$$

equivalently

$$\mathscr{K}er\left(K_\alpha(.)\chi_\omega^*\right) = \{0\}.$$

We now introduce the notion of sensors which plays an important role in the domain of observability, their main role is to collect data on the studied phenomenon.

Definition 7 [11]. A sensor is a couple (Σ, f), where Σ is a non empty subset of the evolution domain Ω, it is called the spatial support of the sensor, and f is the spatial distribution.

A sensor is called zonal if $\Sigma \subset \Omega$ is a subset with strictly positive Lebesgue measure, in this case $f \in L^2(\Sigma)$, $\mathscr{O} = \mathbb{R}$ and $z(t) = Cy(.,t) = \langle f, y(.,t)\rangle_{L^2(\Sigma)}$.

A sensor is called pointwise if $\Sigma = \{b\} \in \Omega$, in this case $f = \delta_b$, where δ_b is the Dirac delta function centered at b, and the output function is written

$$z(t) = \langle \delta_b, y(.,t)\rangle_{L^2(\Omega)} = y(b,t).$$

Definition 8 [13]. A sensor (Σ, f) is called strategic if the system (1) augmented with the output function (2), which is given by means of the sensor (Σ, f) is approximately ω-observable. It is called non strategic if not.

We now present one version of the fractional green's formula [22].
$$\forall \psi \in C^\infty(0,T;L^2(\Omega)),$$

$$\int_0^T \int_\Omega \left[{}^{RL}\mathscr{D}_{0+}^\alpha y(x,t) - Ay(x,t) \right] \psi(x,t)dsdt = \int_\Omega \psi(x,T)\mathscr{I}_{0+}^{1-\alpha}y(x,T)dx$$

$$+ \int_0^T \int_\Omega \left[{}^C\mathscr{D}_{T-}^\alpha \psi(x,t) - A^*\psi(x,t) \right] y(x,t)dxdt - \int_\Omega \psi(x,0) \lim_{t \mapsto 0+} \mathscr{I}_{0+}^{1-\alpha}y(x,t)dx \quad (4)$$

$$+ \int_0^T \int_{\partial\Omega} \frac{\partial y(\varsigma,t)}{\partial v_A} \psi(\varsigma,t)d\varsigma dt - \int_0^T \int_{\partial\Omega} y(\varsigma,t)\frac{\partial \psi(\varsigma,t)}{\partial v_{A^*}}d\varsigma dt.$$

3   HUM Approach

Firstly we define the following set,

$$G = \left\{ g \in L^2(\Omega) \mid g_{|\Omega\setminus\omega} = 0 \right\},$$

this choice of G is not arbitrary, in fact, we are searching for the value of the initial state in a subregion ω without taking into account the residual part (*the value of the initial state in $\Omega \setminus \omega$*), hence it is natural to consider it null as in the definition of G.

For all $\varphi_0 \in G$, we consider the following linear system,

$$\begin{cases} {}^{RL}\mathscr{D}_{0+}^\alpha \varphi(x,t) = A\varphi(x,t) & in \ \Omega \times [0,T], \\[2mm] \varphi(\xi,t) = 0 & on \ \partial\Omega \times [0,T], \\[2mm] \lim_{t \mapsto 0+} \mathscr{I}_{0+}^{1-\alpha} \varphi(x,t) = \varphi_0(x) \ in \ \Omega, \end{cases} \quad (5)$$

the unique mild solution of this system is written,

$$\varphi(x,t) = t^{\alpha-1}R_\alpha(t)\varphi_0(x), \quad \forall (x,t) \in \Omega \times [0,T]. \quad (6)$$

We define on $G \times G$ the following bilinear form,

$$\langle .,. \rangle_G : G \times G \longrightarrow \mathbb{C}$$
$$(\varphi_0,g_0) \longmapsto \int_0^T \langle t^{\alpha-1}CR_\alpha(t)\varphi_0, t^{\alpha-1}CR_\alpha(t)g_0 \rangle_{\mathcal{O}} dt.$$

This form satisfies the properties of conjugate symmetry $\left(\langle \varphi_0, g_0 \rangle_G = \overline{\langle g_0, \varphi_0 \rangle_G} \right)$ and positiveness $\left(\langle \varphi_0, \varphi_0 \rangle_G \geq 0 \right)$.

We give the following proposition.

Proposition 1. *If the system (5) together with (2) is approximately ω-observable then the form $\langle .,. \rangle_G$ defines a scalar product on G.*

Proof. All that remains is to prove the definiteness of $\langle .,.\rangle_G$, $\left(i.e: \langle \varphi_0,\varphi_0\rangle_G = 0 \implies \varphi_0 = 0\right)$, in fact

$$\langle \varphi_0,\varphi_0\rangle_G = \int_0^T \langle t^{\alpha-1}CR_\alpha(t)\varphi_0, t^{\alpha-1}CR_\alpha(t)\varphi_0\rangle_{\mathscr{O}} dt = \int_0^T \|t^{\alpha-1}CR_\alpha(t)\varphi_0\|_{\mathscr{O}}^2 dt,$$

hence

$$\langle \varphi_0,\varphi_0\rangle_G = 0 \implies \|t^{\alpha-1}CR_\alpha(t)\varphi_0\|_{\mathscr{O}}^2 = 0,\ \forall t \in [0,T],$$

then

$$t^{\alpha-1}CR_\alpha(t)\varphi_0 = 0,\ \forall t \in [0,T],$$

which gives

$$t^{\alpha-1}CR_\alpha(t)\chi_\omega^*\chi_\omega\varphi_0 = K_\alpha(t)\chi_\omega^*(\chi_\omega\varphi_0) = 0,\ \forall t \in [0,T],$$

and since the system (5) is approximately ω-observable we have that

$$\chi_\omega\varphi_0 = 0,$$

thus

$$\varphi_0 = 0,\quad \text{in } \omega,$$

and by definition of G, we have

$$\varphi_0 = 0,\quad \text{in } \Omega \setminus \omega,$$

finally

$$\varphi_0 = 0,\quad \text{in } \Omega.$$

$\square$

For the rest of this work, we assume that the system (5) – (2) is approximately ω-observable, hence $\langle .,.\rangle_G$ is a scalar product and we denote by $\|.\|_G := \sqrt{\langle .,.\rangle_G}$ the natural norm on G based upon the scalar product of G.

With the help of the fractional green's formula, see equation(4), we derive the following retrograded system,

$$\begin{cases} {}^CD_{T-}^\alpha\phi(x,t) = A^*\phi(x,t) + C^*C\varphi(.,t) & \text{in } \Omega \times [0,T], \\ \phi(\xi,t) = 0 & \text{on } \partial\Omega \times [0,T], \\ \phi(x,T) = 0 & \text{in } \Omega, \end{cases} \tag{7}$$

which has a unique mild solution $\phi \in C\left(0,T;L^2(\Omega)\right)$, given by the following integral formula,

$$\phi(t) = \int_t^T (\tau-t)^{\alpha-1}R_\alpha^*(\tau-t)C^*C\varphi(\tau)d\tau. \tag{8}$$

if φ_0 is chosen in G such that $C\varphi(t) = z(t)$, then the following system

$$\begin{cases} {}^C D^\alpha_{T^-} \psi(x,t) = A^*\psi(x,t) + C^*z(t) & \text{in } \Omega \times [0,T], \\ \psi(\xi,t) = 0 & \text{on } \partial\Omega \times [0,T], \\ \psi(x,T) = 0 & \text{in } \Omega, \end{cases} \qquad (9)$$

can be seen as the adjoint system of (5).

We define the mapping

$$\Lambda : G \longrightarrow G$$
$$\varphi_0 \longmapsto \chi^*_\omega \chi_\omega (\psi(0)),$$

hence the problem of regional reconstruction is reduced to solving the following equation,

$$\Lambda \varphi_0 = \chi^*_\omega \chi_\omega (\psi(0)). \qquad (10)$$

Remark 3. $\chi^*_\omega \chi_\omega$ is projection operator on G.

We have the following theorem,

Theorem 1. *If the system (5) together with (2) is approximately ω-observable then the equation (10) has a unique solution which corresponds with the initial state in ω.*

Proof. Let's consider $\varphi_0 \in G$, we have :

$$\langle \Lambda \varphi_0, \varphi_0 \rangle_G = \langle \chi^*_\omega \chi_\omega (\psi(0)), \varphi_0 \rangle_G,$$

$$= \langle \psi(0), \varphi_0 \rangle_G,$$

$$= \langle \int_0^T t^{2\alpha-2} R_\alpha(t)^* C^* C R_\alpha(t) \varphi_0 dt, \varphi_0 \rangle_G,$$

$$= \int_0^T \langle t^{\alpha-1} C R_\alpha(t)\varphi_0, t^{\alpha-1} C R_\alpha(t)\varphi_0 \rangle_\mathcal{O} dt,$$

$$= \langle \varphi_0, \varphi_0 \rangle_G$$

$$= \|\varphi_0\|^2_G.$$

Thus Λ is an isomorphism.   □

4   Algorithm

This section is reserved to the proposed algorithm for the reconstruction of the initial state in ω, this leads to some numerical results which will be presented in the next section.

In order to give an algorithm that reconstructs the initial state, we assume that the operator A generates a complete system of eigenfunctions $\{\varphi_i\}_{i\in\mathbb{N}^*}$ on the state space $L^2(\Omega)$ associated with the eigenvalues $\{\lambda_i\}_{i\in\mathbb{N}^*}$.

Note that the family $\{\varphi_i\}_{i\in\mathbb{N}^*}$ is an orthonormal basis of $L^2(\Omega)$.

In this case, R_α and φ are respectively expressed as follows, $\forall w \in L^2(\Omega)$, $\forall (x,t) \in \Omega \times [0,T]$:

$$R_\alpha(t)w = \sum_{i=1}^{+\infty} E_{\alpha,\alpha}\left(\lambda_i t^\alpha\right)\langle w,\varphi_i\rangle_{L^2(\Omega)}\varphi_i(.).$$

$$\varphi(x,t) = \sum_{i=1}^{+\infty} t^{\alpha-1}E_{\alpha,\alpha}\left(\lambda_i t^\alpha\right)\langle \varphi_0,\varphi_i\rangle_{L^2(\Omega)}\varphi_i(x),$$

where, $E_{\alpha,\beta}(t) := \sum_{n=0}^{+\infty}\dfrac{t^n}{\Gamma(n\alpha+\beta)}$, is the two parameter Mittag Lefller function.

The solution of (9) at $t=0$ can be written as follows, $\forall x \in \Omega$,

$$\psi(x,0) = \sum_{i=1}^{+\infty}\int_0^T \tau^{\alpha-1}E_{\alpha,\alpha}\left(\lambda_i\tau^\alpha\right)\langle C^*z(\tau),\varphi_i\rangle_{L^2(\Omega)}d\tau\varphi_i(x)$$

From theorem 1, we have

$$\langle \Lambda\varphi_0,\varphi_0\rangle_G = \|\varphi_0\|_G^2 = \int_0^T \|C\varphi(.,t)\|_{\mathscr{O}}^2\,dt$$

Case 1 : *If the measurements are given by mean of a pointwise sensor* (b,δ_b) :

$$\langle \Lambda\varphi_0,\varphi_0\rangle_G = \sum_{i,j=1}^{+\infty}\int_0^T t^{2\alpha-2}E_{\alpha,\alpha}\left(\lambda_i t^\alpha\right)E_{\alpha,\alpha}\left(\lambda_j t^\alpha\right)dt\,\varphi_i(b)\varphi_j(b)$$
$$\times\langle\varphi_0,\varphi_i\rangle_{L^2(\Omega)}\langle\varphi_0,\varphi_j\rangle_{L^2(\Omega)}.$$

We set

$$\Lambda_{ij} = \int_0^T t^{2\alpha-2}E_{\alpha,\alpha}\left(\lambda_i t^\alpha\right)E_{\alpha,\alpha}\left(\lambda_j t^\alpha\right)dt\,\varphi_i(b)\varphi_j(b),\quad \forall i,j=1,...,\infty.$$

Case 2 : *If the measurements are given by mean of a zonal sensor* (D,f) : $\langle \Lambda\varphi_0,\varphi_0\rangle_G =$
$$\sum_{i,j=1}^{+\infty}\int_0^T t^{2\alpha-2}E_{\alpha,\alpha}\left(\lambda_i t^\alpha\right)E_{\alpha,\alpha}\left(\lambda_j t^\alpha\right)dt\,\langle f,\varphi_i\rangle_{L^2(D)}$$
$$\times\langle f,\varphi_j\rangle_{L^2(D)}\langle\varphi_0,\varphi_i\rangle_{L^2(\Omega)}\langle\varphi_0,\varphi_j\rangle_{L^2(\Omega)}.$$

We set

$$\Lambda_{ij} = \int_0^T t^{2\alpha-2} E_{\alpha,\alpha}\left(\lambda_i t^\alpha\right) E_{\alpha,\alpha}\left(\lambda_j t^\alpha\right) dt \langle f, \varphi_i \rangle_{L^2(D)} \langle f, \varphi_j \rangle_{L^2(D)}, \quad \forall i, j = 1, \dots, \infty.$$

The problem (10) can be written now as

$$AX = b, \quad where \quad A \in \mathcal{M}_{N,N}(\mathbb{C}), \; X \in \mathcal{M}_{N,1}(\mathbb{C}) \; and \; b \in \mathcal{M}_{N,1}(\mathbb{C}), \quad (11)$$

such that,

$$A_{ij} = \Lambda_{ij}, \quad X_i = \langle \varphi_0, \varphi_i \rangle_{L^2(\Omega)} \quad and \quad b_j = \langle \chi_\omega^* \chi_\omega(\psi(0)), \varphi_j \rangle_{L^2(\Omega)}.$$

After resolving the system (11), we obtain the reconstructed initial state. Then we give the following algorithm.

Algorithm

- Initialization of : ε, α, ω, Sensors, y_0.
- Repeat
  - $\to$ Solve (9), and get ψ
  - $\to$ Calculate the components of Λ
  - $\to$ Solve the system (11) and get φ_0.
- Until $\quad \|y_0 - \varphi_0\|_{L^2(\omega)} \le \varepsilon$.

5   Numerical Results

In this section, we show some numerical illustrations of our result. We will present two examples with the same system but with different output functions, the first one will be given with a pointwise senor whereas for the second one, measurements are given by a zonal sensor.

Pointwise Sensor

Let us consider the following time-fractional system,

$$\begin{cases} {}^{RL}\mathscr{D}_{0+}^{0.5} y(x,t) = \dfrac{\partial^2}{\partial x^2} y(x,t) \; in \; [0,1] \times [0,2], \\[2mm] y(0,t) = y(1,t) = 0 \qquad in \; [0,2], \\[2mm] \lim_{t \to 0+} \mathscr{I}_{0+}^{0.5} y(x,t) = y_0(x) \quad in \; [0,1], \end{cases} \qquad (12)$$

The operator $\dfrac{\partial^2}{\partial x^2}$ has a complete set of eigenfunctions $\varphi_i(x) = \sqrt{2}\sin(i\pi x)$ with the corresponding eigenvalues $\lambda_i = -i^2\pi^2$.

The system (12) is augmented by the output equation given by means of a pointwise sensor localized at $b = 0.72$,

$$z(t) = y(b,t), \quad t \in [0,T],$$

we consider the region $\omega =]0.35, 0.65[$ and the initial state (Supposed to be unknown)

$$y_0(x) = 2x(x-1)(2x-1).$$

By applying the proposed algorithm, we obtain Fig. 1.

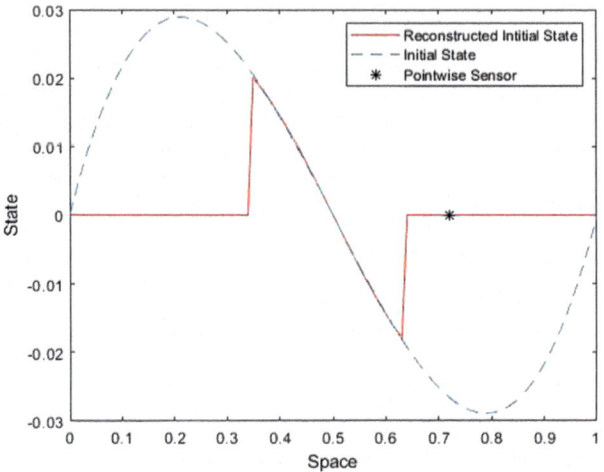

Fig. 1. Initial and estimated initial state in $\omega = [0.35, 0.65]$.

The reconstruction error is $\|y_0 - \varphi_0\|_{L^2(\omega)} = 1.0410 \times 10^{-4}$.

The Fig. 1 shows that the estimated and real initial state are near one another in the subregion ω.

Figure 2, shows the evolution of the reconstruction error in terms of the sensor's location.

Zonal Sensor

We consider, in this example, the same system (12) but with $\alpha = 0.4$ and measurements given by a zonal sensor (D, f) where $D = [0.25, 0.35]$ and $f \equiv 1$. The output function is given as follows

$$z(t) = \langle f, y(.,t) \rangle_{L^2(D)}.$$

The considered subregion is $\omega = [0.2, 0.6]$ and the initial state (Supposedly unknown) is $y_0(x) = (e^x - 1)ln(2-x)$.

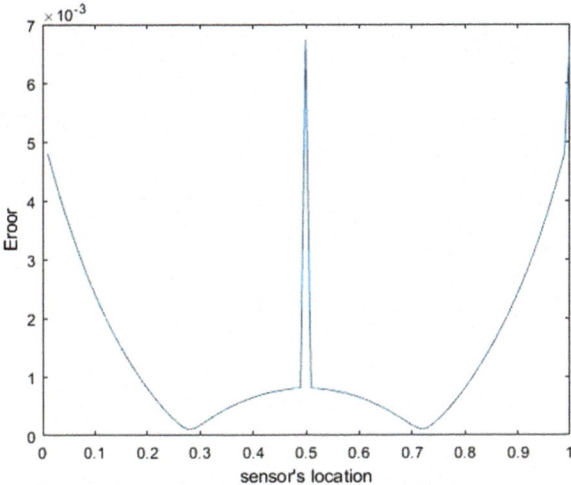

Fig. 2. Evolution of the reconstruction error in function of the sensor's location.

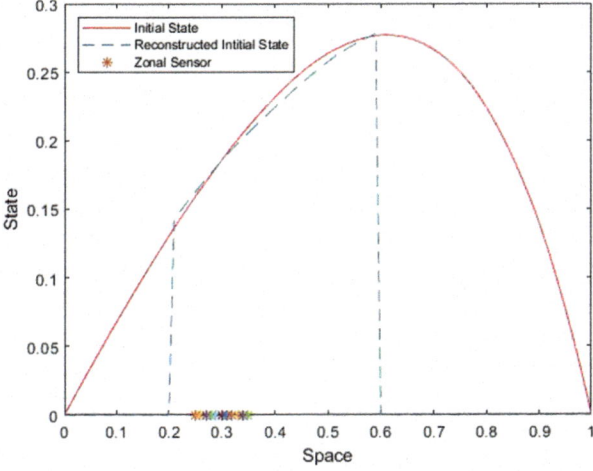

Fig. 3. Initial and estimated initial state in $\omega = [0.2, 0.6]$.

The reconstruction error is $\|y_0 - \varphi_0\|_{L^2(\omega)} = 3.2 \times 10^{-3}$.

We also see, in Fig. 3, that the reconstructed initial state and the initial one are besides each other in the desired subregion ω.

The following table shows the evolution of the reconstruction error in function of the geometric domain of the sensor.

Table 1. Evolution of the reconstruction error in function of the geometric domain of the sensor.

Geometric domain of the sensor	Error
[0.15, 0.25]	1.8×10^{-3}
[0.25, 0.35]	3.2×10^{-3}
[0.35, 0.45]	1.9×10^{-3}
[0.45, 0.55]	4.77×10^{11}
[0.55, 0.65]	3.5×10^{-3}
[0.65, 0.75]	2.9×10^{-3}
[0.75, 0.85]	2.6×10^{-3}
[0.85, 0.95]	3.4×10^{-3}

The same remark about the position of the sensor applies here. We see that if the sensor is placed in $[0.45, 0.55]$ then the sensor is not strategic.

6   Conclusion

In this work, we studied the regional reconstruction of the initial state for a Riemann-Liouville type time-fractional diffusion system, and for that we adopted an extension of the Hilbert uniqueness method, which was introduced by the french mathematicians Jacques-Louis Lions. We supposed the admissibility condition on the observation operator, which is necessary when it is unbounded, and that the considered system is approximately regionally observable, which is also necessary, because we can't reconstruct the initial state, regionally, if the system isn't at least approximately regionally observable. We proposed an algorithm that helps us achieve our goal and at the end we gave two successful numerical results, with satisfying errors of reconstruction, to valid our algorithm. For the future, we are working on the concept of regional observability for semilinear fractional diffusion systems, we also plan on making the algorithm better in order to get a lesser error.

References

1. Afifi, L., El Jai, A.: Strategic sensors and spy sensors. Int. J. Appl. Math. Comput. Sci. **4**(4), 553–573 (1994)
2. Amouroux, M., El Jai, A., Zerrik, E.: Regional observability of distributed systems. Int. J. Syst. Sci. **25**(2), 301–313 (1994)
3. Baleanu, D., Golmankhaneh, A.K., Golmankhaneh, A.K.: On electromagnetic field in fractional space. Nonlinear Anal. Real World Appl. **11**(1), 288–292 (2010)
4. Boutoulout, A., Bourray, H., El Alaoui, F.Z., Benhadid, S.: Regional boundary observability for semi-linear systems approach and simulation. Int. J. Math. Anal. **4**(24), 1153–1173 (2010)
5. Boutoulout, A., Bourray, H., El Alaoui, F.Z.: Regional gradient observability for distributed semilinear parabolic systems. J. Dyn. Control. Syst. **18**(2), 159–179 (2012)

6. Boutoulout, A., Bourray, H., El Alaoui, F.Z.: Boundary gradient observability for semilinear parabolic systems: sectorial approach. Math. Sci. Lett. **2**(1), 45–54 (2013)
7. Boutoulout, A., Bourray, H., El Alaoui, F.Z., Benhadid, S.: Regional observability for distributed semi-linear hyperbolic systems. Int. J. Control. **87**(5), 898–910 (2014)
8. Boutoulout, A., Bourray, H., El Alaoui, F.Z.: Regional boundary observability of semilinear hyperbolic systems: sectorial approach. IMA. J. Math. Control. Inform. **32**(3), 497–513 (2015)
9. Curtain, R.F., Zwart, H.: An Introduction to Infinite-Dimensional Linear Systems Theory. Springer-Verlag, New York (1995)
10. El Alaoui, F. Z.: Regional observability of semilinear systems. Ph.D thesis. Faculty of Sciences. Moulay Ismail University. Meknes (2011)
11. El Jai, A.: Eléments d'analyse et de contrôle des systèmes. Presses Universitaires de Perpignan, Perpignan (2005)
12. El Jai, A., Simon, M.C., Zerrik, E.: Regional observability and sensor structures. Sens. Actuators Phys. **39**(2), 95–102 (1993)
13. Ge, F., Chen, Y., Kou, C.: Regional Analysis of Time-Fractional Diffusion Processes. Springer International Publishing (2018). www.springer.com/gp/book/9783319728957
14. Heymans, N., Podlubny, I.: Physical interpretation of initial conditions for fractional differential equations with Riemann-Liouville fractional derivatives. Rheol. Acta. **45**(5), 765–771 (2006)
15. Hilfer, R.: Applications of Fractional Calculus in Physics. World Scientific Pub. Co., Singapore River Edge N.J (2000)
16. Kalman, R.E.: On the general theory of control systems. In: IFAC Proceedings Volumes, vol. 1 (1), pp. 491–502 (1960)
17. Kilbas, A.A., Srivastava, H.M., Trujillo, J.J.: Theory and Applications of Fractional Differential Equations. Elsevier, Amsterdam (2006)
18. Kumar, D., Singh, J., Al Qurashi, M., Baleanu, D.: A new fractional SIRS-SI malaria disease model with application of vaccines, antimalarial drugs, and spraying. Adv. Differ. Equ. **2019**(1), 278 (2019)
19. Li, Y., Liu, F., Turner, I.W., Li, T.: Time-fractional diffusion equation for signal smoothing. Appl. Math. Comput. **326**, 108–116 (2018)
20. Lions, J.L.: Contrôlabilité exacte, perturbations et stabilisation de systèmes distribués, Paris (1997)
21. Mainardi, F., Mura, A., Pagnini, G.: The M-Wright function in time-fractional diffusion processes: a tutorial survey. Int. J. Differ. Equ. **22**(3), 87–99 (2010)
22. Mophou, G.M.: Optimal control of fractional diffusion equation. Comput. Math. Appl. **61**(1), 68–78 (2011)
23. Podlubny, I.: Fractional Differential Equations: An Introduction to Fractional Derivatives, Fractional Differential Equations, to Methods of Their Solution and Some of Their Applications, 1st edn. Academic Press, Cambridge (1998)
24. Pritchard, A.J., Wirth, A.: Unbounded control and observation systems and their duality. SIAM. J. Control. Optim. **16**(4), 535–545 (1978)
25. Qureshi, S., Yusuf, A., Shaikh, A.A., Inc, M., Baleanu, D.: Fractional modeling of blood ethanol concentration system with real data application. Chaos Interdiscip. J. Nonlinear. Sci. **29**(1), 013143 (2019)
26. Salamon, D.: Infinite dimensional linear systems with unbounded control and observation: a functional analytic approach. Trans. Amer. Math. Soc. **300**, 383–431 (1987)
27. Weiss, G.: Admissible observation operators for linear semigroups. Israel. J. Math. **65**(1), 17–43 (1989)
28. Zerrik, E., Bourray, H., El Jai, A.: Regional observability for semilinear distributed parabolic systems. J. Dyn. Control Syst. **10**(3), 413–430 (2004)

29. Zhou, Y., Zhang, L., Shen, X.H.: Existence of mild solutions for fractional evolution equations. J. Integral Equ. Appl. **25**(4), 557–586 (2013)
30. Zouiten, H., El Alaoui, F.Z., Boutoulout, A.: Regional boundary observability with constraints: a numerical approach. Int. Rev. Automat. Contr. **8**(5), 354–361 (2015)
31. Zouiten, H., Boutoulout, A., Torres, D.: Regional enlarged observability of Caputo fractional differential equations. Discrete. Cont. Dyn-S. **13**(3), 1017–1029 (2018)

Stability Analysis of Fractional Differential Systems Involving Riemann–Liouville Derivative

Hanaa Zitane$^{(\boxtimes)}$, Fatima Zahrae El Alaoui, and Ali Boutoulout

Department of Mathematics, Faculty of Sciences, University of Moulay Ismail,
11201 Meknes, Morocco
hanaa.zit@gmail.com, fzelalaoui2011@yahoo.fr, boutouloutali@yahoo.fr

Abstract. We introduce the stability notion of the fractional differential systems under Riemann–Liouville time derivative of order $\alpha \in (0,1)$, evolving on a spatial domain Ω. Then, we characterize the asymptotic behavior of the state. Also, we present sufficient and necessary conditions to achieve the exponential stability of this important class of systems. Hence, we study the state stabilization of fractional differential systems by means of decomposition method. Several examples and simulations are given to show the applicability of our presented results.

1 Introduction

The fractional calculus history dates back to the Seventeenth century (more precisely to 1695), when the possible meaning of the half order differentiation was discussed by Marquis de L'Hospital and Gottfried Wilhelm Leibniz. Since then, this question has been studied by many well-known mathematicians over the years, such as: Euler, Feller, Fourier, Laurent, Letnikov, Liouville, Grünewald, Riemann and many others. However, the fractional calculus theory has been evolved speedily since the Nineteenth century, mainly as a foundation of several mathematical branches such as, fractional differential systems and fractional geometry. Moreover, during last decades, the investigation, especially, of the fractional differential systems theory was motived by the great development of fractional calculus one. Hence, nowadays, the fractional differential systems have been proved as a powerful tool to characterize various dynamical models and many real world systems. For example, it can be mentioned viscoelastic systems [1], diffusion and some heat transfer process [5], amongst others.

Stability analysis is an interesting concept in differential systems and control theories, as well as in their applications. Indeed, the stability of integer order differential systems was widely studied ([2,13,16]). Many approaches were used to establish several degrees of their stability and stabilization: the asymptotic and exponential stability have been considered using Lyapunov equation [13]. Also, the strong stabilization has been treated by means of Riccati equation [4]. Moreover, the exponential stabilization has been developed using a specific state space and system decomposition [16]. Recently, the stability was introduced

© Springer Nature Switzerland AG 2021
Z. Hammouch et al. (Eds.): SM2A 2019, LNNS 168, pp. 179–193, 2021.
https://doi.org/10.1007/978-3-030-62299-2_13

to fractional calculus [11,14,18–20]. Furthermore, many studies investigated the stability of fractional order systems [17]. For example Qian et al have been developed some stability analytical results for Riemann-Liouville fractional systems of order $\alpha \in (0,1)$, including perturbed systems, linear systems and time-delayed systems [14]. Also, a fractional Lyapunov direct method has been proposed, by Li et al, to study the power law stability and the exponential stability [11]. Moreover, in [6], Ge et al introduced the regional stability notion for Riemann–Liouville linear fractional differential systems, where they charcterized the strong stability using the spectrum properties of the system dynamic and the strong stabilization via decomposition method.

In this work, we shall present some new stability and stabilization theorems for Riemann–Liouville linear fractional differential systems of order $\alpha \in (0,1)$. In details, we formulate the problem and we characterize the asymptotic and exponential stability of such class of systems in Sect. 2. In Sect. 3, we derive the fractional differential systems stabilization using especially decomposition method. Finally, we give a conclusion in the last section which contains a synthesis and some perspectives.

2   Stability of Fractional Differential Systems

In this section we consider, in $\Omega \subset \mathbb{R}^n$ $(n = 1, 2, 3, ..)$, an open bounded subset with a regular boundary $\partial\Omega$, the following fractional diffusion system, defined as

$$\begin{cases} {}^{RL}_{0}D^\alpha_t z(x,t) = Az(x,t), & x \in \Omega, \ t \in]0, +\infty[\\ z(\eta, t) = 0, & \eta \in \partial\Omega, \ t \in]0, +\infty[\\ \lim_{t \longrightarrow 0^+} {}_0I^{1-\alpha}_t z(x,t) = z_0(x), \ x \in \Omega, \end{cases} \tag{1}$$

where $A : D(A) \subset L^2(\Omega) \longrightarrow L^2(\Omega)$ is a linear operator that generates a C_0-semi-group $(S(t))_{t \geq 0}$ [3] on $L^2(\Omega)$, ${}^{RL}_{0}D^\alpha_t$ and ${}_0I^\alpha_t$ denote, respectively, the Riemann-Liouville derivative and integral of order $\alpha \in (0,1)$ [10], that are given by

$$_0I^\alpha_t z(.,t) = \Gamma(\alpha)^{-1} \int_0^t (t-v)^{\alpha-1} z(.,v) \, dv$$

and

$$^{RL}_{0}D^\alpha_t z(.,t) = \frac{d}{dt} {}_0I^{1-\alpha}_t z(.,t),$$

with $\Gamma(\alpha) = \int_0^{+\infty} y^{\alpha-1} e^{-y} \, dy$ is the Gamma function.

The mild solution $z \in C(0, T, L^2(\Omega))$ of system (1) [7] is defined as

$$z(.,t) = H_\alpha(t)z_0(.) = t^{\alpha-1} K_\alpha(t) z_0(.), \tag{2}$$

where

$$K_\alpha(t) = \alpha \int_0^{+\infty} \xi \phi_\alpha(\xi) S(t^\alpha \xi) \, d\xi \tag{3}$$

and

$$\phi_\alpha(\xi) = \alpha^{-1}\xi^{-1-\frac{1}{\alpha}}P_\alpha(\xi^{-\frac{1}{\alpha}}), \tag{4}$$

with

$$P_\alpha(\xi) = \pi^{-1}\sum_{n=1}^{+\infty}(-1)^n\frac{\Gamma(n\alpha+1)}{n!}\xi^{\alpha n-1}\sin(n\pi\alpha), \ \xi \in (0,\infty).$$

Remark 1. $P_\alpha(.)$ represents the function of the probability density.

First, we state some stability definitions.

Definition 1. System (1) is said to be

- Exponentially stable, if for all $z_0 \in L^2(\Omega)$ there exist Q and $\sigma > 0$ satisfying

$$\|z(.,t)\| \leq Qe^{-\sigma t}\|z_0\|, \ \forall t \geq 0.$$

- Strongly stable, if for all $z_0 \in L^2(\Omega)$ the corresponding solution $z(.,t)$ of (1) fulfills

$$\lim_{t\longrightarrow+\infty}\|z(.,t)\| = 0.$$

Remark 2. The exponential stability implies the strong one.

In the following theorem we present the link between the strong stability of the fractional differential system (1) and the spectrum properties of its dynamic A .
Let us introduce the sets

$$\sigma^1(A) = \left\{\lambda \in \sigma(A) : |arg(\lambda)| \leq \frac{\alpha\pi}{2}\right\}$$

and

$$\sigma^2(A) = \left\{\lambda \in \sigma(A) : |arg(\lambda)| > \frac{\alpha\pi}{2}\right\},$$

where $\sigma(A)$ indicates the points spectrum of the operator A.

Theorem 1. *Let $(\lambda_n)_{n\geq 1}$ and $(\chi_n)_{n\geq 1}$ be the eigenvalues and the corresponding eigenfunctions of the operator A, with $(\chi_n)_{n\geq 1}$ form an orthonormal basis on $L^2(\Omega)$. If $\sigma^1(A) = \varnothing$ and $\forall\lambda_n \in \sigma^2(A)$, $n = 1, 2, ...$, there exists $\varepsilon > 0$ satisfying $\lambda_n \leq -\varepsilon$, then the system (1) is strongly stable in Ω.*

Proof. For $z_0 \in L^2(\Omega)$, the solution of (1) [7] can be written as

$$z(.,t) = t^{\alpha-1}\sum_{n=1}^{+\infty}E_{\alpha,\alpha}(\lambda_n t^\alpha)\langle z_0, \chi_n\rangle\chi_n(.), \ \forall z_0 \in L^2(\Omega), \tag{5}$$

where

$$E_{\alpha,\alpha}(\lambda_n t^\alpha) = \sum_{k=0}^{+\infty}\frac{(\lambda_n t^\alpha)^k}{\Gamma(\alpha k+\alpha)}.$$

From (5), one has

$$\|z(.,t)\|^2 = t^{2(\alpha-1)} \sum_{n=1}^{+\infty} (E_{\alpha,\alpha}(\lambda_n t^\alpha))^2 \langle z_0, \chi_n \rangle^2.$$

Also, using the fact that $\sigma^1(A) = \varnothing$ and $\lambda_n \leq -\varepsilon$ for all $\lambda_n \in \sigma^2(A)$, and the Mittag-Leffler function $E_{\alpha,\alpha}(-x)$, $x \geq 0$, is completely monotonic [15], yields

$$\|z(.,t)\| \leq t^{\alpha-1} E_{\alpha,\alpha}(-\varepsilon t^\alpha)\|z_0\|.$$

It follows, since $|E_{\alpha,\alpha}(-\varepsilon t^\alpha)| \leq 1$ for $\alpha \in (0,1)$ [8], that

$$\|z(.,t)\| \longrightarrow 0 \ \ as \ \ t \longrightarrow +\infty.$$

Example 1. Let's consider the sub-diffusion system

$$\begin{cases} {}_0^{RL}D_t^{0.6} z(x,t) = \dfrac{\partial^2}{\partial x^2} z(x,t), & x \in \Omega, t \in \,]0,+\infty[\\ z(\eta,t) = 0, & \eta \in \partial\Omega, t \in \,]0,+\infty[\\ \lim_{t \longrightarrow 0^+} {}_0 I_t^{0.4} z(x,t) = x^3(x-1), & x \in \Omega, \end{cases} \quad (6)$$

with $\Omega = \,]0,1[$. According to (6), we get that the dynamic $A = \dfrac{\partial^2}{\partial x^2}$, with the eigenvalues being

$$\lambda_n = -n^2\pi^2, \ n \geq 1 \quad (7)$$

and the corresponding eigenfunctions being

$$\chi_n(x) = \sqrt{2}\sin(n\pi x), \ n \geq 1.$$

The solution of system (6) is defined by

$$z(x,t) = t^{0.4} \sum_{n=1}^{+\infty} E_{0.6}(\lambda_n t^{0.6}) \langle z_0, \chi_n \rangle \chi_n(x).$$

One has, for all $n \geq 1$, that

$$|arg(\lambda_n)| = \pi > \frac{\alpha\pi}{2} = \frac{3\pi}{10},$$

which implies that $\sigma^1(A) = \varnothing$ and $\sigma^2(A) = \{-n^2\pi^2, n \geq 1\}$.

Also, from (7), one has that

$$\lambda_n \leq -\pi^2, \ n \geq 1.$$

so, for all $\lambda_n \in \sigma^2(A)$, there exist $\varepsilon = \pi^2 > 0$ such that

$$\lambda_n \leq -\varepsilon, \ n \geq 1.$$

Hence, by applying the above Theorem, we get the strong stability of system (6) as it is illustrated by Fig. 1.

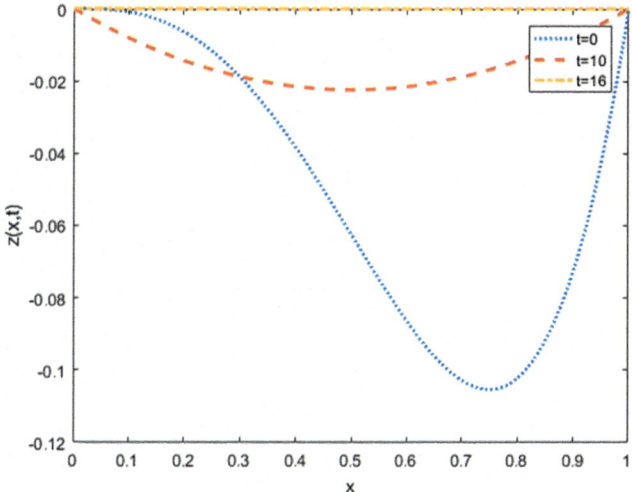

Fig. 1. The state $z(x, t)$ behavior of system (6) at $t = 0$, $t = 10$, $t = 16$.

We shall use the next lemma to study the exponential stability for system (1).

Lemma 1. *Suppose there exists a function $R(.) \in L^2(0, +\infty; \mathbb{R}^+)$ fulfilling*

$$\|H_\alpha(t + \tau)z\| \leq R(t)\|H_\alpha(\tau)z\|, \forall t, \tau \geq 0, \tag{8}$$

for all $z \in L^2(\Omega)$, then the operators $(H_\alpha(t))_{t\geq 0}$ are uniformly bounded.

Proof. To show the boundedness of $(H_\alpha(t))_{t\geq 0}$, we prove that

$$\sup_{t\geq 0}\|H_\alpha(t)z\| < \infty, \ \forall z \in L^2(\Omega).$$

Otherwise, there exists a sequence $(t_1 + r_n)$, $t_1 > 0$ and $r_n \longrightarrow +\infty$ with

$$\|H_\alpha(t_1 + r_n)z\| \longrightarrow +\infty \ as \ n \longrightarrow +\infty. \tag{9}$$

From the following relation

$$\int_0^{+\infty} \|H_\alpha(\tau + r_n)z\|^2 \, d\tau = \int_{r_n}^{+\infty} \|H_\alpha(\tau)z\|^2 \, d\tau, \ \underset{n \longrightarrow +\infty}{\longrightarrow} 0,$$

and by Fatou's Lemma, it follows that

$$\lim_{n \longrightarrow +\infty} \inf \|H_\alpha(\tau + r_n)z\| = 0,$$

almost everywhere $0 \leq \tau < +\infty$.

Thus, for some $\tau_0 < t_1$ we can find a subsequence r_{n_p} such that

$$\lim_{p \longrightarrow +\infty} \|H_\alpha(\tau_0 + r_{n_p})z\| = 0. \tag{10}$$

Moreover, by virtue of (8), one has

$$\|H_\alpha(t_1 + r_{n_p})z\| \le R(t_1 - \tau_0)\|H_\alpha(\tau_0 + r_{n_p})z\|. \tag{11}$$

Then, combining (11) and (10), one obtains

$$\|H_\alpha(t_1 + r_{n_p})z\| \longrightarrow 0 \ as \ p \longrightarrow +\infty, \tag{12}$$

which is absurd. Hence, using the principale of the uniform boundedness, we get the stated result.

Theorem 2. *Assume that the operator* $(H_\alpha(t))_{t\ge 0}$ *satisfies the condition* (8) *and the inequality*

$$\|H_\alpha(t + \tau)z\| \le \|H_\alpha(t)z\|.\|H_\alpha(\tau)z\|, \quad \forall t, \tau \ge 0, \tag{13}$$

holds for all $z \in L^2(\Omega)$, *then the system* (1) *is exponentially stable, if and only if*

$$\int_0^{+\infty} \|H_\alpha(t)z\|^2 \, \mathrm{d}t < \infty, \ \forall z \in L^2(\Omega). \tag{14}$$

Proof. Let us show that

$$\sigma_0 = \lim_{t \longrightarrow +\infty} \frac{\ln \|H_\alpha(t)\|}{t}. \tag{15}$$

We have, for all $t \ge 0$, the following relation

$$\begin{aligned} t\|H_\alpha(t)z\|^2 &= \int_0^t \|H_\alpha(t)z\|^2 \, \mathrm{d}\tau \\ &= \int_0^t \|H_\alpha(\tau + t - \tau)z\|^2 \, \mathrm{d}\tau. \end{aligned}$$

Using (13), yields

$$t\|H_\alpha(t)z\|^2 \le \int_0^t \|H_\alpha(t - \tau)z\|^2 \|H_\alpha(\tau)z\|^2 \, \mathrm{d}\tau.$$

Since the operator $H_\alpha(t)$ is bounded for all $t \ge 0$, and by virtue of (14), one gets

$$t\|H_\alpha(t)z\|^2 \le \xi\|z\|^2, \ for \ some \ \xi > 0,$$

moreover, for t sufficiently large, yields

$$\|H_\alpha(t)\| < 1,$$

hence, there exists $t_1 > 0$ satisfying

$$\ln \|H_\alpha(t)\| < 0,$$

for all $t \geq t_1$. Then,

$$\sigma_0 = \inf_{t \geq 0} \frac{\ln \|H_\alpha(t)\|}{t} < 0.$$

Furthermore, let $S_p = \sup_{t \in]0, t_1]} \|H_\alpha(t)\|$ with $t_1 > 0$ is fixed. Thus, for every $t > t_1$ we may find an integer $\beta \geq 0$ such that $\beta t_1 \leq t \leq (\beta + 1)t_1$.

From (13), yields

$$\|H_\alpha(t)\| = \|H_\alpha(\beta t_1 + (t - \beta t_1))\|$$
$$\leq \|H_\alpha(\beta t_1)\| \|H_\alpha(t - \beta t_1)\|,$$

which implies that

$$\frac{\ln \|H_\alpha(t)\|}{t} \leq \frac{\ln \|H_\alpha(\beta t_1)\|}{t} + \frac{\ln \|H_\alpha(t - \beta t_1)\|}{t},$$

using again condition (13), we obtain that

$$\frac{\ln \|H_\alpha(t)\|}{t} \leq \frac{\beta t_1}{t} \frac{\ln \|H_\alpha(t_1)\|}{t_1} + \frac{\ln \|S_p\|}{t},$$

taking into acount that t_1 is arbitrary and $\dfrac{\beta t_1}{t} \leq 1$, it follows

$$\limsup_{t \longrightarrow +\infty} \frac{\ln \|H_\alpha(t)\|}{t} \leq \inf_{t > 0} \frac{\ln \|H_\alpha(t)\|}{t} \leq \liminf_{t \longrightarrow +\infty} \frac{\ln \|H_\alpha(t)\|}{t},$$

which implies that (15) is satisfied.

Then, for all $\sigma \in]0, -\sigma_0]$, there exists $Q > 0$ such that

$$\|H_\alpha(t)z\| \leq Q e^{-\sigma t} \|z\|,$$

for all $z \in L^2(\Omega)$ and $t \geq 0$.

The converse implication of the theorem is immediate.

Remark 3. When $\alpha = 1$, we retrieve the exponential stability result established in [2].

3   Stabilization of Fractional Differential Systems

In this section we investigate the strong stabilization of time fractional differential systems under Riemann-Liouville derivative of order $\alpha \in (0, 1)$, described by

$$\begin{cases} {}^{RL}_{0}D^\alpha_t z(x, t) = A z(x, t) + B u(x, t), & x \in \Omega, \ t \in]0, +\infty[\\ z(\eta, t) = 0, & \eta \in \partial\Omega, \ t \in]0, +\infty[\\ \lim_{t \longrightarrow 0^+} {}_0 I^{1-\alpha}_t z(x, t) = z_0(x), & x \in \Omega, \end{cases} \quad (16)$$

where the operator A is defined as in system (1), the operator B is linear and bounded from X into $L^2(\Omega)$, with $\Omega \subset \mathbb{R}^n$ is an open bounded subset and X is a Hilbert space of controls, and $u \in L^2(0, +\infty, X)$.

Definition 2. The system (16) is said to be strongly stabilizable if there exists a bounded operator $K \in \mathcal{L}(L^2(\Omega), X)$ such that the system

$$\begin{cases} {}_{0}^{RL}D_t^\alpha z(x,t) = (A + BK)z(x,t), & x \in \Omega, \ t \in]0, +\infty[\\ z(\eta, t) = 0, & \eta \in \partial\Omega, \ t \in]0, +\infty[\\ \lim_{t \longrightarrow 0^+} {}_0 I_t^{1-\alpha} z(x,t) = z_0(x), & x \in \Omega \end{cases} \quad (17)$$

is strongly stable in Ω.

The solution of system (17) is defined by

$$z(.,t) = t^{\alpha-1} K_\alpha^K(t) z_0(.),$$

with

$$K_\alpha^K(t) = \alpha \int_0^{+\infty} \xi \phi_\alpha(\xi) S^K(t^\alpha \xi) \, d\xi,$$

where $\phi_\alpha(.)$ is given by (4) and $(S^K(t))_{t \geq 0}$ is the semi-group generated by $A + BK$.

3.1 Characterization of Stabilization

We have the following theorem.

Theorem 3. *Let $(\lambda_n^K)_{n \geq 1}$ and $(\chi_n^K)_{n \geq 1}$ be the eigenvalues and the corresponding eigenfunctions of the operator $A + BK$, with $(\chi_n^K)_{n \geq 1}$ form an orthonormal basis on $L^2(\Omega)$. If $\sigma^1(A + BK) = \varnothing$ and $\forall \lambda_n^K \in \sigma^2(A + BK)$, $n = 1, 2, ...,$ there exists $\varepsilon > 0$ satisfying $\lambda_n^K \leq -\varepsilon$, then the system (16) is strongly stabilizable in Ω by the control*

$$u(x,t) = Kz(x,t). \quad (18)$$

Proof. The system (1) admits a unique mild solution [7] given by

$$z(.,t) = t^{\alpha-1} \sum_{n=1}^{+\infty} E_{\alpha,\alpha}(\lambda_n^K t^\alpha)\langle z_0, \chi_n^K \rangle \chi_n^K(.), \ \forall z_0 \in L^2(\Omega),$$

It follows

$$\|z(.,t)\|^2 = t^{2(\alpha-1)} \sum_{n=1}^{+\infty} (E_{\alpha,\alpha}(\lambda_n^K t^\alpha))^2 \langle z_0, \chi_n \rangle^2.$$

Also, Using the fact that $\sigma^1(A) = \varnothing$ and $\lambda_n^K \leq -\varepsilon$ for all $\lambda_n^K \in \sigma^2(A)$, and the Mittag-Leffler function $E_{\alpha,\alpha}(-x)$, $x \geq 0$, is completely monotonic [15], yields

$$\|z(.,t)\| \leq t^{\alpha-1} E_{\alpha,\alpha}(-\varepsilon t^\alpha)\|z_0\|.$$

Using the fact that $|E_{\alpha,\alpha}(-\varepsilon t^\alpha)| \leq 1$ for $\alpha \in (0,1)$ [8], it follows that

$$\|z(.,t)\| \longrightarrow 0 \ as \ t \longrightarrow +\infty,$$

which means that the system (16) is strongly stabilizable by the feedback control $u(x,t) = Kz(x,t)$.

Example 2. Let us consider the fractional diffusion system

$$
\begin{cases}
{}^{RL}_{0}D^{0.8}_{t}z(x,t) = Az(x,t) + Bu(t), & x \in \Omega, \ t \in \,]0,+\infty[\\
z(0,t) = z(\pi,t) = 0, & t \in \,]0,+\infty[\\
\lim_{t \longrightarrow 0^{+}} {}_{0}I^{0.2}_{t}z(x,t) = x\sin(x), & x \in \Omega,
\end{cases}
\tag{19}
$$

where $\Omega = \,]0,\pi[$, the operator $Az = z + \dfrac{1}{4\pi^2}\dfrac{\partial^2 z}{\partial x^2}$ and the control operator $B = I$. The eigenvalues of A are defined by

$$
\lambda_n = 1 - \frac{n^2}{4}, \ n \geq 1
\tag{20}
$$

and the corresponding eigenfunctions are given by

$$
\chi_n(x) = \sqrt{2}\sin(n\pi x), \ n \geq 1.
$$

System (19) is unstable since $\lambda_1, \lambda_2 \geq 0$.

Applying the control (18), with $K = -I$, to system (19). One has, the operator $A + BK = \dfrac{1}{4\pi^2}\dfrac{\partial^2}{\partial x^2}$, with the eigenvalues being

$$
\lambda^K_n = -\frac{n^2}{4}, \ n \geq 1
\tag{21}
$$

and the corresponding eigenfunctions being $\chi^K_n(x) = \chi_n(x), \ n \geq 1$.

Moreover, one has

$$
z(x,t) = t^{0.2}\sum_{n=1}^{+\infty} E_{0.8,0.8}(\lambda^K_n t^{0.8})\langle z_0, \chi^K_n \rangle \chi^K_n(x).
$$

From (21), we have

$$
|arg(\lambda^K_n)| = \pi > \frac{\alpha\pi}{2} = \frac{2\pi}{5},
$$

yields $\sigma^1(A + BK) = \varnothing$ and $\sigma^2(A + BK) = \left\{-\dfrac{n^2}{4}, n \geq 1\right\}$.

Also, from (21), one has that

$$
\lambda^K_n \leq -\frac{1}{4}, \ n \geq 1.
$$

Consequently, from Theorem 3, we conclude that the system (19) is strongly stabilizable in Ω by the beedback control (18). Numerical illustration is given in Fig. 2.

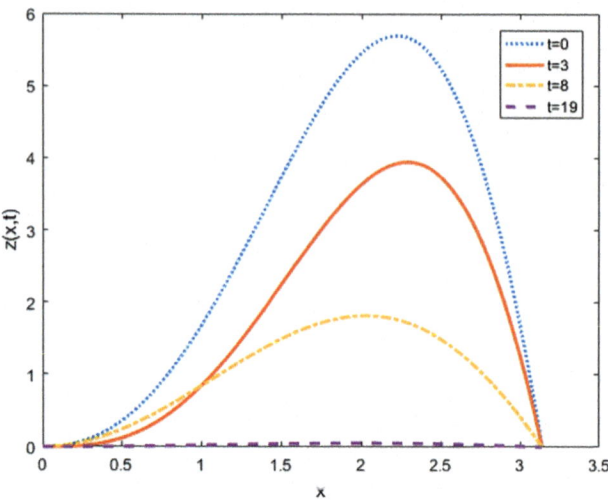

Fig. 2. The state $z(x, t)$ behavior of system (19) at $t = 0$, $t = 3$, $t = 8$ and $t = 19$.

3.2   Decomposition Approach

In the following, we propose an approach characterizing a feedback control that guarantees the stabilization of system (16). We suppose that A is a self adjoint operator with compact resolvent on $H = L^2(\Omega)$. So, the eigenvalues $(\lambda_n)_{n \geq 1}$ of A are real (which can be numbered in decreasing order in such away that $\lambda_n \xrightarrow[n \to +\infty]{} -\infty$) and there are at most finitely many nonnegative eigenvalues $(\lambda_n)_{1 \leq n \leq l}$, each with finite-dimensional eigenspace, such that $\lambda_n \geq -\delta$, for some $\delta > 0$. Yields $\sigma(A)$ can be decomposed as

$$\sigma(A) = \sigma_s(A) \cup \sigma_u(A), \tag{22}$$

with $\sigma_s(A)$ and $\sigma_u(A)$ defined as

$$\sigma_s(A) = \{\lambda_n \leq -\delta, \quad n = l + 1, l + 2...\},$$

$$\sigma_u(A) = \{\lambda_n \geq \delta, \quad n = 1, 2, ..., l\}.$$

Since the eigenvectors $(\chi_n)_{n \geq 1}$ associated to the eigenvalues $(\lambda_n)_{n \geq 1}$ forms a complete and orthonormal basis in H [16], then one has the following decomposition of the state space

$$H = H_s \oplus H_u,$$

with

$$H_s = (I - P)H = Vect\{\chi_{l+1}, \chi_{l+2}, ...\}$$

and

$$H_u = PH = Vect\{\chi_1, \chi_2, ..., \chi_l\},$$

where the operator $P \in L(H)$ represents the projection one [9].

Furthermore, the decomposition of system (16) may be described by

$$
\begin{cases}
{}_0^{RL}D_t^\alpha z_s(x,t) = A_s z_s(x,t) + (I-P)Bu(x,t), & x \in \Omega,\ t \in]0,+\infty[\\
z_s(\eta,t) = 0, & \eta \in \partial\Omega,\ t \in]0,+\infty[\\
\lim_{t \longrightarrow 0^+} z_s(x,t) = z_{0s}(x) = (I-P)z_0(x), & x \in \Omega \\
z_s = (I-P)z, & z \in L^2(\Omega)
\end{cases}
\tag{23}
$$

and

$$
\begin{cases}
{}_0^{RL}D_t^\alpha z_u(x,t) = A_u z_u(x,t) + PBu(x,t), & x \in \Omega,\ t \in]0,+\infty[\\
z_u(\eta,t) = 0, & \eta \in \partial\Omega,\ t \in]0,+\infty[\\
\lim_{t \longrightarrow 0^+} z_u(x,t) = z_{0u}(x) = Pz_0(x), & x \in \Omega \\
z_u = Pz, & z \in L^2(\Omega),
\end{cases}
\tag{24}
$$

where A_s and A_u define the restrictions of A on H_s and H_u respectively, with

$$
\begin{cases}
\sigma(A_s) = \sigma_s(A) \\
\sigma(A_u) = \sigma_u(A)
\end{cases}
$$

and the operator A_u is bounded on H_u.

For $\alpha = 1$ case, in [16], it has been shown that if A_s satisfies the following spectrum growth condition

$$
\lim_{t \longrightarrow +\infty} \frac{\|S_s(t)\|}{t} = sup(Re(\sigma(A_s))),
$$

then the stabilization of system (16) boils down to the stabilization of (24).

The following theorem gives an extension of this result to $\alpha \in (0,1)$ case.

Theorem 4. *Let the spectrum $\sigma(A)$ of A satisfies the above spectrum decomposition assumption (22) and $\sigma(A_s) \subset \sigma^2(A)$. If the system (24) is strongly stabilizable by the control*

$$
u(x,t) = K_u z_u(x,t),
\tag{25}
$$

where $K_u \in L(H,U)$ with

$$
\|z_u(.,t)\| \leq M\ t^{-\mu},
\tag{26}
$$

for some $\mu > 0$ and $M > 0$, then the system (16) is strongly stabilizable using the feedback control (25).

Proof. One has that the system (23) admits a unique mild solution [7] given by

$$
\begin{aligned}
z_s(.,t) = {}& t^{\alpha-1} \sum_{n=l+1}^{+\infty} E_{\alpha,\alpha}(\lambda_n t^\alpha)\langle z_{0s}, \chi_n\rangle \chi_n(.) \\
& + \sum_{n=l+1}^{+\infty} \int_0^t (t-\tau)^{\alpha-1} E_{\alpha,\alpha}(\lambda_n(t-\tau)^\alpha)\langle (I-P)Bu(.,\tau), \chi_n\rangle \chi_n(.)\, d\tau.
\end{aligned}
\tag{27}
$$

From the spectrum decomposition relation (22), we have that $\lambda_n \leq -\delta$, for all $n \geq l+1$, then using the completely monotonic property [15] of the Mittag-Leffler function $E_{\alpha,\alpha}(-x)$, $x \geq 0$, yields

$$E_{\alpha,\alpha}(\lambda_n t^\alpha) \leq E_{\alpha,\alpha}(-\delta t^\alpha), \ \forall \, n \geq l+1, \tag{28}$$

and

$$E_{\alpha,\alpha}(\lambda_n (t-\tau)^\alpha) \leq E_{\alpha,\alpha}(-\delta(t-\tau)^\alpha), \ \forall \, n \geq l+1. \tag{29}$$

Replacing (28) and (29) in (27) and applying the control $u(x,t) = K_u z_u(x,t)$ into (23), one obtains

$$\|z_s(.,t)\| \leq t^{\alpha-1} E_{\alpha,\alpha}(-\delta t^\alpha) \|z_{0s}\|$$
$$+ C_p \int_0^t (t-\tau)^{\alpha-1} \tau^{-\mu} E_{\alpha,\alpha}(-\delta(t-\tau)^\alpha) \, d\tau,$$

with $C_p = M\Gamma(1-\mu)\|K_u\|\|I - P\|\|B\|$. It implies

$$\|z_s(.,t)\| \leq \frac{E_{\alpha,\alpha}(-\delta t^\alpha)}{t^{1-\alpha}} \|z_{0s}\| + C_p \sum_{k=1}^{+\infty} \int_0^t \frac{(-\delta)^k (t-s)^{\alpha k + \alpha - 1} s^{-\mu} \, ds}{\Gamma(\alpha k + \alpha)}$$
$$\leq \frac{E_{\alpha,\alpha}(-\delta t^\alpha)}{t^{1-\alpha}} \|z_{0s}\| + C_p \sum_{k=1}^{+\infty} \frac{(-\delta)^k t^{\alpha k + \alpha - \mu}}{\Gamma(\alpha k + \alpha - \mu - 1)} \Gamma(1-\mu)$$
$$\leq \frac{E_{\alpha,\alpha}(-\delta t^\alpha)}{t^{1-\alpha}} \|z_{0s}\| + C_p t^{\alpha-\mu} E_{\alpha,\alpha-\mu+1}(-\delta t^\alpha).$$

Then, since $\sigma(A_s) \subset \sigma^2(A)$, it follows

$$\|z_s(.,t)\| \leq \frac{\omega_1}{t^{1-\alpha}(1+\delta t^\alpha)} \|z_{s0}\| + C_p \frac{\omega_2 t^{\alpha-\mu}}{1+\delta t^\alpha}, \omega_1, \ \omega_2 > 0,$$

which leads to

$$\lim_{t \longrightarrow +\infty} \|z_s(.,t)\| = 0. \tag{30}$$

On the other hand, taking into acount (26) and that control (25) strongly stabilizes system (24), one gets

$$\lim_{t \longrightarrow +\infty} \|z_u(.,t)\| = 0. \tag{31}$$

Hence, from the relation

$$\|z(.,t)\| \leq \|z_s(.,t)\| + \|z_u(.,t)\|, \tag{32}$$

it follows, by using (31) and (30), that $\lim_{t \longrightarrow +\infty} \|z(.,t)\| = 0$, which achieves the proof.

Example 3. Let's consider $\Omega = \,]0, 2[$ and the fractional diffusion system

$$
\begin{cases}
{}^{RL}_0 D_t^{0.4} z(x,t) = \dfrac{2}{9\pi^2} \dfrac{\partial^2 z}{\partial x^2} z(x,t) + \delta z(x,t) + Bu(t), \; x \in \Omega, \; t \in]0, +\infty[\\
z(0,t) = z(2,t) = 0, & t \in]0, +\infty[\\
\lim_{t \to 0^+} {}_0 I_t^{0.6} z(x,t) = z_0(x), & x \in \Omega,
\end{cases}
\tag{33}
$$

where the operator $Az = \dfrac{2}{9\pi^2} \dfrac{\partial^2 z}{\partial x^2} + \delta z$, with $\delta = 2$ and

$$
D(A) = \{ z \in L^2(0,2), z(0,t) = z(2,t) = 0, (\forall\, t > 0)\}
$$

is self-adjoint and the control operator $B = \delta I$.

The eigenvalues and the eigenfunctions of A are given by

$$
\begin{cases}
\lambda_n = 2 - \frac{2n^2}{9}, & n \geq 1 \\
\chi_n(x) = \sqrt{2}\sin(n\pi x), & n \geq 1.
\end{cases}
$$

One has $\sigma(A) = \{\dfrac{16}{9}, \dfrac{4}{3}, 0\} \cup \{\lambda_n, \quad n = 4,5,...\}$ which satisfies the spectrum decomposition assumption (22) with

$$
\sigma_s(A) = \{ 2 - \frac{2n^2}{9}, \quad n = 4,5,...\},
$$

$$
\sigma_u(A) = \{\frac{16}{9}, \frac{4}{3}, 0\},
$$

and $\sigma_s(A) \subset \sigma^2(A)$ because of $|arg(2 - \dfrac{2n^2}{9})| = \pi > \dfrac{0.4\pi}{2}$, for all $n \geq 4$.

The eigenvectors $(\chi_n)_{n \geq 1}$ associated to the eigenvalues $(\lambda_n)_{n \geq 1}$ forms a complete basis on $L^2(\Omega)$, thus the system (33) may be decomposed into sub-systems (23) and (24) with

$$
A_s z_s = \sum_{n=4}^{+\infty} \lambda_n \langle z_s, \chi_n \rangle \chi_n, \forall\, z_s \in H_s = Vect\{\chi_n, n \geq 4\}
$$

$$
A_u z_u = \sum_{n=1}^{3} \lambda_n \langle z_u, \chi_n \rangle \chi_n, \forall\, z_u \in H_u = Vect\{\chi_1, \chi_2, \chi_3\}.
\tag{34}
$$

On the other hand, system (33) is unstable since $\lambda_1, \lambda_2, \lambda_3 \geq 0$.

So, applying the control (18), with $K = -I$, to the unstable part of system (33) with (34). It follows that the operator $A_u + B_u K = \dfrac{2}{9\pi^2} \dfrac{\partial^2}{\partial x^2}$, with the eigenvalues being

$$
\lambda_n^K = -\frac{2n^2}{9}, \; n \geq 1
$$

and the corresponding eigenfunctions being $\chi_n^K(x) = \chi_n(x)$, $n \geq 1$.

Yields (33) is strongly stabilizable. Indeed, for all $\lambda_n^K \in \sigma_u(A)$, one has that

$$|arg(\lambda_n^K)| = |arg(-\frac{2n^2}{9})| = \pi > \frac{0.4\pi}{2}, \ 1 \leq n \leq 3$$

and

$$\lambda_n^K < -\frac{2}{9}, \ 1 \leq n \leq 3.$$

Moreover, one has

$$z_u(x,t) = t^{0.6} \sum_{n=4}^{+\infty} E_{0.4,0,4}(\lambda_n^K t^{0.4})\langle z_0, \chi_n^K \rangle \chi_n^K(x).$$

Using the fact that the Mittag-Leffler function $E_{\alpha,\alpha}(-x)$, $x \geq 0$, is completely monotonic [15], it follows that

$$\|z_u(.,t)\| \leq \|t^{0.6} \sum_{n=1}^{+\infty} E_{0.4,0,4}(-\frac{2}{9}t^{0.4})\langle z_{u0}, \chi_n^K \rangle \chi_n^K \|.$$

Also, by considering the fact that $|E_{\alpha,\alpha}(-\varepsilon t^\alpha)| \leq 1$ for $\alpha \in (0,1)$ [8] yields

$$\|z_u(.,t)\| \leq t^{0.6}\|z_{u0}\|,$$

which means that (26) holds with $M = \|z_{u0}\|$ and $\mu = 0.6$.

Hence, all the contitions of Theorem 4 are satisfied. Thus system (33) is strongly stabilizable by $u(x,t) = -2z(x,t)$.

4   Conclusion

The present paper deals with the concepts of the stability and stabilization of the state for Riemann–Liouville time fractional differential system of order $\alpha \in (0,1)$. We investigated several interesting strong stability criterion's. Also, we explored the exponential stability. Furthermore, the decomposition method is utilized to derive the stabilization of fractional differential systems. Hence, we presented different examples with some simultations to illustrate the applicability of the estabilished theorems. We claim that our developed results can be useful to analyse and control the behaviour of several real world phenomena such as some heat transfer processes and anomalous sub-diffusion ones.

The problem of the state gradient stability of fractional time differential systems of order $\alpha \in (0,1)$ could be considered as our future work. Various questions are still open, for example, extending the presented results here to a class of complex linear fractional systems and studying the stabilization of fractional semilinear systems as well as nonlinear ones in general, which are closer to real applications.

References

1. Bagley, R., Calico, R.: Fractional order state equations for the control of viscoelastically damped structures. J. Guid. Control Dyn. **14**(91), 304–311 (1991)
2. Curtain, R.F., Zwart, H.J.: An Introduction to Infinite Dimensional Linear Systems Theory. Springer-Verlag, New York (1995)
3. Engel, K.J., Nagel, R.: One-Parameter Semigroups for Linear Evolution Equations. Graduate Texts in Mathematics 194. Springer-Verlag, New York (2000)
4. Balakrishnan, A.V.: Strong stabilizability and the steady state Riccati equation. Appl. Math. Optim. **7**(1), 335–345 (1981)
5. Gabano, J.D., Poinot, T.: Fractional modelling and identification of thermal systems. Sign. Process **91**, 531–41 (2011)
6. Ge, F., Chen, Y., Kou, C.: Regional Analysis of Time-Fractional Diffusion Processes. Springer, Cham, Switzerland (2018)
7. Ge, F., Chen, Y., Kou, C.: Regional controllability of anomalous diffusion generated by the time fractional diffusion equations. In: ASME IDETC/CIE 2015 August, Boston, pp. 2–5 (2015)
8. Joshi, S., Mittal, E., Pandey, R.M.: On Euler type integrals involving extended Mittag-Leffler functions. Bol. Soc. Parana. Mat **38**, 125–134 (2020)
9. Kato, T.: Perturbation Theory for Linear Operators. Springer-Verlag, New York (1980)
10. Jun, Y.X.: General Fractional Derivatives: Theory. Methods and Applications. CRC Press, Boca Raton (2019)
11. Li, Y., Chen, Y.Q., Podlubny, I.: Stability of fractional-order nonlinear dynamic systems: Lyapunov direct method and generalized Mittag-Leffler stability. Comput. Math. Appl. **59**, 1810–1821 (2010)
12. Mainardi, F., Paradisi, P., Gorenflo, R.: Probability distributions generated by fractional diffusion equations. arXiv:0704.0320 (2007)
13. Pritchard, A.J., Zabczyk, J.: Stability and stabilizability of infinite dimensional systems. SIAM Rev. **23**(1), 25–51 (1981)
14. Qian, D.L., et al.: Stability analysis of fractional differential system with Riemann Liouville derivative. Math. Comput. Model. **52**, 862–874 (2010)
15. Schneider, W.R.: Completely monotone generalized Mittag-Leffler functions. Expo. Math. **14**, 3–16 (1996)
16. Triggiani, R.: On the stabilizability problem in Banach space. J. Math. Anal. Appl. **52**(3), 383–403 (1975)
17. Zitane, H., Boutoulout, A., Torres, D.F.M.: The stability and stabilization of infinite dimensional Caputo-time fractional differential linear systems. Mathematics **8**, 353 (2020)
18. Zitane, H., Larhrissi, R., Boutoulout, A.: On the fractional output stabilization for a class of infinite dimensional linear system. In: Recent Advances in Modeling, Analysis and Systems Control: Theoretical Aspects and Applications, pp. 241–259 (2020)
19. Zitane, H., Larhrissi, R., Boutoulout, A.: Fractional output stabilization for a class of bilinear distributed systems. Rend. Circ. Mat. Palermo, II. Ser. **69**(3), 737–752 (2020)
20. Zitane, H., Larhrissi, R., Boutoulout, A.: Riemann Liouville fractional spatial derivative stabilization of bilinear distributed systems. J. Appl. Nonlinear Dyn. **8**(3), 447–461 (2019)

Deformed Joint Free Distributions of Semicircular Elements Induced by Multi Orthogonal Projections

Ilwoo Cho[✉]

Department of of Mathematics and Statistics, 421 Ambrose Hall,
Saint Ambrose University, 518 W. Locust St., Davenport, IA 52803, USA
choilwoo@sau.edu

Abstract. In this paper, we consider (i) how to establish semicircular elements $\{U_k\}_{k=1}^N$ induced by N-many mutually orthogonal projections $\{q_k\}_{k=1}^N$, for $N \in (\mathbb{N} \setminus \{1\}) \cup \{\infty\}$, and the corresponding free product Banach $*$-probability space $\mathbb{L}_Q^{(N)}$ generated by $\{U_k\}_{k=1}^N$, (ii) the free-distributional data on $\mathbb{L}_Q^{(N)}$, (iii) certain $*$-homomorphisms on $\mathbb{L}_Q^{(N)}$, and (iv) how the $*$-homomophisms of (iii) deform the original free-distributional data of (ii).

Keywords: Free probability · Projections · (Weighted-)semicircular elements · Banach $*$-probability spaces · Integer-shifts · Restricted-integer-shifts

1 Introduction

In this paper, we study certain $*$-*homomorphisms* acting on a *free product Banach $*$-algebra* $\mathbb{L}_Q^{(N)}$ generated by mutually free, N-many *semicircular elements* $\mathcal{S}^{(N)} = \{U_k\}_{k=1}^N$, induced by mutually orthogonal N-many *projections* $\mathbf{Q}_o = \{q_k\}_{k=1}^N$, for

$$N \in \mathbb{N}_{>1}^\infty \overset{def}{=} (\mathbb{N} \setminus \{1\}) \cup \{\infty\},$$

where $\infty = |\mathbb{N}|$. Especially, we consider the cases where such $*$-homomorphisms are constructed by some shifting processes on the index set

$$\{1, ..., N\}$$

of $\mathcal{S}^{(N)}$, or of $\mathbf{Q}_o$. The main results show how our $*$-homomorphisms deform the free probability on $\mathbb{L}_Q^{(N)}$.

1991 Mathematics Subject Classification. 46L10, 46L40, 46L53, 46L54, 47L15, 47L30, 47L55.

Z. Hammouch et al. (Eds.): SM2A 2019, LNNS 168, pp. 194–232, 2021.
https://doi.org/10.1007/978-3-030-62299-2_14

1.1 Motivations

In earlier works (e.g., [1,4,6,12,15,19–21]), semicircular elements are constructed-and-studied in topological $*$-probability spaces (e.g., C^*-probability spaces, or W^*-probability spaces, or Banach $*$-probability spaces, etc.). Different from them, the construction of semicircular elements here is motivated by that of weighted-semicircular elements in the Banach $*$-probability spaces of [5] and [8] from an analysis on the p-adic number fields $\mathbb{Q}_p$, for primes p.

By mimicking the weighted-semicircularity of [5] and [8], a construction of (weighted-)semicircular elements from arbitrary mutually orthogonal $|\mathbb{Z}|$-many projections in fixed C^*-probability spaces is introduced, and the corresponding (weighted-)semicircular law(s) is (are) considered in [6] (See short Sects. 3 through Sect. 5 below). Independently, free distributions of free reduced words in mutually free, multi semicircular elements were characterized, estimated, and asymptotically estimated by their joint free moments in [7] (See Sect. 6.2 below).

Based on the main results of [6] and [7], the free product Banach $*$-algebra $\mathbb{L}_Q^{(N)}$ generated by the free semicircular family $\mathcal{S}^{(N)} = \{U_k\}_{k=1}^N$, which is induced by the family $\mathbf{Q}_o = \{q_k\}_{k=1}^N$ of mutually orthogonal projections $q_1, ..., q_N$ of an arbitrary C^*-probability space (A_o, ψ_o), is considered as a Banach $*$-subalgebra of the Banach $*$-probability space $\mathbb{L}_Q$, generated by mutually free, $|\mathbb{Z}|$-many semicircular elements. Then the free-distributional data on $\mathbb{L}_Q^{(N)}$ would be characterized naturally (e.g., [7]). And then, we define-and-study a certain type of $*$-homomorphisms on $\mathbb{L}_Q^{(N)}$. In particular, we are interested in how these morphisms on $\mathbb{L}_Q^{(N)}$ affect the original free-probabilistic information on $\mathbb{L}_Q^{(N)}$.

1.2 Overview

In short Sects. 2, 3, 4 and 5, we introduce backgrounds of our works briefly. In Sect. 6, we construct an operator algebra $\mathbb{L}_Q$ generated by our semicircular elements under free product, and free-distributional data on $\mathbb{L}_Q$ are considered.

In Sect. 7, certain shifting processes on $\mathbb{Z}$ are defined, and the corresponding $*$-isomorphisms are established on $\mathbb{L}_Q$. It is shown that such $*$-isomorphisms form a subgroup $\mathfrak{B}$ of the automorphism group $Aut(\mathbb{L}_Q)$ of $\mathbb{L}_Q$; and, it is isomorphic to the infinite cyclic abelian group $(\mathbb{Z}, +)$ as groups. Interestingly, our (weighted-)semicircularity on $\mathbb{L}_Q$ is preserved by the action of $\mathfrak{B}$, implying that the action of $\mathfrak{B}$ preserves the free probability on $\mathbb{L}_Q$.

In Sect. 8, arbitrarily given N-many mutually orthogonal projections of a C^*-algebra are fixed for $N \in \mathbb{N}_{>1}^\infty$, and we study how they induce the corresponding free semicircular family $\mathcal{S}^{(N)}$, and show this family generates the Banach $*$-probability space $\mathbb{L}_Q^{(N)}$. Especially, $\mathbb{L}_Q^{(N)}$ can be understood as a free-probabilistic sub-structure of $\mathbb{L}_Q$ of Sect. 7. By restricting the action of $\mathfrak{B}$ on $\mathbb{L}_Q$ to that on $\mathbb{L}_Q^{(N)}$, it is proven that this restricted action of $\mathfrak{B}$ distorts the original free probability "on $\mathbb{L}_Q^{(N)}$." Such distortions are characterized.

2   Preliminaries

For fundamental free probability theory, e.g., see [17,19], and the citations therein. *Free probability* is the noncommutative operator-algebraic analogue of classical *measure theory* (including *probability theory*) and *statistical analysis*. It is not only an important branch of functional analysis (e.g., [2–4,7,12,14,15]), but also an interesting application in related fields (e.g., [5,6,8,13,16,20,21]).

We here use combinatorial approach [17] of Speicher. *Joint free moments* and *joint free cumulants* of operators will be computed, and the (free-probabilistic) free product (of [17] and [19]) will be used without detailed definitions and backgrounds.

3   The Banach *-Algebra $\mathfrak{L}_Q$

Let $(\mathcal{B}, \varphi)$ be a topological *-probability space (a C^*-probability space, or a W^*-probability space, or a Banach *-probability space, etc.), where $\mathcal{B}$ is a topological *-algebra (a C^*-algebra, resp., a W^*-algebra, resp., a Banach *-algebra, etc.), and φ is a bounded linear functional on $\mathcal{B}$.

An operator $a \in \mathcal{B}$ is said to be a *free random variable*, if we understand it as an element of $(\mathcal{B}, \varphi)$. A free random variable $a \in (\mathcal{B}, \varphi)$ is said to be *self-adjoint*, if the operator a is self-adjoint in $\mathcal{B}$ in the sense that $a^* = a$ in $\mathcal{B}$, where a^* is the *adjoint of a* (e.g., [11]).

Definition 3.1. *A self-adjoint free random variable a is weighted-semicircular in $(\mathcal{B}, \varphi)$ with the weight $t_0 \in \mathbb{C}^\times = \mathbb{C} \setminus \{0\}$ (or, in short, t_0-semicircular), if*

$$k_n^{\mathcal{B}}(a, ..., a) = \begin{cases} k_2^{\mathcal{B}}(a, \, a) = t_0 & \text{if } n = 2 \\ 0 & \text{otherwise,} \end{cases} \tag{3.1}$$

for all $n \in \mathbb{N}$, where $k_\bullet^{\mathcal{B}}(...)$ is the free cumulant on $\mathcal{B}$ in terms of φ under the Möbius inversion of [17].

If $t_0 = 1$ in (3.1), the 1-semicircular element a is said to be semicircular in $(\mathcal{B}, \varphi)$. i.e., a is semicircular in $(\mathcal{B}, \varphi)$, if

$$k_n(a, ..., a) = \begin{cases} 1 & \text{if } n = 2 \\ 0 & \text{otherwise,} \end{cases} \tag{3.2}$$

for all $n \in \mathbb{N}$.

By the *Möbius inversion* of [17], the weighted-semicircularity (3.1) is re-characterized as follows: a self-adjoint operator a is t_0-semicircular in $(\mathcal{B}, \varphi)$, if and only if

$$\varphi(a^n) = \omega_n \left(t_0^{\frac{n}{2}} c_{\frac{n}{2}} \right), \tag{3.3}$$

where

$$\omega_n \overset{def}{=} \begin{cases} 1 & \text{if } n \text{ is even} \\ 0 & \text{if } n \text{ is odd,} \end{cases}$$

for all $n \in \mathbb{N}$, and c_k are the k-th *Catalan numbers*,

$$c_k = \frac{1}{k+1}\binom{2k}{k} = \frac{1}{k+1}\frac{(2k)!}{k!(2k-k)!} = \frac{(2k)!}{k!(k+1)!},$$

for all $k \in \mathbb{N}_0 = \mathbb{N} \cup \{0\}$.

So, by (3.3), a free random variable a is semicircular in $(\mathcal{B}, \varphi)$, if and only if a is 1-semicircular in $(\mathcal{B}, \varphi)$, if and only if

$$\varphi(a^n) = \omega_n \, c_{\frac{n}{2}}, \tag{3.4}$$

for all $n \in \mathbb{N}$.

From below, we use the t_0-semicircularity (3.1) (or the semicircularity (3.2)) and its characterization (3.3) (resp., (3.4)) alternatively.

If a is a self-adjoint free random variable of $(\mathcal{B}, \varphi)$, then

$$\text{the } free\,moments \; \{\varphi(a^n)\}_{n=1}^\infty,$$

and

$$\text{the } free\,cumulants \; \{k_n^{\mathcal{B}}(a, ..., a)\}_{n=1}^\infty$$

provide equivalent free-distributional data of a in $(\mathcal{B}, \varphi)$ (e.g., [17]). Indeed, the *Möbius inversion* makes us have

$$\varphi(a^n) = \sum_{\pi \in NC(n)} \left(\underset{V \in \pi}{\Pi} \, k_{|V|}^{\mathcal{B}}(a, \, ..., \, a) \right),$$

and

$$k_n^{\mathcal{B}}(a, ..., a) = \sum_{\pi \in NC(n)} \left(\underset{V \in \theta}{\Pi} \, \varphi(a^{|V|}) \right) \mu(\pi, 1_n),$$

where $NC(n)$ is the *lattice* of all *noncrossing partitions* over $\{1, ..., n\}$, and "$V \in \pi$" means "V is a *block* of π," and where $\mu(\pi, 1_n)$ is the Möbius functional values at the interval $[\pi, 1_n]$ in $NC(n)$, where

$$1_n = \{(1, ..., n)\}$$

is the maximal element of the lattice $NC(n)$ having a single block $(1, ..., n)$.

We now fix a C^*-probability space (A, ψ), where A is a C^*-algebra, and assume that A contains mutually orthogonal, $|\mathbb{Z}|$-many projections $\{q_j\}_{j \in \mathbb{Z}}$, i.e.,

$$q_j^* = q_j = q_j^2 \text{ in } A, \text{ for all } j \in \mathbb{Z}, \tag{3.5}$$

and

$$q_i q_j = \delta_{i,j} q_j \text{ in } A, \text{ for all } i, j \in \mathbb{Z},$$

where δ is the *Kronecker delta*. Remark that there do exist such C^*-probabilistic structures naturally (e.g., [5,8,11]), or artificially (e.g., [6]).

Fix the family,

$$\mathbf{Q} = \{q_j : j \in \mathbb{Z}\} \text{ in } A, \tag{3.6}$$

of mutually orthogonal projections q_j's of (3.5).

And let Q be the C^*-*subalgebra of* A generated by the family $\mathbf{Q}$ of (3.6),

$$Q \stackrel{def}{=} C^*(\mathbf{Q}) \subseteq A, \tag{3.7}$$

where $C^*(Y)$ are the C^*-subalgebras generated by the subsets $Y \cup Y^*$ of A, where

$$Y^* = \{y^* : y \in Y\} \text{ in } A.$$

Proposition 3.1. *If Q is the C^*-subalgebra (3.7) of A, then*

$$Q \stackrel{*\text{-}iso}{=} \bigoplus_{j \in \mathbb{Z}} (\mathbb{C} \cdot q_j) \stackrel{*\text{-}iso}{=} \mathbb{C}^{\oplus |\mathbb{Z}|}, \tag{3.8}$$

in A, where $\oplus$ is the direct product of C^*-algebras.

Proof. The characterization (3.8) is proven by the mutual-orthogonality (3.5) of $\mathbf{Q}$. ∎

Define the linear functionals ψ_j on the C^*-algebra Q by

$$\psi_j(q_i) = \delta_{ij} \psi(q_j), \text{ for all } i \in \mathbb{Z}, \tag{3.9}$$

for all $j \in \mathbb{Z}$, where ψ is the linear functional of (A, ψ). These linear functionals $\{\psi_j\}_{j \in \mathbb{Z}}$ of (3.9) are well-defined on Q by (3.8).

Assumption. Let (A, ψ) be a fixed C^*-probability space, and let Q be the C^*-subalgebra (3.7) of A. From below, we assume

$$\psi(q_j) \neq 0 \text{ in } \mathbb{C}^\times = \mathbb{C} \setminus \{1\}, \text{ for all } j \in \mathbb{Z},$$

where q_j are projections in the generating family $\mathbf{Q}$ of (3.6).   □

Then, as an independent C^*-algebra, the C^*-subalgebra Q of A forms C^*-probability spaces (Q, ψ_j), where ψ_j are the linear functionals (3.9) on Q, for all $j \in \mathbb{Z}$. We call them, *the j-th C^*-probability spaces of Q in (A, ψ)*, for all $j \in \mathbb{Z}$.

Now, define bounded linear transformations $\mathbf{c}$ and $\mathbf{a}$ acting on the C^*-algebra Q, by linear morphisms satisfying

$$\mathbf{c}(q_j) = q_{j+1}, \text{ and } \mathbf{a}(q_j) = q_{j-1}, \tag{3.10}$$

for all $j \in \mathbb{Z}$. Then $\mathbf{c}$ and $\mathbf{a}$ are well-defined operators "acting on Q" by (3.8). These are understood to be *Banach-space operators* in the *operator space* $B(Q)$, consisting of all bounded linear transformations on Q, by understanding Q as a Banach space under its C^*-norm topology (e.g., [9]).

Definition 3.2. *The Banach-space operators* **c** *and* **a** *of (3.10) are said to be the* creation, *respectively, the* annihilation *on Q. Define*

$$\mathbf{l} = \mathbf{c} + \mathbf{a} \ on \ Q. \tag{3.11}$$

We call this Banach-space operator **l** *of (3.11), the* radial operator *on Q.*

Now, define a subspace $\mathfrak{L}$ of $B(Q)$ by

$$\mathfrak{L} \overset{def}{=} \overline{\mathbb{C}[\{\mathbf{l}\}]}^{\|\cdot\|}, \tag{3.12}$$

equipped with the operator norm,

$$\|T\| = \sup\{\|Tq\|_Q : \|q\|_Q = 1\}, \tag{3.13}$$

where $\|.\|_Q$ is the C^*-*norm on* Q, where $\overline{Z}^{\|\cdot\|}$ are the *operator-norm closures* of subsets $Z \subseteq B(Q)$ (e.g., [9]). By (3.12), this subspace $\mathfrak{L}$ forms a Banach algebra in the vector space $B(Q)$.

On this Banach algebra $\mathfrak{L}$ of (3.12), define an operation $(*)$ by

$$\left(\sum_{n=0}^{\infty} t_n \mathbf{l}^n\right)^* = \sum_{n=0}^{\infty} \overline{t_n}\mathbf{l}^n \ in \ \mathfrak{L}, \tag{3.14}$$

where $\overline{z}$ are the *conjugates of* $z \in \mathbb{C}$.

Then the operation (3.13) is a well-defined *adjoint* on $\mathfrak{L}$ (See [6]), and hence, every element of $\mathfrak{L}$ is *adjointable* (in the sense of [9]) in $B(Q)$. So, the Banach algebra $\mathfrak{L}$ of (3.12) forms a *Banach $*$-algebra* with the adjoint (3.13) in $B(Q)$. We call this Banach $*$-algebra $\mathfrak{L}$, the *radial (Banach $*$-)algebra on Q.*

Construct now the *tensor product Banach $*$-algebra* $\mathfrak{L}_Q$,

$$\mathfrak{L}_Q = \mathfrak{L} \otimes_{\mathbb{C}} Q, \tag{3.15}$$

where $\otimes_{\mathbb{C}}$ is the tensor product of Banach $*$-algebras, where $\mathfrak{L}$ is the radial algebra (3.12).

Definition 3.3. *The Banach $*$-algebra* $\mathfrak{L}_Q$ *of (3.14) is called the* radial projection *(Banach $*$-)algebra on Q.*

4   Weighted-Semicircular Elements

In this section, we construct weighted-semicircular elements induced by the family $\mathbf{Q}$ of (3.6) in the radial projection algebra $\mathfrak{L}_Q$ of (3.14). Let (Q, ψ_j) be the j-th C^*-probability spaces of Q in (A, ψ), where ψ_j are the linear functionals of (3.9), for all $j \in \mathbb{Z}$.

Note that, if

$$u_j \overset{def}{=} \mathbf{1} \otimes q_j \in \mathfrak{L}_Q, \ \text{for all} \ j \in \mathbb{Z}, \tag{4.1}$$

then

$$u_j^n = (1 \otimes q_j)^n = 1^n \otimes q_j, \text{ for all } n \in \mathbb{N},$$

for $j \in \mathbb{Z}$. i.e., such operators $\{u_j\}_{j \in \mathbb{Z}}$ generate $\mathfrak{L}_Q$, by (3.8), (3.12) and (3.14).

By (4.1), one can define a linear functional φ_j on $\mathfrak{L}_Q$ by a morphism satisfying that

$$\varphi_j \left((1 \otimes q_i)^n \right) \overset{def}{=} \psi_j \left(1^n(q_i) \right) \tag{4.2}$$

for all $n \in \mathbb{N}$, for all $i, j \in \mathbb{Z}$.

By the well-definedness of the linear functionals $\{\varphi_j\}_{j \in \mathbb{Z}}$ of (4.2), the Banach $*$-algebra $\mathfrak{L}_Q$ forms well-defined Banach $*$-probability spaces,

$$\left(\mathfrak{L}_Q, \varphi_j \right), \text{ for all } j \in \mathbb{Z}. \tag{4.3}$$

If $\mathbf{c}$ and $\mathbf{a}$ are the creation, respectively, the annihilation on Q of (3.10), then

$$\mathbf{c}\mathbf{a} = 1_Q = \mathbf{a}\mathbf{c} \text{ , the identity operator on } Q,$$

in $B(Q)$. So, one has

$$\mathbf{c}^{n_1} \mathbf{a}^{n_2} = \mathbf{a}^{n_2} \mathbf{c}^{n_1}, \forall n_1, n_2 \in \mathbb{N}. \tag{4.4}$$

By (4.4), we have

$$1^n = (\mathbf{c} + \mathbf{a})^n = \sum_{k=0}^{n} \binom{n}{k} \mathbf{c}^k \mathbf{a}^{n-k}, \tag{4.5}$$

for all $n \in \mathbb{N}$, where

$$\binom{n}{k} = \frac{n!}{k!(n-k)!}, \text{ for all } k \leq n \in \mathbb{N}_0.$$

By (4.5), for any $n \in \mathbb{N}$,

$$1^{2n-1} = \sum_{k=0}^{2n-1} \binom{2n-1}{k} \mathbf{c}^k \mathbf{a}^{n-k}, \tag{4.6}$$

and

$$1^{2n} = \sum_{k=0}^{2n} \binom{2n}{k} \mathbf{c}^k \mathbf{a}^{n-k} = \binom{2n}{n} \mathbf{c}^n \mathbf{a}^n + [\text{ Rest terms}] \tag{4.7}$$

(e.g., see [6] for details).

Proposition 4.1. *Let* 1 *be the radial operator on* Q. *Then*

(4.8) 1^{2n-1} *does not contain the nonzero* 1_Q *-summand.*

(4.9) 1^{2n} *contains the nonzero* 1_Q *-summand,* $\binom{2n}{n} \cdot 1_Q$.

Proof. The statements (4.8) and (4.9) are shown by (4.6) and (4.7), respectively. ∎

Since

$$u_j^n = (1 \otimes q_j)^n = 1^n \otimes q_j,$$

one has

$$\varphi_j \left(u_j^{2n-1} \right) = \psi_j \left(1^{2n-1} (q_j) \right) = 0, \tag{4.10}$$

for all $n \in \mathbb{N}$, by (4.8).

Also, we have

$$\varphi_j \left(u_j^{2n} \right) = \psi_j \left(1^{2n} (q_j) \right) = \psi_j \left(\binom{2n}{n} q_j + [\text{Rest terms}] \right)$$

$$= \binom{2n}{n} \psi_j (q_j) = \binom{2n}{n} \psi (q_j),$$

by (4.7) and (4.9). i.e.,

$$\varphi_j \left(u_j^{2n} \right) = \binom{2n}{n} \psi (q_j), \text{ for all } n \in \mathbb{N}. \tag{4.11}$$

Thus, by (4.10) and (4.11), the following free-distributional data are obtained.

Proposition 4.2. *Fix $j \in \mathbb{Z}$, and let $u_k = 1 \otimes q_k$ be the k-th generating operators of $(\mathfrak{L}_Q, \varphi_j)$, for all $k \in \mathbb{Z}$. Then*

$$\varphi_j \left(u_k^n \right) = \delta_{j,k} \omega_n \left(\left(\frac{n}{2} + 1 \right) \psi (q_j) \right) c_{\frac{n}{2}}, \tag{4.12}$$

where ω_n and $c_{\frac{n}{2}}$ are in the sense of (3.3) for all $k \in \mathbb{Z}$, and $n \in \mathbb{N}$.

Proof. By (4.10) and (4.11), one can get that: if u_j is the j-th generating operator of $\mathfrak{L}_Q$, then

$$\varphi_j \left(u_j^{2n-1} \right) = 0,$$

and

$$\varphi_j \left(u_j^{2n} \right) = \binom{2n}{n} \psi (q_j) = \left(\frac{n+1}{n+1} \right) \binom{2n}{n} \psi (q_j)$$

$$= ((n+1)\psi (q_j)) \left(\frac{1}{n+1} \binom{2n}{n} \right)$$

$$= ((n+1)\psi (q_j)) c_n,$$

for all $n \in \mathbb{N}$.

If $k \neq j$ in $\mathbb{Z}$, and u_k is the k-th generating operator of $\mathfrak{L}_Q$, then

$$\varphi_j \left(u_k^n \right) = 0, \text{ for all } n \in \mathbb{N},$$

by (3.9) and (4.2). ∎

Based on (4.12), define the linear morphisms,

$$E_{j,Q} : \mathfrak{L}_Q \to \mathfrak{L}_Q,$$

by linear transformations satisfying

$$E_{j,Q}\left(u_i^n\right) \overset{def}{=} \begin{cases} \frac{\psi(q_j)^{n-1}}{\left(\left[\frac{n}{2}\right]+1\right)}\, u_j^n & \text{if } i = j \\[2ex] 0_{\mathfrak{L}_Q}, \text{ the zero operator of } \mathfrak{L}_Q & \text{otherwise,} \end{cases} \tag{4.13}$$

for all $n \in \mathbb{N}$, $i, j \in \mathbb{Z}$, where $\left[\frac{n}{2}\right]$ means the *minimal integer* greater than or equal to $\frac{n}{2}$.

The linear transformations $E_{j,Q}$ of (4.13) are well-defined on $\mathfrak{L}_Q$ by the cyclicity (3.12) of a tensor factor $\mathfrak{L}$ of $\mathfrak{L}_Q$, and by the structure theorem (3.8) of the other tensor factor Q of $\mathfrak{L}_Q$, by (3.14).

Now, define the linear functionals τ_j on $\mathfrak{L}_Q$ by

$$\tau_j \overset{def}{=} \varphi_j \circ E_{j,Q} \text{ on } \mathfrak{L}_Q, \text{ for all } j \in \mathbb{Z}, \tag{4.14}$$

where $E_{j,Q}$ are in the sense of (4.13).

Definition 4.1. *The Banach $*$-probability spaces,*

$$\mathfrak{L}_Q(j) \overset{denote}{=} \left(\mathfrak{L}_Q, \tau_j\right), \tag{4.15}$$

are called the j-th (free) filter of $\mathfrak{L}_Q$, for all $j \in \mathbb{Z}$.

Observe on the j-th filter $\mathfrak{L}_Q(j)$ of (4.15) that:

$$\tau_j\left(u_j^n\right) = \varphi_j\left(E_{j,Q}\left(u_j^n\right)\right)$$

$$= \varphi_j\left(\frac{\psi(q_j)^{n-1}}{\left(\left[\frac{n}{2}\right]+1\right)}\left(u_j^n\right)\right) = \frac{\psi(q_j)^{n-1}}{\left(\left[\frac{n}{2}\right]+1\right)}\varphi_j\left(u_j^n\right)$$

$$= \frac{\psi(q_j)^{n-1}}{\left(\left[\frac{n}{2}\right]+1\right)}\omega_n\left(\left(\frac{n}{2}+1\right)\psi\left(q_j\right)\right)c_{\frac{n}{2}},$$

by (4.12), i.e.,

$$\tau_j\left(u_j^n\right) = \omega_n\psi(q_j)^n c_{\frac{n}{2}}, \tag{4.16}$$

where ω_n and $c_{\frac{n}{2}}$ are in the sense of (3.3), for all $n \in \mathbb{N}$, for $j \in \mathbb{Z}$.

Lemma 4.3. *Let $\mathfrak{L}_Q(j) = \left(\mathfrak{L}_Q, \tau_j\right)$ be the j-th filter of $\mathfrak{L}_Q$, for $j \in \mathbb{Z}$. Then*

$$\tau_j\left(u_i^n\right) = \delta_{j,i}\left(\omega_n\psi(q_j)^n c_{\frac{n}{2}}\right), \tag{4.17}$$

for all $n \in \mathbb{N}$, for all $i \in \mathbb{Z}$.

Proof. If $i = j$ in $\mathbb{Z}$, then the formula (4.17) holds by (4.16). Meanwhile, if $i \neq j$ in $\mathbb{Z}$, then $\tau_j (u_i^n) = 0$, by (4.2) and (4.13). Therefore, the formula (4.17) holds for all $i \in \mathbb{Z}$. ■

The following theorem is proven by (4.17).

Theorem 4.4. *Let $\mathfrak{L}_Q(j)$ be the j-th filter of $\mathfrak{L}_Q$, for $j \in \mathbb{Z}$. Then the j-th generating operator (4.1) of $\mathfrak{L}_Q$ is $\psi(q_j)^2$-semicircular in $\mathfrak{L}_Q(j)$. Meanwhile, all other k-th generating operators u_k of $\mathfrak{L}_Q$ have the zero free distribution on $\mathfrak{L}_Q(j)$, for all $k \neq j$ in $\mathbb{Z}$.*

Proof. Note that the generating operators u_k are self-adjoint in $\mathfrak{L}_Q$, since

$$u_k^* = (\mathbf{1} \otimes q_k)^* = \mathbf{1} \otimes q_k = u_k$$

for all $k \in \mathbb{Z}$, by (3.13).

If u_j is the j-th generating operator of $\mathfrak{L}_Q$, then

$$\tau_j \left(u_j^n \right) = \omega_n \left(\psi \left(q_j \right)^2 \right)^{\frac{n}{2}} c_{\frac{n}{2}},$$

for all $n \in \mathbb{N}$, by (4.17). Therefore, by (3.3), u_j is $\psi(q_j)^2$-semicircular in $\mathfrak{L}_Q(j)$.

Now, suppose $k \neq j$ in $\mathbb{Z}$, and take the generating operator u_k of $\mathfrak{L}_Q(j)$. By the self-adjointness of u_k, the free distribution of u_k is characterized by the free-moment sequence,

$$(\tau (u_k^n))_{n=1}^{\infty} = (0, 0, 0, 0, ...),$$

by (4.17). So, u_k has the zero free distribution on $\mathfrak{L}_Q(j)$, whenever $k \neq j$. ■

By using the Möbius inversion of [17], one can obtain that: if $k_\bullet^j(...)$ is the free cumulant on $\mathfrak{L}_Q$ in terms of a linear functional τ_j, then

$$k_n^j \left(\underbrace{u_k,\ u_k,\ ......,\ u_k}_{n\text{-times}} \right) = \begin{cases} \delta_{j,k}\psi(q_j)^2 & \text{if } n = 2 \\ 0 & \text{otherwise,} \end{cases} \tag{4.18}$$

for all $n \in \mathbb{N}$, for all $j, k \in \mathbb{Z}$ (e.g., see [6] for details).

5   Semicircular Elements

Let $\mathfrak{L}_Q(j)$ be the j-th filter for $j \in \mathbb{Z}$. Then, for a fixed $j \in \mathbb{Z}$, the j-th generating operator u_j is $\psi(q_j)^2$-semicircular in $\mathfrak{L}_Q(j)$, since

$$\tau_j \left(u_j^n \right) = \omega_n \psi(q_j)^n c_{\frac{n}{2}},$$

equivalently,

$$k_n^j \left(u_j, ..., u_j \right) = \begin{cases} \psi(q_j)^2 & \text{if } n = 2 \\ 0 & \text{otherwise,} \end{cases} \tag{5.1}$$

for all $n \in \mathbb{N}$, by (4.17) and (4.18).

Theorem 5.1. Let $U_j = \frac{1}{\psi(q_j)} u_j$ in the j-th filter $\mathfrak{L}_Q(j)$ for $j \in \mathbb{Z}$, where u_j is the j-th generating operator of $\mathfrak{L}_Q$. If

$$\psi(q_j) \in \mathbb{R}^\times = \mathbb{R} \setminus \{0\} \text{ in } \mathbb{C}^\times, \tag{5.2}$$

then U_j is semicircular in $\mathfrak{L}_Q(j)$, for $j \in \mathbb{Z}$.

Proof. Fix $j \in \mathbb{Z}$, and assume the condition (5.2) holds. Then

$$U_j^* = \left(\frac{1}{\psi(q_j)} u_j \right)^* = U_j,$$

in $\mathfrak{L}_Q$, because u_j is self-adjoint. Consider now that

$$\tau_j \left(U_j^n \right) = \frac{1}{\psi(q_j)^n} \tau \left(u_j^n \right)$$
$$= \frac{1}{\psi(q_j)^n} \left(\omega_n \psi(q_j)^n c_{\frac{n}{2}} \right) = \omega_n c_{\frac{n}{2}}, \tag{5.3}$$

for all $n \in \mathbb{N}$.

Therefore, under (5.2), the self-adjoint free random variable U_j is semicircular in $\mathfrak{L}_Q(j)$, by (3.4) and (5.3). ∎

Assumption. For convenience, we assume from below that

$$\psi(q_j) \in \mathbb{R}^\times \text{ in } \mathbb{C}, \text{ for } q_j \in \mathbf{Q},$$

for all $j \in \mathbb{Z}$. □

6 The Free Filterization $\mathfrak{L}_Q(\mathbb{Z})$

In this section, we construct the free product Banach $*$-probability space $\mathfrak{L}_Q(\mathbb{Z})$ of the free filters $\{\mathfrak{L}_Q(j)\}_{j \in \mathbb{Z}}$, and the corresponding sub-structure $\mathbb{L}_Q = (\mathbb{L}_Q, \tau)$ generated by a free semicircular family

$$\{U_j \in \mathfrak{L}_Q(j) : j \in \mathbb{Z}\},$$

and study free-distributional information on $\mathbb{L}_Q$.

6.1 The Semicircular Filterization $\mathbb{L}_Q$

As before, let (A, ψ) be the fixed C^*-probability space containing a family $\mathbf{Q} = \{q_j\}_{j \in \mathbb{Z}}$ of mutually orthogonal projections, satisfying

$$\psi(q_j) \in \mathbb{R}^\times, \text{ for all } j \in \mathbb{Z},$$

and let $\mathfrak{L}_Q(j)$ be the j-th filters of Q, for all $j \in \mathbb{Z}$.
From the system,

$$\{\mathfrak{L}_Q(j) : j \in \mathbb{Z}\},$$

define the *free product Banach* $*$-*probability space* $\mathfrak{L}_Q(\mathbb{Z})$ by

$$\mathfrak{L}_Q(\mathbb{Z}) \overset{denote}{=} (\mathfrak{L}_Q(\mathbb{Z}),\ \tau) \\ \overset{def}{=} \underset{j \in \mathbb{Z}}{\star} \mathfrak{L}_Q(j) = \left(\underset{j \in \mathbb{Z}}{\star} \mathfrak{L}_{Q,j},\ \underset{j \in \mathbb{Z}}{\star} \tau_j \right), \tag{6.1.1}$$

with

$$\mathfrak{L}_Q(\mathbb{Z}) = \underset{j \in \mathbb{Z}}{\star} \mathfrak{L}_{Q,j}\ ,\ \text{with}\ \mathfrak{L}_{Q,j} = \mathfrak{L}_Q, \forall j \in \mathbb{Z},$$

and

$$\tau = \underset{j \in \mathbb{Z}}{\star} \tau_j\ \text{on}\ \ \mathfrak{L}_Q(\mathbb{Z}).$$

For more about free-probabilistic free product, see [17] and [19].

Definition 6.1. *The free product Banach $*$-probability space $\mathfrak{L}_Q(\mathbb{Z})$ of (6.1.1) is said to be the free filterization of $Q \subset (A,\ \psi)$.*

Define now two subsets $\mathcal{X}$ and $\mathcal{S}$ of $\mathfrak{L}_Q(\mathbb{Z})$ by

$$\mathcal{X} = \{u_j \in \mathfrak{L}_Q(j) : j \in \mathbb{Z}\}, \tag{6.1.2}$$

and

$$\mathcal{S} = \{U_j = \frac{1}{\psi(q_j)} u_j \in \mathfrak{L}_Q(j) : j \in \mathbb{Z}\},$$

where u_j are the operators (4.1), for all $j \in \mathbb{Z}$.

A subset $\mathcal{Y}$ of a topological $*$-probability space $(\mathcal{B},\ \varphi)$ is said to be a *free family*, if all elements of $\mathcal{Y}$ are mutually free in $(\mathcal{B},\ \varphi)$. And, a free family $\mathcal{Y}$ is called a *free (weighted-)semicircular family* in $(\mathcal{B},\ \varphi)$, if all elements of $\mathcal{Y}$ are (weighted-)semicircular in $(\mathcal{B},\ \varphi)$. (e.g., [5] and [19]).

Theorem 6.1. *Let $\mathcal{X}$ and $\mathcal{S}$ be the families of (6.1.2) in $\mathfrak{L}_Q(\mathbb{Z})$.*

(6.1.3) $\mathcal{X}$ is a free weighted-semicircular family in $\mathfrak{L}_Q(\mathbb{Z})$.

(6.1.4) $\mathcal{S}$ is a free semicircular family in $\mathfrak{L}_Q(\mathbb{Z})$.

Proof. Let $\mathcal{X}$ be in the family of (6.1.2) in $\mathfrak{L}_Q(\mathbb{Z})$. By (6.1.1), all elements u_j of $\mathcal{X}$ are from mutually distinct free blocks $\mathfrak{L}_Q(j)$ for all $j \in \mathbb{Z}$, and hence, they are mutually free in $\mathfrak{L}_Q(\mathbb{Z})$. Thus, the subset $\mathcal{X}$ forms a free family in $\mathfrak{L}_Q(\mathbb{Z})$. Moreover, the powers $u_j^n \in \mathfrak{L}_Q(\mathbb{Z})$ of $u_j \in \mathcal{X}$ are contained in the same free block $\mathfrak{L}_Q(j)$ as free reduced words with their lengths-1 of $\mathfrak{L}_Q(\mathbb{Z})$, for all $n \in \mathbb{N}$, implying that

$$\tau\left(u_j^n\right) = \tau_j\left(u_j^n\right) = \omega_n \psi(q_j)^n c_{\frac{n}{2}}, \forall n \in \mathbb{N},$$

by (5.1). Therefore, the statement (6.1.3) holds.

Similarly, the family $\mathcal{S}$ of (6.1.2) is a free family in $\mathfrak{L}_Q(\mathbb{Z})$, because

$$U_j = \frac{1}{\psi(q_j)} u_j = U_j^*, \text{ for all } j \in \mathbb{Z},$$

and the family $\mathcal{X}$ is a free family in $\mathfrak{L}_Q(\mathbb{Z})$. So, the semicircularity (5.4) of U_j's shows that the statement (6.1.4) holds.   ∎

By (6.1.3) and (6.1.4), the "j-th" generating operators u_j of the free blocks $\mathfrak{L}_Q(j)$, and their powers $u_j^n \in \mathfrak{L}_Q(j)$ provide nonzero free-distributional data on the free filterization $\mathfrak{L}_Q(\mathbb{Z})$. In particular, the free (reduced) words in $\mathcal{X} \cup \mathcal{S}$ (under operator-multiplication on $\mathfrak{L}_Q(\mathbb{Z})$) have non-vanishing free distributions on $\mathfrak{L}_Q(\mathbb{Z})$.

Definition 6.2. *In the free filterization $\mathfrak{L}_Q(\mathbb{Z})$, define a Banach $*$-subalgebra $\mathbb{L}_Q$ of $\mathfrak{L}_Q(\mathbb{Z})$ by*

$$\mathbb{L}_Q \overset{def}{=} \overline{\mathbb{C}[\mathcal{X}]}, \tag{6.1.5}$$

where $\mathcal{X}$ is the free weighted-semicircular family (6.1.3) of $\mathfrak{L}_Q(\mathbb{Z})$, and $\overline{Y}$ are the topological closures of the subsets Y of $\mathfrak{L}_Q(\mathbb{Z})$. Construct the Banach $*$-probability space,

$$\mathbb{L}_Q \overset{denote}{=} \left(\mathbb{L}_Q, \ \tau = \tau \mid_{\mathbb{L}_Q}\right), \tag{6.1.6}$$

in $\mathfrak{L}_Q(\mathbb{Z}) = (\mathfrak{L}_Q(\mathbb{Z}), \tau)$. We call $\mathbb{L}_Q$ of (6.1.5) or (6.1.6), the semicircular (free-sub-)filterization of $\mathfrak{L}_Q(\mathbb{Z})$.

By the definitions (6.1.5) and (6.1.6), one obtains the following structure theorem.

Theorem 6.2. *Let $\mathbb{L}_Q$ be the semicircular filterization (6.1.5). Then*

$$\mathbb{L}_Q = \overline{\mathbb{C}[\mathcal{S}]} \overset{*-iso}{=} \underset{j \in \mathbb{Z}}{\star} \overline{\mathbb{C}[\{u_j\}]} \overset{*-iso}{=} \mathbb{C}\left[\underset{j \in \mathbb{Z}}{\star} \{u_j\}\right], \tag{6.1.7}$$

in $\mathfrak{L}_Q(\mathbb{Z})$, where "$\overset{*-iso}{=}$" means "being Banach-$*$-isomorphic," and where $(\star)$ in the first $*$-isomorphic relation of (6.1.7) means the free-probabilistic free product of [17] and [19], and $(\star)$ in the second $*$-isomorphic relation of (6.1.7) is the pure-algebraic free product inducing noncommutative free words in $\mathcal{X}$.

Proof. Set-theoretically, one has

$$\mathcal{X} = \{\psi(q_j) U_j \in \mathfrak{L}_Q(j) : j \in \mathbb{Z}\}$$

in the free filterization $\mathfrak{L}_Q(\mathbb{Z})$, where $U_j \in \mathcal{S}$ are the semicircular elements of (6.1.4). Therefore,

$$\overline{\mathbb{C}[\mathcal{X}]} = \overline{\mathbb{C}[\mathcal{S}]} \text{ in } \mathfrak{L}_Q(\mathbb{Z}),$$

i.e., the equality $(=)$ of (6.1.7) holds.

By the definition (6.1.5) of $\mathbb{L}_Q$, it is generated by the free family $\mathcal{X}$, and hence, the first $*$-isomorphic relation of (6.1.7) holds in $\mathfrak{L}_Q(\mathbb{Z})$ by (6.1.1) and (6.1.6).

Since

$$\mathbb{L}_Q \overset{*\text{-iso}}{=} \underset{j \in \mathbb{Z}}{\star} \overline{\mathbb{C}[\{u_j\}]} \text{ in } \mathfrak{L}_Q(\mathbb{Z}),$$

every element T of $\mathbb{L}_Q$ is a limit of linear combinations of free reduced words in $\mathcal{X}$. Note that all (pure-algebraic) free words in $\mathcal{X}$ have their unique free-reduced-word forms as their operator-product in $\mathfrak{L}_Q(\mathbb{Z})$ (e.g., [17] and [19]). Therefore, the second $*$-isomorphic relation of (6.1.7) holds. ∎

6.2   Free-Distributional Data Induced by Semicircular Elements

In this section, we consider general free-distributional data on $\mathbb{L}_Q$. In particular, we are interested in joint free moments of $\mathcal{S}$ in $\mathbb{L}_Q$. Throughout this section, let (B, φ) be an arbitrarily fixed topological $*$-probability space, and suppose there are N-many semicircular elements $x_1, ..., x_N$ in (B, φ), for $N \in \mathbb{N} \setminus \{1\}$. Assume further that they are free from each other in (B, φ).

By the self-adjointness of these semicircular elements $x_1, ..., x_N \in (B, \varphi)$, the free distribution, say

$$\rho \overset{denote}{=} \rho_{x_1,...,x_N},$$

of them are characterized by the joint free-moments

$$\overset{\infty}{\underset{n=1}{\cup}} \left(\underset{(i_1,...,i_n) \in \{1,...,N\}^n}{\cup} \{\varphi(x_{i_1} x_{i_2}...x_{i_n})\} \right) \tag{6.2.1}$$

(e.g., [17]). i.e., the free distribution ρ of (6.2.1), is characterized by the free-moments,

$$\overset{N}{\underset{l=1}{\cup}} \{\varphi(x_l^n)\}_{n=1}^{\infty}, \tag{6.2.2}$$

and the "mixed" free-moments,

$$\overset{\infty}{\underset{s=2}{\cup}} \left\{ \varphi\left(x_{i_1}^{n_1} x_{i_2}^{n_2}...x_{i_s}^{n_s}\right) \, \middle| \, \begin{array}{l} (i_1,...,i_s) \in \{1,...,N\}^s \\ \text{are mixed in } \{1,...,N\}, \\ \text{for all } n_1,...,n_s \in \mathbb{N} \end{array} \right\}, \tag{6.2.3}$$

by (6.2.1). To characterize the free distribution ρ, we consider the free-distributional data (6.2.2) and (6.2.3), independently.

Corollary 6.3. *The free-distributional data (6.2.2) of ρ are determined by the semicircularity. i.e.,*

$$\varphi\left(x_l^n\right) = \omega_n c_{\frac{n}{2}}, \ \text{for all } n \in \mathbb{N}, \tag{6.2.4}$$

for all $l = 1, \dots, N$.

Proof. The formula (6.2.4) is obtained by the semicircularity (3.4). ∎

Let's concentrate on the free-distributional data (6.2.3) of the free distribution ρ. For any $s \in \mathbb{N} \setminus \{1\}$, we fix an s-tuple I_s,

$$I_s \overset{denote}{=} (i_1, \dots, i_s) \in \{1, \dots, N\}^s, \tag{6.2.5}$$

which is mixed in $\{1, \dots, N\}$ in the sense that there exists at least one entry i_{k_0} in I_s such that $i_{k_0} \neq i_l$, for some $l \neq k_0$ in $\{1, \dots, s\}$.
For example,

$$I_8 = (1, 1, 3, 2, 4, 2, 2, 1),$$

in $\{1, 2, 3, 4, 5\}^8$.
From the sequence I_s of (6.2.5), define a set

$$[I_s] = \{i_1, i_2, \dots, i_s\}, \tag{6.2.6}$$

without considering repetition. For instance, if I_8 is as above, then

$$[I_8] = \{i_1, i_2, \dots, i_8\},$$

with

$$i_1 = i_2 = i_8 = 1,$$
$$i_4 = i_6 = i_7 = 2,$$
$$i_3 = 3, \ \text{and } i_5 = 4.$$

i.e., all 1's in I_8 are regarded as distinct elements i_1, i_2 and i_8 in the set $[I_8]$.
From the set $[I_s]$ of (6.2.6), define a unique "noncrossing" partition $\pi_{(I_s)}$ of the non-crossing-partition lattice $NC\left([I_s]\right)$ over $[I_s]$, such that (i) starting from the very first entry i_1, construct the block V_1 of $\pi_{(I_s)}$, satisfying

$$V_1 = \left(i_{j_1} = i_1, \ i_{j_2}, \ \dots, \ i_{j_{|V_1|}}\right) \in \pi_{(I_s)}, \tag{6.2.7}$$

$\Longleftrightarrow$

$$\exists k \in \{1, \dots, N\}, \ \text{s.t.}, \ i_{j_1} = i_{j_2} = \dots = i_{j_{|V|}} = k,$$

and then do the same process to the very next entry other than $i_{j_1}, \dots, i_{j_{|V_1|}}$, step-by-step, until such processes end, (ii) such a partition $\pi_{(I_s)}$ of (i) is "maximal" in $NC\left([I_s]\right)$ (e.g., [17]).

For example, if I_8 and $[I_8]$ are as above, then there exists a noncrossing partition

$$\pi_{(I_8)} = \{(i_1, i_2, i_8), (i_3), (i_4, i_6, i_7), (i_5)\}$$
$$= \{(1, 1, 1), (3), (2, 2, 2), (4)\},$$

in $NC([I_8])$, satisfying the above conditions (i) and (ii). In this case,

$$V_1 = \{i_1, i_2, i_8\} = \{1, 1, 1\}, \text{ as in (6.2.7).}$$

Denote the noncrossing partition $\pi_{(I_s)} \in NC([I_s])$ of (6.2.6) by

$$\pi_{(I_s)} = \{V_1, ..., V_t\},$$

where $t \leq s$ and $V_k \in \pi_{(I_s)}$ are the blocks of (ii), satisfying (i), for $k = 1, ..., t$. Then the partition $\pi_{(I_s)}$ is the joint partition,

$$\pi_{(I_s)} = 1_{|V_1|} \vee 1_{|V_2|} \vee ... \vee 1_{|V_t|}, \tag{6.2.8}$$

where $1_{|V_k|}$ are the maximal partitions of $NC(V_k)$, for all $k = 1, ..., t$, by regarding V_k as discrete sets.

Let I_s be in the sense of (6.2.5), and let $x_{i_1}, ..., x_{i_s}$ be the corresponding semicircular elements of (B, φ) induced by I_s, without considering repetition in the set $\{x_1, ..., x_N\}$ of our fixed mutually free, N-many semicircular elements of (B, φ). And then, define a free random variable $X[I_s]$ by

$$X[I_s] \overset{def}{=} \overset{s}{\underset{l=1}{\Pi}} x_{i_l} \in (B, \varphi). \tag{6.2.9}$$

If $X[I_s]$ is a free random variable (6.2.9), then

$$\varphi(X[I_s]) = \sum_{\pi \in NC([I_s])} k_\pi$$

by the Möbius inversion of [17], where k_π are the partition-depending free cumulants of [17],

$$k_\pi = \underset{V \in \pi}{\Pi} k_V,$$

where k_V is the block-depending free cumulants of [17], and hence, it goes to

$$= \sum_{\pi \in NC([I_s]), \, \pi \leq \pi_{(I_s)}} k_\pi$$

by the mutual-freeness of $x_1, ..., x_N$ in (B, φ)

$$= \sum_{(\theta_1, ..., \theta_t) \in NC(V_1) \times ... \times NC(V_t)} k_{\theta_1 \vee ... \vee \theta_t}$$

by (6.2.8)

$$
= \sum_{(\theta_1, \ldots, \theta_t) \in NC_2(V_1) \times \ldots \times NC_2(V_t)} k_{\theta_1 \vee \ldots \vee \theta_t}
$$

$$
= \sum_{(\theta_1, \ldots, \theta_t) \in NC_2(V_1) \times \ldots \times NC_2(V_t)} \left(\prod_{l=1}^{t} k_{\theta_l} \right), \tag{6.2.10}
$$

by the semicircularity (3.2) of $x_{i_1}, \ldots, x_{i_s}$ in (B, φ), where $NC_2(X)$ is the subset,

$$
NC_2(X) = \{\pi \in NC(X) : \forall V \in \pi, |V| = 2\}, \tag{6.2.11}
$$

of the noncrossing-partition lattice $NC(X)$ over sets X.

By (6.2.10), (6.2.11) and (3.2), if there is at least one $k_0 \in \{1, \ldots, t\}$, such that $|V_{k_0}|$ is odd in $\mathbb{N}$ (or equivalently, if s is odd), then

$$
\varphi\left(X[I_s]\right) = 0,
$$

where $X[I_s]$ is in the sense of (6.2.9).

Meanwhile, if

$$
|V_k| \in 2\mathbb{N}, \text{ for all } k = 1, \ldots, t, \tag{6.2.12}
$$

where $2\mathbb{N} = \{2n : n \in \mathbb{N}\}$, then the formula (6.2.10) is nonzero.

More precisely, if the condition (6.2.12) is satisfied, then the summands $k_{\theta_1 \vee \ldots \vee \theta_t}$ of (6.2.10) satisfy that

$$
k_{\theta_1 \vee \ldots \vee \theta_t} = \prod_{V \in \theta_1 \vee \ldots \vee \theta_t} k_V = \prod_{V \in \theta_1 \vee \ldots \vee \theta_t} \left(\prod_{i=1}^{t} 1^{\#(\theta_i)} \right) = 1, \tag{6.2.13}
$$

by the semicircularity (3.2), where $\#(\theta_i)$ are the number of blocks of θ_i, for all $i = 1, \ldots, t$. Therefore, if the condition (6.2.12) holds, then

$$
\varphi\left(X[I_s]\right) = \sum_{(\theta_1, \ldots, \theta_t) \in NC_2(V_1) \times \ldots \times NC_2(V_t)} 1
$$

$$
= |NC_2(V_1) \times \ldots \times NC_2(V_t)|, \tag{6.2.14}
$$

by (6.2.10) and (6.2.13), where $|Y|$ are the cardinalities of sets Y.

Theorem 6.4. *Let I_s be a mixed s-tuple (6.2.5), and let $X[I_s] = \prod_{l=1}^{s} x_{i_l}$ be the corresponding free random variable (6.2.9) of (B, φ). If*

$$
\pi_{(I_s)} = 1_{|V_1|} \vee \ldots \vee 1_{|V_t|},
$$

in the sense of (6.2.7) and (6.2.8), then

$$
\varphi\left(X[I_s]\right) = \begin{cases} \prod_{i=1}^{t} c_{\frac{|V_i|}{2}} & \begin{array}{l} \text{if } |V_k| \in 2\mathbb{N}, \\ \text{for all } k = 1, \ldots, t \end{array} \\ \\ 0 & \text{otherwise,} \end{cases} \tag{6.2.15}
$$

where c_k are the k-th Catalan numbers for all $k \in \mathbb{N}_0$.

Proof. Under hypothesis,
$$\varphi\left(X[I_s]\right)$$

$$= \begin{cases} |NC_2\left(V_1\right) \times \ldots \times NC_2\left(V_t\right)| & \begin{array}{l} \text{if } |V_k| \in 2\mathbb{N}, \\ \text{for all } k = 1, \ldots, t \end{array} \\ \\ 0 & \text{otherwise,} \end{cases}$$

by (6.2.14).
Recall that, for every countable set X, with $|X| \in 2\mathbb{N}$, the set

$$NC_2\left(X\right) = \{\theta \in NC(X) : \forall V \in \theta, |V| = 2\}$$

is equipotent (or bijective) to the noncrossing-partition lattice $NC\left(\frac{|X|}{2}\right)$ over $\{1, \ldots, \frac{|X|}{2}\}$ (e.g., [5] and [8]). i.e., if $|V_k| \in 2\mathbb{N}$, then

$$|NC_2\left(V_k\right)| = \left|NC\left(\frac{|V_k|}{2}\right)\right|, \tag{6.2.16}$$

for all $k = 1, \ldots, t$. So, we have
$$\varphi\left(X[I_s]\right)$$

$$= \begin{cases} \left|NC\left(\frac{|V_1|}{2}\right) \times \ldots \times NC\left(\frac{|V_t|}{2}\right)\right| & \begin{array}{l} \text{if } |V_k| \in 2\mathbb{N}, \\ \text{for all } k = 1, \ldots, t \end{array} \\ \\ 0 & \text{otherwise,} \end{cases}$$

$$= \begin{cases} \prod_{l=1}^{t} c_{\frac{|V_l|}{2}} & \text{if } |V_l| \in 2\mathbb{N}, \text{ for all } l = 1, \ldots, t \\ \\ 0 & \text{otherwise,} \end{cases} \tag{6.2.17}$$

by (6.2.16), because $|NC(X)| = c_{|X|}$, for all finite sets X (e.g., [7,8,14,17]). Therefore, the formula (6.2.15) holds by (6.2.17). ∎

Remark 6.1. *The more combinatorial computational techniques, and the refined results of (6.2.15) are considered "analytically" in [7], including direct estimations, and asymptotic estimations of (6.2.15). However, in this paper, the free-distributional data (6.2.15) is enough for our purposes. The importance here is that the free-distributional data induced by mutually free, multi semicircular elements are dictated by the semicircularity (3.2), by (6.2.4) and (6.2.15).*

We provide some examples before finishing this section.

Example 6.1. *(1) Let x_1, x_2, x_3, x_4 be mutually free semicircular elements of (B, φ), and let*

$$W = x_1^2 x_2^4 x_1^2 x_3^2 \in (B, \varphi)$$

be a free reduced word with its length-4. Then one can take

$$I_W = (1,\ 1,\ 2,\ 2,\ 2,\ 2,\ 1,\ 1,\ 3,\ 3) \overset{let}{=} (i_1, ..., i_{10}),$$

and

$$\pi_{(I_W)} = \{(i_1, i_2, i_7, i_8), (i_3, i_4, i_5, i_6), (i_9, i_{10})\},$$

with

$$V_1 = \{i_1,\ i_2,\ i_7,\ i_8\} = \{1,\ 1,\ 1,\ 1\},$$
$$V_2 = \{i_3,\ i_4,\ i_5,\ i_6\} = \{2,\ 2,\ 2,\ 2\},$$

and

$$V_3 = \{i_7,\ i_8\} = \{3,\ 3\}.$$

Therefore,

$$\varphi(W) = c_{\frac{4}{2}} c_{\frac{4}{2}} c_{\frac{2}{2}} = c_2^2 c_1 = 4.$$

(2) Meanwhile, if $W = x_1^2 x_3 x_2 x_4 x_2^2 x_1 \in (B,\ \varphi)$, then

$$I_W = (1,\ 1,\ 3,\ 2,\ 4,\ 2,\ 2,\ 1),$$

and

$$\pi_{(I_8)} = \{(i_1,\ i_2,\ i_8),\ (i_3),\ (i_4,\ i_6,\ i_7),\ (i_5)\}$$
$$= \{(1,\ 1,\ 1),\ (3),\ (2,\ 2,\ 2),\ (4)\}$$

satisfying that

$$\varphi(X[I_8]) = \varphi\left(x_1^2 x_3 x_2 x_4 x_2^2 x_1\right) = c_{\frac{3}{2}} c_{\frac{1}{2}} c_{\frac{3}{2}} c_{\frac{1}{2}} = 0,$$

by (6.2.15).
(3) Now, let $W = x_1^4 x_2^4 x_1^6 x_2^4 \in (B,\ \varphi)$. Then one can take

$$I_W = (i_1, i_2, ..., i_{18}),$$

having

$$\pi_{(I_{18})} = \{V_1, V_2, V_3\},$$

with

$$V_1 = \{i_1, i_2, i_3, i_4, i_9, i_{10}, i_{11}, i_{12}, i_{13}, i_{14}\},$$

and

$$V_2 = \{i_5, i_6, i_7, i_8\}, V_3 = \{i_{15}, i_{16}, i_{17}, i_{18}\}.$$

(Here, since all entries of V_2 and V_3 are identical to 2, one may/can be tempted to make a block

$$\{i_5, i_6, i_7, i_8, i_{15}, i_{16}, i_{17}, i_{18}\},$$

but, in such a case, this block has crossing with V_1, disobeying the conditions (i) and (ii)!)
Therefore,

$$\varphi\left(x_1^4 x_2^4 x_1^6 x_3^4\right) = c_{\frac{10}{2}} c_{\frac{4}{2}} c_{\frac{4}{2}} = 168,$$

by (6.2.15).

6.3   Free-Distributional Data on $\mathbb{L}_Q$

Let $\mathbb{L}_Q$ be our semicircular filterization (6.1.5) of the free filterization $\mathfrak{L}_Q(\mathbb{Z})$, generated by the free semicircular family $\mathcal{S}$ of (6.1.4). By the structure theorem (6.1.7), all free random variables of $\mathbb{L}_Q$ are the limits of linear combinations of free reduced words, formed by

$$W = \prod_{l=1}^{N} U_{j_l}^{n_l}, \text{ for } U_{j_l} \in \mathcal{S}, \forall l = 1, ..., N, \tag{6.3.1}$$

in $\mathcal{S}$, for all $N \in \mathbb{N}$, where $n_1, ..., n_N \in \mathbb{N}$, and the N-tuple $(j_1, ..., j_N)$ is alternating in $\mathbb{Z}$.

Theorem 6.5. *Let W be a free reduced word (6.3.1) of $\mathbb{L}_Q$ in $\mathcal{S}$.*

(6.3.2) *If $j_1 = j_2 = ... = j_N$ in (6.3.1), then $\tau(W)$ is characterized by (6.2.4).*

(6.3.3) *If $(j_1, ..., j_N)$ is mixed in (6.3.1), then $\tau(W)$ is determined by (6.2.15).*

Proof. Note that all semicircular elements of any topological $*$-probability spaces have the same free distribution, "the" semicircular law, characterized by the free-moment sequence

$$(0, c_1, 0, c_2, 0, c_3, 0, c_4, ...),$$

equivalently, the free-cumulant sequence

$$(0, 1, 0, 0, 0, 0, ...),$$

by (3.2) and (3.4), where c_k are the k-th Catalan numbers for all $k \in \mathbb{N}$.
By this universality of the semicircular law (or, by the identically-free-distributedness of all semicircular elements in terms of [19]), the statements (6.3.2) and (6.3.3) are shown by (6.2.4) and (6.2.15), respectively.   ∎

The above theorem characterizes the free-distributional data on the semicircular filterization $\mathbb{L}_Q$, in terms of joint free moments of generating semicircular elements of $\mathcal{S}$, by (6.3.2) and (6.3.3).

7   Integer-Shifts on $\mathbb{L}_Q$

In this section, let (A, ψ) be the fixed C^*-probability space containing a family $\mathbf{Q} = \{q_j\}_{j\in\mathbb{Z}}$ of mutually-orthogonal projections q_j's having

$$\psi(q_j) \in \mathbb{R}^{\times}, \text{ for all } j \in \mathbb{Z},$$

and let $\mathbb{L}_Q$ be the semicircular filterization.

7.1   ($\pm$)-Shifts on $\mathbb{Z}$

Let $\mathbb{Z}$ be the set of all integers. Define functions h_+ and h_- on $\mathbb{Z}$ by

$$h_+(j) = j + 1, \tag{7.1.1}$$

and

$$h_-(j) = j - 1,$$

for all $j \in \mathbb{Z}$. By the definition (7.1.1), these two functions $h_{\pm}$ are well-defined bijections on $\mathbb{Z}$, satisfying $h_+^{-1} = h_-$, where f^{-1} mean the functional inverses of invertible functions f.

Then, for these bijections $h_{\pm}$ of (7.1.1), one can construct the bijections $h_{\pm}^{(n)}$ on $\mathbb{Z}$,

$$h_{\pm}^{(n)} = \underbrace{h_{\pm} \circ h_{\pm} \circ \cdots \circ h_{\pm}}_{n\text{-times}}, \tag{7.1.2}$$

for all $n \in \mathbb{N}$, with identities, $h_{\pm}^{(1)} = h_{\pm}$, where $(\circ)$ is the usual functional composition. It is easy to check that

$$h_{\pm}^{(n)}(j) = j \pm n, \text{ for all } j \in \mathbb{Z},$$

for all $n \in \mathbb{N}$.

Definition 7.1. *We call the functions $h_{\pm}^{(n)}$ of (7.1.2), the n-($\pm$)-shifts on $\mathbb{Z}$, for $n \in \mathbb{N}$.*

7.2   Integer-Shifts on $\mathbb{L}_Q$

Let $h_{\pm}^{(n)}$ be n-($\pm$)-shifts of (7.1.2) on $\mathbb{Z}$, for all $n \in \mathbb{N}$. Define now a multiplicative bounded linear transformation $\beta_{\pm}$ on $\mathbb{L}_Q$ by the morphisms satisfying that:

$$\beta_{\pm}(U_j) = U_{h_{\pm}(j)}, \tag{7.2.1}$$

for $U_j \in \mathcal{S}$, for all $j \in \mathbb{Z}$, where $\mathcal{S}$ is our free semicircular family (6.1.4) of $\mathfrak{L}_Q(\mathbb{Z})$, generating $\mathbb{L}_Q$.

By (6.1.6) and (6.1.7), the above multiplicative linear transformation $\beta_{\pm}$ of (7.2.1) is well-defined on $\mathbb{L}_Q$.

Lemma 7.1. *Let* $Y = \prod_{l=1}^{N} U_{j_l}^{n_l} \in \mathbb{L}_Q$, *for* $U_{j_1}, ..., U_{j_N} \in \mathcal{S}$, *and* $n_1, ..., n_N \in \mathbb{N}$, *for* $N \in \mathbb{N}$. *Then*

$$\beta_{\pm}(Y) = \prod_{l=1}^{N} U_{h_{\pm}(j_l)}^{n_l}. \tag{7.2.2}$$

Proof. If $Y \in \mathbb{L}_Q$ be as above, then, by the multiplicativity of $\beta_{\pm}$, one has that

$$\beta_{\pm}(Y) = \prod_{l=1}^{N} \beta_{\pm}\left(U_{j_l}^{n_l}\right) = \prod_{l=1}^{N} \left(\beta_{\pm}\left(U_{j_l}\right)\right)^{n_l} = \prod_{l=1}^{N} U_{h_{\pm}(j_l)}^{n_l}.$$

So, the formula (7.2.2) holds. ∎

Let $u_{j_1}, ..., u_{j_N} \in \mathcal{X}$ be weighted-semicircular elements of $\mathbb{L}_Q$, for $N \in \mathbb{N}$, where $\mathcal{X}$ is the free weighted-semicircular family (6.1.3), generating $\mathbb{L}_Q$, and let

$$X = \prod_{l=1}^{N} u_{j_l}^{n_l}, \text{ for } n_1, ..., n_N \in \mathbb{N}.$$

Then one has

$$\beta_{\pm}(X) = \beta_{\pm}\left(\left(\prod_{l=1}^{N} \psi(q_{j_l})^{n_l}\right)\left(\prod_{l=1}^{N} U_{j_l}^{n_l}\right)\right)$$

since

$$U_{j_l} = \frac{1}{\psi(q_{j_l})} u_{j_l} \in \mathcal{S} \Longleftrightarrow u_{j_l} = \psi(q_{j_l})U_{j_l} \in \mathcal{X},$$

so, the above formula goes to

$$= \left(\prod_{l=1}^{N} \psi(q_{j_l})^{n_l}\right) \beta_{\pm}\left(\prod_{l=1}^{N} U_{j_l}^{n_l}\right)$$

$$= \left(\prod_{l=1}^{N} \psi(q_{j_l})^{n_l}\right)\left(\prod_{l=1}^{N} U_{h_{\pm}(j_l)}^{n_l}\right), \tag{7.2.2'}$$

by (7.2.2)

By (7.2.2)', one can see that the freeness on $\mathbb{L}_Q$ is preserved to that on $\mathbb{L}_Q$.

Theorem 7.2. *The multiplicative linear transformations* $\beta_{\pm}$ *of (7.2.1) are* ∗*-isomorphisms on* $\mathbb{L}_Q$.

Proof. By (6.1.5) and (6.1.6), each element of the semicircular filterization $\mathbb{L}_Q$ is a limit of linear combinations of free reduced words in the generating free semicircular family $\mathcal{S}$. So, we focus on free reduced words of $\mathbb{L}_Q$ in $\mathcal{S}$.

Let $(j_1, ..., j_N)$ be an alternating N-tuple in $\mathbb{Z}$ for $N \in \mathbb{N}$, and let

$$Y = \prod_{l=1}^{N} U_{j_l}^{n_l}, \text{ for } n_1, ..., n_N \in \mathbb{N}.$$

be a free reduced word with its length-N in $\mathbb{L}_Q$ by (6.1.7).
Then, by (7.2.2),

$$\beta_\pm(Y) = \prod_{l=1}^{N} U_{h_\pm(j_l)}^{n_l}, \qquad (7.2.3)$$

are free reduced words with their lengths-N in $\mathbb{L}_Q$, where $h_\pm$ are the ($\pm$)-shifts (7.1.1) on $\mathbb{Z}$, since

$$(h_\pm(j_1), \ ..., \ h_\pm(j_N)) = (j_1 \pm 1, ..., j_N \pm 1)$$

are alternating in $\mathbb{Z}$, too.
Also, if Y is as above, then

$$\beta_\pm(Y^*) = \beta_\pm\left(\prod_{l=1}^{N} U_{j_{N-l+1}}^{n_{N-l+1}} \right)$$

by the self-adjointness of $U_{j_1}, ..., U_{j_N}$

$$= \prod_{l=1}^{N} U_{h_\pm(j_{N-l+1})}^{n_{N-l+1}}$$

by

$$= \left(\prod_{l=1}^{N} U_{h_\pm(j_l)}^{n_l} \right)^* = (\beta_\pm(Y))^*,$$

showing that

$$\beta_\pm(S^*) = (\beta_\pm(S))^*, \text{ for all } S \in \mathbb{L}_Q. \qquad (7.2.4)$$

By (7.2.4), the bijective bounded multiplicative linear transformations $\beta_\pm$ of (7.2.1) are well-defined $*$-isomorphisms on $\mathbb{L}_Q$. ∎

The above theorem illustrates that the ($\pm$)-shifts $h_\pm$ of (7.1.1) on $\mathbb{Z}$ induce the $*$-isomorphisms $\beta_\pm$ of (7.2.2) on $\mathbb{L}_Q$.

Definition 7.2. *The $*$-isomorphisms $\beta_\pm$ of (7.2.1) are said to be ($\pm$)-integer-shift(-$*$-isomorphism)s on $\mathbb{L}_Q$.*

These two $*$-isomorphisms $\beta_\pm$ satisfy the following identity relation on $\mathbb{L}_Q$.

Lemma 7.3. *The ($\pm$)-integer shifts $\beta_\pm$ satisfy*

$$\beta_+\beta_- = 1_{\mathbb{L}_Q} = \beta_-\beta_+ \text{ on } \mathbb{L}_Q, \qquad (7.2.5)$$

where $1_{\mathbb{L}_Q}$ is the identity map on $\mathbb{L}_Q$.

Proof. It is enough to consider the cases where we have free reduced words formed by

$$Y = \prod_{l=1}^{N} U_{j_l}^{n_l} \text{ of } \mathbb{L}_Q, \text{ for } n_1, ..., n_N \in \mathbb{N},$$

for $N \in \mathbb{N}$, where $(j_1, ..., j_N)$ is alternating in $\mathbb{Z}$. Observe that

$$\beta_+\beta_-(Y) = \beta_+\left(\prod_{l=1}^{N} U_{h_-(j_l)}^{n_l} \right) = \beta_+\left(\prod_{l=1}^{N} U_{j_l-1}^{n_l} \right)$$

$$= \prod_{l=1}^{N} U_{h_+(j_l-1)}^{n_l} = \prod_{l=1}^{N} U_{j_l-1+1}^{n_l} = Y.$$

Similarly,

$$\beta_- \beta_+(Y) = Y, \text{ in } \mathbb{L}_Q.$$

So, for any arbitrary operators $S \in \mathbb{L}_Q$,

$$\beta_+ \beta_-(S) = S = \beta_- \beta_+(S), \text{ in } \mathbb{L}_Q.$$

Thus, the relation (7.2.5) holds. ∎

From the $(\pm)$-shifts $\beta_\pm$ on $\mathbb{L}_Q$, construct the $*$-isomorphisms $\beta_\pm^n$,

$$\beta_\pm^n = \underbrace{\beta_\pm \beta_\pm \cdots\cdots \beta_\pm}_{n\text{-times}} \text{ on } \mathbb{L}_Q, \tag{7.2.6}$$

for all $n \in \mathbb{N}_0 = \mathbb{N} \cup \{0\}$, with axiomatization:

$$\beta_+^0 = 1_{\mathbb{L}_Q} = \beta_-^0.$$

Since $\beta_\pm$ and $1_{\mathbb{L}_Q}$ are well-defined $*$-isomorphisms, the morphisms $\beta_\pm^n$ of (7.2.9) are well-defined $*$-isomorphisms on $\mathbb{L}_Q$, too, for all $n \in \mathbb{N}_0$.

Definition 7.3. *Let $\beta_\pm^n$ be the $*$-isomorphisms (7.2.6) on the semicircular filterization $\mathbb{L}_Q$, for all $n \in \mathbb{N}_0$, with axiomatization $\beta_\pm^0 = 1_{\mathbb{L}_Q}$. Then we call them the n-$(\pm)$-(integer-)shifts on $\mathbb{L}_Q$, for all $n \in \mathbb{N}_0$.*

By (7.2.5) and (7.2.6), we obtain the following relation on the set $\{\beta_\pm^n : n \in \mathbb{N}_0\}$.

Lemma 7.4. *Let $\beta_\pm^n$ be the n-$(\pm)$-shifts on $\mathbb{L}_Q$, for $n \in \mathbb{N}_0$. Then*

$$\beta_+^{n_1} \beta_-^{n_2} = \beta_-^{n_2} \beta_+^{n_1} = \begin{cases} 1_{\mathbb{L}_Q} & \text{if } n_1 = n_2 \\ \beta_+^{n_1 - n_2} & \text{if } n_1 > n_2 \\ \beta_-^{n_2 - n_1} & \text{if } n_1 < n_2, \end{cases} \tag{7.2.7}$$

on $\mathbb{L}_Q$, for all $n_1, n_2 \in \mathbb{N}_0$. And

$$\beta_+^{n_1} \beta_+^{n_2} = \beta_+^{n_1 + n_2}, \text{ and } \beta_-^{n_1} \beta_-^{n_2} = \beta_-^{n_1 + n_2}, \tag{7.2.8}$$

on $\mathbb{L}_Q$, for all $n_1, n_2 \in \mathbb{N}_0$.

Proof. The formulas (7.2.7) and (7.2.8) are proven by the straightforward computations. ∎

The above relations (7.2.7) and (7.2.8) can be re-expressed as follows;

$$\beta_{e_1}^{n_1} \beta_{e_2}^{n_2} = \beta_{e_2}^{n_2} \beta_{e_1}^{n_1} = \beta_{sgn(e_1 n_1 + e_2 n_2)}^{|e_1 n_1 + e_2 n_2|} \text{ on } \mathbb{L}_Q, \tag{7.2.9}$$

with

$$sgn(e_1 n_1 + e_2 n_2) = \begin{cases} + \text{ if } e_1 n_1 + e_2 n_2 \geq 0 \\ - \text{ if } e_1 n_1 + e_2 n_2 \leq 0, \end{cases}$$

for all $e_1, e_2 \in \{\pm\}$, and $n_1, n_2 \in \mathbb{N}_0$, where sgn is the *sign map*,

$$sgn(j) = \begin{cases} + \text{ if } j \geq 0 \\ - \text{ if } j < 0, \end{cases}$$

for all $j \in \mathbb{Z}$, and $|.|$ is the *absolute value* on $\mathbb{Z}$.

Now, let

$$\mathfrak{B} = \{\beta_\pm^n\}_{n \in \mathbb{N}_0}. \tag{7.2.10}$$

Then it is a subset of the *automorphism group*,

$$Aut(\mathbb{L}_Q) = \left(\left\{ \alpha : \mathbb{L}_Q \to \mathbb{L}_Q \;\middle|\; \begin{array}{c} \alpha \text{ are} \\ * \text{-isomorphisms} \\ \text{on } \mathbb{L}_Q \end{array} \right\}, \cdot \right), \tag{7.2.11}$$

of all $*$-isomorphisms on $\mathbb{L}_Q$, where the operation $(\cdot)$ is the product of $*$-isomorphisms.

By (7.2.10), this system $\mathfrak{B}$ is clearly a "subset" of the automorphism group $Aut(\mathbb{L}_Q)$ of (7.2.11). Note that, by (7.2.9),

$$\left(\beta_{e_1}^{n_1} \beta_{e_2}^{n_2} \right) \beta_{e_3}^{n_3} = \beta_{sgn(e_1 n_1 e_2 n_2)}^{|e_1 n_1 e_2 n_2|} \beta_{e_3}^{n_3} = \beta_{sgn(e_1 n_1 e_2 n_2 e_3 n_3)}^{|e_1 n_1 e_2 n_2 e_3 n_3|}$$

$$= \beta_{e_1}^{n_1} \beta_{sng(e_2 n_2 e_3 n_3)}^{|e_2 n_2 e_3 n_3|} = \beta_{e_1}^{n_1} \left(\beta_{e_2}^{n_2} \beta_{e_3}^{n_3} \right), \tag{7.2.12}$$

on $\mathbb{L}_Q$, for all $e_1, e_2, e_3 \in \{\pm\}$, $n_1, n_2, n_3 \in \mathbb{N}_0$.

Theorem 7.5. *Let $\mathfrak{B}$ be the subset (7.2.10) of $Aut(\mathbb{L}_Q)$. Then*

$$(7.2.13) \qquad \mathfrak{B} \text{ is a subgroup of } Aut(\mathbb{L}_Q).$$

Proof. Let $\mathfrak{B}$ be the set (7.2.10). By (7.2.9), the operation $(\cdot)$ is closed on $\mathfrak{B}$. By (7.2.12), this operation is associative on $\mathfrak{B}$.

Since

$$\beta_+^0 = 1_{\mathbb{L}_Q} = \beta_-^0 \in \mathfrak{B},$$

and

$$\beta_e^n \cdot 1_{\mathbb{L}_Q} = \beta_e^n = 1_{\mathbb{L}_Q} \cdot \beta_e^n \text{ on } \mathbb{L}_Q,$$

for all $e \in \{\pm\}$, and $n \in \mathbb{N}_0$, the set $\mathfrak{B}$ contains the group-identity $1_{\mathbb{L}_Q}$ of $Aut(\mathbb{L}_Q)$.

By (7.2.7), all elements $\beta_\pm^n \in \mathfrak{B}$ have their unique $(\cdot)$-inverses $\beta_\mp^n \in \mathfrak{B}$, such that

$$\beta_+^n \beta_-^n = 1_{\mathbb{L}_Q} = \beta_-^n \beta_+^n \text{ on } \mathbb{L}_Q,$$

for all $n \in \mathbb{N}_0$,

Therefore, the system $\mathfrak{B}$ is a subgroup of $Aut(\mathbb{L}_Q)$. ∎

As a subgroup of $Aut(\mathbb{L}_Q)$, the group $\mathfrak{B}$ satisfies the following algebraic property.

Theorem 7.6. *Let $\mathfrak{B}$ be the subgroup (7.2.10) of $Aut(\mathbb{L}_Q)$. Then*

$$\mathfrak{B} \overset{Group}{=} (\mathbb{Z}, +), \tag{7.14}$$

where "$\overset{Group}{=}$" means "being group-isomorphic," where $(\mathbb{Z}, +)$ is the infinite cyclic abelian group.

Proof. Define a function $\Phi : \mathbb{Z} \to \mathfrak{B}$ by

$$\Phi : j \in \mathbb{Z} \longmapsto \beta_{sgn(j)}^{|j|} \in \mathfrak{B}, \tag{7.2.15}$$

with correspondence

$$0 \in \mathbb{Z} \longmapsto 1_{\mathbb{L}_Q} = \beta_{\pm}^{0} \in \mathfrak{B}.$$

Then this function Φ of (7.2.15) is a well-defined bijection from $\mathbb{Z}$ onto $\mathfrak{B}$, by (7.2.6). And it satisfies that

$$\begin{aligned}
\Phi(j_1 + j_2) = \beta_{sgn(j_1+j_2)}^{|j_1+j_2|} &= \beta_{sgn(j_1)}^{|j_1|} \beta_{sgn(j_2)}^{|j_2|} \\
&= \Phi(j_1)\Phi(j_2),
\end{aligned} \tag{7.2.16}$$

in $\mathfrak{B}$, for all $j_1, j_2 \in \mathbb{Z}$.

So, the bijection Φ is a group-homomorphism by (7.2.16), i.e., the relation (7.14) holds. ∎

The above theorem characterizes the algebraic structure of the group $\mathfrak{B} = \{\beta_{\pm}^n\}_{n \in \mathbb{N}_0}$ in the automorphism group $Aut(\mathbb{L}_Q)$.

Definition 7.4. *The subgroup $\mathfrak{B}$ of the automorphism group $Aut(\mathbb{L}_Q)$ is called the integer-shift (sub)group (of $Aut(\mathbb{L}_Q)$ acting) on $\mathbb{L}_Q$.*

7.3 Free Distributions on $\mathbb{L}_Q$ Affected by $\mathfrak{B}$

Let $\mathfrak{B}$ be the integer-shift group (7.2.10) acting on the semicircular filterization $\mathbb{L}_Q$ of Q. We here consider how the action of our $*$-isomorphisms $\beta_{\pm}^n \in \mathfrak{B}$ affects the original free-distributional data on the semicircular filterization $\mathbb{L}_Q$.

Take an arbitrary free reduced word,

$$Y = \prod_{l=1}^{N} U_{j_l}^{n_l} \, in \mathbb{L}_Q \tag{7.3.1}$$

where $U_{j_l} = \frac{1}{\psi(q_{j_l})} u_{j_l} \in \mathcal{S}$ are the generating semicircular elements of $\mathbb{L}_Q$, where $u_{j_l} \in \mathcal{X}$ are the $\psi(q_j)^2$-semicircular elements, for all $l = 1, ..., N$, for $N \in \mathbb{N}$, and where the N-tuple $(j_1, ..., j_N)$ is alternating in $\mathbb{Z}$, and $n_1, ..., n_N \in \mathbb{N}$.

Theorem 7.7. *Let Y be a free reduced word (7.3.1) of $\mathbb{L}_Q$ in S. Then*

$$\tau\left(\beta_e^k(Y)\right) = \tau(Y), \tag{7.3.2}$$

for all $e \in \{\pm\}$ and $k \in \mathbb{N}_0$.

Proof. First assume that $N = 1$, and hence, $Y = U_{j_1}^{n_1}$ in $\mathbb{L}_Q$. Then, by the semicircularity of U_{j_1}, $U_{j_1 ek} \in S$ in $\mathbb{L}_Q$, one has that

$$\tau\left(\beta_e^k(Y)\right) = \tau\left(U_{jek}^{n_1}\right) = \omega_{n_1} c_{\frac{n_1}{2}} = \tau\left(U_{j_1}^{n_1}\right), \tag{7.3.3}$$

for all $\beta_e^k \in \mathfrak{B}$.

Assume now that $N > 1$ in $\mathbb{N}$, and the free reduced word $Y \in \mathbb{L}_Q$ with its length-N is in the sense of (7.3.1). Note that the image

$$\beta_e^k(Y) = \prod_{l=1}^{N} U_{j_l ek}^{n_k}$$

is again a free reduced word with the same length-N in $\mathbb{L}_Q$, for all $\beta_e^k \in \mathfrak{B}$. Now, let I_s be the s-tuple of (6.2.5), satisfying

$$Y = X[I_s] \text{ in } \mathbb{L}_Q, \text{ for some } s \geq N,$$

where $X[I_s]$ is in the sense of (6.2.9). Similarly, let $I_{s'}$ be the s'-tuple of (6.2.5), satisfying

$$\beta_e^k(Y) = X[I_{s'}] \text{ in } \mathbb{L}_Q,$$

for $\beta_e^k \in \mathfrak{B}$, where $X[I_{s'}]$ is in the sense of (6.2.9).

Then, since Y and $\beta_e^k(Y)$ are the same-length free reduced words having the same free-ness structure, one has not only

$$s = s' \text{ in } \mathbb{N},$$

but also

$$\pi(I_s) = \pi(I_{s'}) \text{ in } NC\left([I_s]\right) \stackrel{\text{lattice}}{=} NC\left([I_{s'}]\right),$$

where $\pi(I_s)$ and $\pi(I_{s'})$ are the noncrossing partitions of (6.2.7), and "$\stackrel{\text{lattice}}{=}$" means "being lattice-isomorphic." (Recall that if X and Y are finite discrete sets, then $NC(X)$ and $NC(Y)$ are lattice-isomorphic, if and only if $|X| = |Y|$ in $\mathbb{N}$).

By the semicircularity (7.3.3) of $U_{j_1}, ..., U_{j_N}, U_{j_1 ek}, ..., U_{j_N ek} \in S$ in $\mathbb{L}_Q$, we have

$$\tau\left(\beta_e^k(Y)\right) = \tau\left(X[I_{s'}]\right) = \tau\left(X[I_s]\right) = \tau(Y),$$

by (6.2.15) or (6.3.3), for all $\beta_e^k \in \mathfrak{B}$.

Therefore, the statement (7.3.2) holds. ∎

The above theorem shows how the original free distributional data on the semicircular filterization $\mathbb{L}_Q$ is affected by the group-action of the integer-shifts of $\mathfrak{B}$ on $\mathbb{L}_Q$. i.e., $\mathfrak{B}$ preserves the free probability on $\mathbb{L}_Q$, by (7.3.2).

8   Semicircular Elements Induced by Multi Projections

Now, we have all ingredients for studying our main interests. In this section, we show that if there are N-many mutually orthogonal projections in an arbitrary C^*-probability space, then there exists a corresponding free semicircular family $\mathcal{S}^{(N)}$, induced by the projections in a certain free product Banach $*$-probability space $\mathbb{L}_Q^{(N)}$, for any

$$N \in \mathbb{N}_{>1}^\infty = (\mathbb{N} \setminus \{1\}) \cup \{\infty\}.$$

And then, consider certain $*$-homomorphisms acting on $\mathbb{L}_Q^{(N)}$, and study how they deform the free probability on $\mathbb{L}_Q^{(N)}$.

8.1   A Free Semicircular Family $\mathcal{S}^{(N)}$ Induced by N-many Projections

Let (A_o, ψ_o) be a C^*-probability space containing its N-many mutually orthogonal projections

$$\mathbf{Q}_o = \{q_1^o, ..., q_N^o\} \tag{8.1.1}$$

for $N \in \mathbb{N}_{>1}^\infty$, and let

$$Q_o = C^*(\mathbf{Q}_o) \subseteq A_o \tag{8.1.2}$$

be the C^*-subalgebra of A generated by the family $\mathbf{Q}_o$ of (8.1.1).
Suppose

$$\psi_o(q_k^o) \in \mathbb{R}^\times \text{ in } \mathbb{C}, \forall k = 1, ..., N. \tag{8.1.3}$$

Proposition 8.1. *Let Q_o be the C^*-subalgebra (8.1.2). Then*

$$Q_o \overset{*\text{-}iso}{=} \overset{N}{\underset{l=1}{\oplus}} (\mathbb{C} \cdot q_l^o) \overset{*\text{-}iso}{=} \mathbb{C}^{\oplus N}. \tag{8.1.4}$$

Proof. Since the generating set $\mathbf{Q}_o$ of Q_o consists of mutually orthogonal N-many projections q_1^o, ..., q_N^o, the structure theorem (8.1.4) is immediately proven.   ∎

Suppose there is a C^*-probability space (A, ψ) containing a family $\mathbf{Q} = \{q_j\}_{j \in \mathbb{Z}}$ of mutually orthogonal $|\mathbb{Z}|$-many projections q_j's, satisfying

$$\psi(q_j) \in \mathbb{R}^\times \text{ in } \mathbb{C}, \text{ for all } j \in \mathbb{Z}.$$

Assume further that there exist projections $q_{j_1}, ..., q_{j_N} \in \mathbf{Q}$, such that

$$\psi(q_{j_l}) = \psi_o(q_l^o) \text{ in } \mathbb{R}^\times, \tag{8.1.5}$$

for all $l = 1, ..., N$, where ψ_o is the linear functional on the C^*-algebra A_o, satisfying the condition (8.1.3).

For convenience, we re-index the subfamily

$$\{q_{j_1}, ..., q_{j_N}\} \text{ of } \mathbf{Q} \tag{8.1.6}$$

by

$$\{q_1, ..., q_N\} \text{ in } \mathbf{Q},$$

without loss of generality, from below.

Theorem 8.2. *Let Q_o be the C^*-subalgebra (8.1.2) of a fixed C^*-probability space (A_o, ψ_o), generated by the family $\mathbf{Q}_o$ of (8.1.1). Then, under (8.1.5) and (8.1.6), there exists a Banach $*$-subalgebra*

$$\mathbb{L}_Q^{(N)} \overset{*-iso}{=} \overset{N}{\underset{l=1}{\star}} \overline{\mathbb{C}[\{U_l\}]}$$

of the semicircular filterization $\mathbb{L}_Q$ of (6.1.5).

Proof. Let $Q = C^*(\mathbf{Q})$ be the C^*-subalgebra (3.7) of the C^*-probability space (A, ψ) satisfying (8.1.5), under the re-indexing process (8.1.6). First, define a linear morphism

$$\Psi : Q_o \to Q$$

by

$$\Psi \left(\sum_{l=1}^{N} t_l q_l^o \right) \overset{def}{=} \sum_{l=1}^{N} t_l q_l + \sum_{j \in \mathbb{Z} \setminus \{1,...,N\}} 0 \cdot q_j.$$

Then it is an injective $*$-homomorphism from Q_o to Q, by (3.8) and (8.1.4). Thus, one can construct the semicircular elements

$$U_l = \mathbf{1} \otimes q_l = \mathbf{1} \otimes \Psi(q_l^o) \in \mathbb{L}_Q, \tag{8.1.7}$$

in the free semicircular family $\mathcal{S}$ of (6.1.4), generating the semicircular filterization $\mathbb{L}_Q$, for all $l = 1, ..., N$.

By the structure theorem (6.1.7) of $\mathbb{L}_Q$, one can define the Banach $*$-subalgebra

$$\mathbb{L}_Q^{(N)} \overset{def}{=} \overline{\mathbb{C}[\{U_1, ..., U_N\}]}$$
$$\overset{*-iso}{=} \overline{\mathbb{C}[\{\mathbf{1} \otimes \Psi(q_l^o) : l = 1, ..., N\}]}$$
$$\overset{*-iso}{=} \overset{N}{\underset{l=1}{\star}} \overline{\mathbb{C}[\{\mathbf{1} \otimes \Psi(q_j^o)\}]} = \overset{N}{\underset{l=1}{\star}} \overline{\mathbb{C}[\{U_l\}]} \tag{8.1.8}$$

of $\mathbb{L}_Q$, by (8.1.7).

It shows that: if the condition (8.1.5) is satisfied under (8.1.6), then the family $\mathbf{Q}_o$ of (8.1.1) induces Banach $*$-probability space,

$$\mathbb{L}_Q^{(N)} = \left(\mathbb{L}_Q^{(N)}, \ \tau = \tau \mid_{\mathbb{L}_Q^{(N)}} \right),$$

generated by the free semicircular family

$$\mathcal{S}^{(N)} = \{ U_l = \mathbf{1} \otimes \Psi(q_l^o) \}_{l=1}^N, \tag{8.1.9}$$

as a free-probabilistic sub-structure of the semicircular filterization $\mathbb{L}_Q$. ∎

The above theorem shows that if the conditions (8.1.3) and (8.1.5) are satisfied under the re-indexing process (8.1.6), then the family $\mathbf{Q}_o$ of (8.1.1) induces a Banach $*$-probability space $\mathbb{L}_Q^{(N)}$ of (8.1.8) generated by the free semicircular family $\mathcal{S}^{(N)}$ of (8.1.9), as an embedded free-probabilistic sub-structure of the semicircular filterization $\mathbb{L}_Q$ of (6.1.5).

Remark 8.1. *As we briefly discussed in* [6], *whenever such a family $\mathbf{Q}_o$ of (8.1.1) in a C^*-probability space (A_o, ψ_o) is fixed, in fact, one can construct the corresponding C^*-probability space (A, ψ), having its family $\mathbf{Q}$ of mutually orthogonal $|\mathbb{Z}|$-many projections, artificially-but-naturally.*

Assume first that $N < \infty$ in $\mathbb{N}_{>1}^\infty$. If $Q_o = C^*(\mathbf{Q}_o)$ is the C^*-subalgebra (8.1.2) of A_o, satisfying (8.1.4), then one can construct the direct product C^*-algebra Q,

$$Q = \bigoplus_{k \in \mathbb{Z}} Q_{o,k} \overset{*-iso}{=} \mathbb{C}^{\oplus |\mathbb{Z}|}, \ \ with \ Q_{o,k} = Q_o,$$

equipped with its linear functional ψ,

$$\psi = \bigoplus_{k \in \mathbb{Z}} \psi_{o,k} \ on \ Q, \ \ with \ \psi_{o,k} = \psi_o.$$

So, in the above C^*-probability space (Q, ψ) (or a C^*-probability space (A, ψ) with $A \supseteq Q$), there exist infinitely many projections q, such that

$$\psi(q) = \psi_o(q_l), \ for \ some \ l \in \{1, ..., N\}.$$

Assume now that $N = \infty$ in $\mathbb{N}_{>1}^\infty$, and let

$$Q_o = C^*(\mathbf{Q}_o) = C^* \left(\{q_1^o, \ q_2^o, \ q_3^o, \ ...\} \right).$$

Then, for convenience, by the canonical re-indexing, one can let

$$Q_o = C^* \left(\{q_0^o, \ q_1^o, \ q_2^o, \ ...\} \right),$$

by identifying q_i^o with q_{i-1}^o, for all $i \in \mathbb{N}$.
Now, consider a subfamily $\mathbf{Q}_o'$ of $\mathbf{Q}_o$,

$$\mathbf{Q}_o' = \{q_{-1}^o, q_{-2}^o, ...\} = \{q_0^o, q_1^o, q_2^o, ...\} \setminus \{q_0^o\},$$

with identity:

$$q^o_{-k} = q^o_k, \text{ for all } k \in \mathbb{N},$$

of the re-indexed family $\mathbf{Q}_o$, *and construct*

$$Q' = C^* \left(\mathbf{Q}'_o \right).$$

Then we have the direct product C^*-*algebra*

$$Q = Q_o \oplus Q'_o \overset{*-iso}{=} \mathbb{C}^{\oplus |\mathbb{Z}|},$$

equipped with its linear functional

$$\psi = \psi_o \oplus \psi'_o, \text{ with } \psi'_o = \psi_o \mid_{Q'_o},$$

satisfying

$$\psi \left(q^o_0 \right) = \psi_o(q^o_0), \text{ and } \psi \left(q^o_{\pm n} \right) = \psi_o \left(q^o_n \right),$$

for all $n \in \mathbb{N}$.

Therefore, in fact, whenever such a family $\mathbf{Q}_o$ *of (8.1.1) is fixed in a* C^*-*probability space* (A_o, ψ_o), *there does exist a family* $\mathbf{Q}$ *of mutually orthogonal* $|\mathbb{Z}|$-*many projections in a* C^*-*probability space* (Q, ψ) *(or* (A, ψ) *with* $A \supseteq Q$), *such that* Q *automatically satisfies (8.1.5) (and (8.1.6)).*

By the above remark, the following corollary is regarded as a re-statement of the above theorem.

Corollary 8.3. *Let* (A_o, ψ_o) *be an arbitrary* C^*-*probability space containing mutually orthogonal* N-*many projections* $q_1, ..., q_N$, *satisfying (8.1.3), for* $N \in \mathbb{N}^\infty_{>1}$. *Then there exists a free semicircular family* $\mathcal{S}^{(N)}$ *induced by* $\{q_k\}^N_{k=1}$ *in a certain Banach* $*$-*probability space* $\mathbb{L}^{(N)}_Q$.

Proof. The proof is done by (8.1.7), (8.1.8), (8.1.9), and the very above remark. ∎

8.2   Restricted Integer-Shifts on $\mathbb{L}^{(N)}_Q$

Let (A_o, ψ_o) be a C^*-probability space containing the family $\mathbf{Q}_o$ of (8.1.1), satisfying (8.1.3). In Sect. 8.1, we showed that such a family $\mathbf{Q}_o$ of mutually orthogonal N-many projections induces the free semicircular family $\mathcal{S}^{(N)}$ of (8.1.9) in a free product Banach $*$-probability space

$$\mathbb{L}^{(N)}_Q = (\mathbb{L}^{(N)}_Q, \tau), \text{ with } \tau = \tau \mid_{\mathbb{L}^{(N)}_Q},$$

of (8.1.8), for $N \in \mathbb{N}^\infty_{>1}$. Moreover, this Banach $*$-probability space is understood as a free-probabilistic sub-structure of the semicircular filterization $\mathbb{L}_Q$ of (6.1.5).

Since the integer-shift group

$$\mathfrak{B} = \{\beta_e^k \in Aut(\mathbb{L}_Q) : e \in \{\pm\}, k \in \mathbb{N}_0\}$$

of (7.2.13) acts on $\mathbb{L}_Q$ naturally, as an infinite cyclic abelian group, one can restrict the action of $\mathfrak{B}$ on $\mathbb{L}_Q$ to that on $\mathbb{L}_Q^{(N)}$.

Lemma 8.4. *Let $\beta_e^k \in \mathfrak{B}$ be an integer-shift on $\mathbb{L}_Q$, and $U_l \in \mathcal{S}^{(N)}$, a semicircular element, generating $\mathbb{L}_Q^{(N)}$, for $l = 1, ..., N$, and suppose $\beta_e^k \mid_{\mathbb{L}_Q^{(N)}}$ is the restriction of β_e^k on $\mathbb{L}_Q^{(N)}$, also denoted simply by β_e^k. If*

$$N \in \mathbb{N}_{>1} = \mathbb{N}_{>1}^\infty \setminus \{\infty\},$$

then

$$\beta_e^k(U_l) = \begin{cases} U_{lek} & \text{if } 1 \leq lek \leq N \\ O & \text{otherwise,} \end{cases} \qquad (8.2.1)$$

in $\mathbb{L}_Q^{(N)}$, where O is the zero element of $\mathbb{L}_Q^{(N)}$.
Meanwhile, if $N = \infty$ in $\mathbb{N}_{>1}^\infty$, then

$$\beta_e^k(U_l) = \begin{cases} U_{l+k} & \text{if } e = + \\ U_{l-k} & \text{if } e = -, \text{ and } l > k \\ O & \text{if } e = -, \text{ and } l \leq k, \end{cases} \qquad (8.2.2)$$

in $\mathbb{L}_Q^{(N)}$.

Proof. First, assume that $N < \infty$ in $\mathbb{N}_{>1}^\infty$, i.e., $N \in \mathbb{N}_{>1}$, and fix $l \in \{1, ..., N\}$ arbitrarily, and let $\beta_e^k \in \mathfrak{B}$ be an integer-shift on $\mathbb{L}_Q$. Let's restrict β_e^k on $\mathbb{L}_Q^{(N)}$, i.e.,

$$\beta_e^k \overset{denote}{=} \beta_e^k \mid_{\mathbb{L}_Q^{(N)}} \text{ on } \mathbb{L}_Q^{(N)}.$$

Then, for a semicircular elements $U_l \in \mathcal{S}^{(N)}$, generating $\mathbb{L}_Q^{(N)}$, one has that: if $e = +$, then

$$\beta_e^k(U_l) = \begin{cases} U_{l+k} & \text{if } l + k \leq N \\ O & \text{if } l + k > N; \end{cases} \qquad (8.2.3)$$

and if $e = -$, then

$$\beta_e^k(U_l) = \begin{cases} U_{l-k} & \text{if } l - k \geq 1 \\ O & \text{if } l - k < 1, \end{cases} \qquad (8.2.4)$$

in $\mathbb{L}_Q^{(N)}$.
By (8.2.3) and (8.2.4), one obtains that

$$\beta_e^k(U_l) = \begin{cases} U_{lek} & \text{if } 1 \leq lek \leq N \\ O & \text{otherwise,} \end{cases} \qquad (8.2.5)$$

in $\mathbb{L}_Q^{(N)}$. Therefore, the formula (8.2.1) holds by (8.2.5).

Now, assume that $N = \infty$ in $\mathbb{N}_{>1}^\infty$. Then, the restricted action of $\beta_e^k \in \mathfrak{B}$ on $\mathbb{L}_Q^{(N)}$ satisfies that: if $e = +$, then

$$\beta_e^k (U_l) = U_{l+k};\tag{8.2.6}$$

if $e = -$, then

$$\beta_e^k(U_l) = \begin{cases} U_{l-k} & \text{if } l - k \geq 1 \\ O & \text{if } l - k < 1, \end{cases}\tag{8.2.7}$$

in $\mathbb{L}_Q^{(N)}$. Therefore, the formula (8.2.2) is shown by (8.2.6) and (8.2.7). ■

The above lemma not only show how the restricted action of the integer-shift group $\mathfrak{B}$ on $\mathbb{L}_Q^{(N)}$ acts on the free generator set $\mathcal{S}^{(N)}$, but also demonstrates that the restrictions of integer-shifts are no longer $*$-isomorphisms on $\mathbb{L}_Q^{(N)}$, in general.

Let B be an arbitrary topological $*$-algebra. Then the $(*-)homomorphism$ semigroup $Hom(B)$ is defined to be the semigroup (under composition)

$$Hom(B) = \{f : f \text{ is a } * \text{-homomorphism on } B\}.$$

Since the zero map on B is contained in $Hom(B)$, it cannot be a group (under composition), however, it forms a well-defined semigroup.

Notation. From below, we denote the family of restricted integer-shifts on $\mathbb{L}_Q^{(N)}$ by $\mathfrak{B}^{(N)}$, i.e.,

$$\mathfrak{B}^{(N)} = \left\{ \beta_e^k \mid_{\mathbb{L}_Q^{(N)}} \ \middle| \ \begin{array}{l} \beta_e^k \in \mathfrak{B}, \text{ with} \\ e \in \{\pm\}, \ k \in \mathbb{N}_0 \end{array} \right\}.\tag{8.2.8}$$

Also, for convenience, we denote the restrictions $\beta_e^k \mid_{\mathbb{L}_Q^{(N)}} \in \mathfrak{B}^{(N)}$, simply by β_e^k, as above.   □

Lemma 8.5. *Let $\mathfrak{B}^{(N)}$ be the set (8.2.8) of the restricted integer-shifts on $\mathbb{L}_Q^{(N)}$. Then*

$$\mathfrak{B}^{(N)} \subseteq Hom\left(\mathbb{L}_Q^{(N)}\right),\tag{8.2.9}$$

equivalently, every element $\beta_e^k \in \mathfrak{B}^{(N)}$ is a $*$-homomorphism on $\mathbb{L}_Q^{(N)}$.

Proof. First, assume that $N \in \mathbb{N}_{>1}$ in $\mathbb{N}_{>1}^\infty$. If $\beta_e^k \in \mathfrak{B}^{(N)}$ satisfies

$$lek < 1, \text{ or } lek > N, \forall l = 1, ..., N,\tag{8.2.10}$$

then such a restricted integer-shift β_e^k is identified with the zero $*$-homomorphism $0_Q^{(N)}$ on $\mathbb{L}_Q^{(N)}$, i.e.,

$$\beta_e^k(T) = 0_Q^{(N)}(T) = O \text{ in } \mathbb{L}_Q^{(N)},$$

for all $T \in \mathbb{L}_Q^{(N)}$, by (8.1.8), (8.2.1) and (8.2.10).

And hence, all elements β_e^k of $\mathfrak{B}^{(N)}$ satisfying (8.2.10) are identified with $0_Q^{(N)}$, i.e.,

$$\beta_e^k = 0_Q^{(N)} \in Hom\left(\mathbb{L}_Q^{(N)}\right). \tag{8.2.11}$$

Suppose that $\beta_e^k \in \mathfrak{B}^{(N)}$ satisfies

$$1 \leq lek \leq N, \tag{8.2.12}$$

for some $l \in \{1, ..., N\}$. Then, by (8.2.1), the morphism β_e^k is a well-defined $*$-homomorphism on $\mathbb{L}_Q^{(N)}$, since it is the restriction of a $*$-isomorphism on $\mathbb{L}_Q \supseteq \mathbb{L}_Q^{(N)}$. i.e.,

$$\beta_e^k \in Hom(\mathbb{L}_Q^{(N)}), \tag{8.2.13}$$

under (8.2.12).

So, if $N \in \mathbb{N}_{>1}$, then

$$\mathfrak{B}^{(N)} \subseteq Hom\left(\mathbb{L}_Q^{(N)}\right), \tag{8.2.14}$$

by (8.2.11) and (8.2.13).

Assume now that $N = \infty$ in $\mathbb{N}_{>1}^\infty$. If $\beta_+^k \in \mathfrak{B}^{(N)}$, then

$$\beta_+^k \in Hom\left(\mathbb{L}_Q^{(N)}\right); \tag{8.2.15}$$

if $\beta_-^k \in \mathfrak{B}^{(N)}$, then

$$\beta_-^k \in Hom\left(\mathbb{L}_Q^{(N)}\right), \tag{8.2.16}$$

by (8.2.2), because there are infinitely many semicircular elements $\{U_l\}_{l=1}^\infty$ generating $\mathbb{L}_Q^{(N)}$. However, in this case,

$$\beta_e^k \neq 0_Q^{(N)} inHom\left(\mathbb{L}_Q^{(N)}\right), \forall k \in \mathbb{N}_0, e \in \{\pm\}, \tag{8.2.17}$$

different from (8.2.11). Indeed, for any arbitrarily fixed $\beta_-^k \in \mathfrak{B}^{(N)}$, there always exists $n > k$ in $\mathbb{N}$, such that

$$\beta_-^k(U_n) = U_{n-k} \neq O \text{ in } \mathbb{L}_Q.$$

Therefore, if $N = \infty$ in $\mathbb{N}_{>1}^\infty$, then

$$\mathfrak{B}^{(N)} \subseteq Hom\left(\mathbb{L}_Q^{(N)}\right) \setminus \{0_Q^{(N)}\}, \tag{8.2.18}$$

by (8.2.15), (8.2.16) and (8.2.17).

In conclusion, if $\mathfrak{B}^{(N)}$ is the family (8.2.8) of the restricted integer-shifts on $\mathbb{L}_Q^{(N)}$, then

$$\mathfrak{B}^{(N)} \subseteq Hom\left(\mathbb{L}_Q^{(N)}\right),$$

by (8.2.14) and (8.2.18), for all $N \in \mathbb{N}_{>1}^{\infty}$.   ∎

The relation (8.2.9) shows that all restricted integer-shifts of the family $\mathfrak{B}^{(N)}$ of (8.2.8) are well-defined $*$-homomorphisms on $\mathbb{L}_Q^{(N)}$. However, they cannot be $*$-isomorphisms in general on $\mathbb{L}_Q^{(N)}$. Also, from the proof of (8.2.9), one can realize that the size of $\mathfrak{B}^{(N)}$ can be much smaller than the original integer-shift group $\mathfrak{B}$, especially when $N < \infty$ in $\mathbb{N}_{>1}^{\infty}$. Also, the proof shows that

$$N < \infty \Longleftrightarrow 0_Q^{(N)} \in \mathfrak{B}^{(N)}.$$

Definition 8.1. *Let $\mathfrak{B}^{(N)}$ be in the sense of (8.2.8). We call $\mathfrak{B}^{(N)}$, the restricted(-integer)-shift family on $\mathbb{L}_Q^{(N)}$, for $N \in \mathbb{N}_{>1}^{\infty}$.*

Now, consider an algebraic property of the restricted-shift family $\mathfrak{B}^{(N)}$ in the homomorphism semigroup $Hom\left(\mathbb{L}_Q^{(N)}\right)$. Recall that the integer-shift group $\mathfrak{B}$ is a group acting on $\mathbb{L}_Q$ in $Aut\left(\mathbb{L}_Q\right)$. How about the restricted-shift family $\mathfrak{B}^{(N)}$ in $Hom\left(\mathbb{L}_Q^{(N)}\right)$? This question can be answered by (8.2.1) and (8.2.2).

If the subset $\mathfrak{B}^{(N)}$ were an algebraic structure embedded in $Hom\left(\mathbb{L}_Q^{(N)}\right)$, then the following relation should hold;

$$\beta_{e_1}^{k_1}, \beta_{e_2}^{k_2} \in \mathfrak{B}^{(N)} \Longrightarrow \beta_{e_1}^{k_1}\beta_{e_2}^{k_2} \in \mathfrak{B}^{(N)}. \tag{8.2.19}$$

For instance, if $\mathfrak{B}^{(N)}$ were an algebraic structure, and if $\beta_e^k \in \mathfrak{B}^{(N)}$, then

$$\beta_+^k \beta_-^k, \beta_-^k \beta_+^k \text{ should be in } \mathfrak{B}^{(N)}.$$

Observe now that, for a generating semicircular element $U_k \in \mathcal{S}^{(N)}$ of $\mathbb{L}_Q^{(N)}$, with

$$k \geq 1, \text{ with } 1 \leq 2k \leq N,$$

in $\{1, ..., N\}$, one has that

$$\beta_-^k \beta_+^k (U_k) = \beta_-^k \left(\beta_+^k (U_k)\right) = \beta_-^k (U_{k+k})$$
$$= \beta_-^k (U_{2k}) = U_{2k-k} = U_k, \tag{8.2.20}$$

meanwhile, (8.2.20)

$$\beta_+^k \beta_-^k (U_k) = \beta_+^k (U_{k-k}) = \beta_+^k (O) = O,$$

in $\mathbb{L}_Q^{(N)}$.

The relation (8.2.20) illustrates that two $*$-homomorphisms $\beta_+^k \beta_-^k$, and $\beta_-^k \beta_+^k$ are distinct $*$-homomorphisms in $Hom\left(\mathbb{L}_Q^{(N)}\right)$, i.e.,

$$\beta_+^k \beta_-^k \neq \beta_-^k \beta_+^k \text{ in } Hom\left(\mathbb{L}_Q^{(N)}\right). \tag{8.2.21}$$

The relation (8.2.21) shows that there does "not" exist $\beta_e^n = \beta_e^n \mid_{\mathbb{L}_Q^{(N)}} \in \mathfrak{B}^{(N)}$, such that either

$$\beta_e^n = \beta_+^k \beta_-^k, \text{ or } \beta_e^n = \beta_-^k \beta_+^k, \text{ in } \mathfrak{B}^{(N)},$$

by (8.2.8), i.e.,

$$\beta_+^k \beta_-^k \neq \beta_-^k \beta_+^k \notin \mathfrak{B}^{(N)}, \tag{8.2.22}$$

$\Longleftrightarrow$

$$\beta_+^k \beta_-^k \neq \beta_-^k \beta_+^k \in Hom\left(\mathbb{L}_Q^{(N)}\right) \setminus \mathfrak{B}^{(N)}.$$

i.e., the relation (8.2.19) does not hold on $\mathfrak{B}^{(N)}$, equivalently, the multiplication on $*$-homomorphisms is not closed (or, well-defined) on $\mathfrak{B}^{(N)}$.

Theorem 8.6. *The restricted-shift family $\mathfrak{B}^{(N)}$ of (8.2.8) is a subset of $Hom\left(\mathbb{L}_Q^{(N)}\right)$, but it is not an algebraic sub-structure of $Hom\left(\mathbb{L}_Q^{(N)}\right)$.*

Proof. By (8.2.9), the restricted-shift family $\mathfrak{B}^{(N)}$ is a well-defined subset of the homomorphism semigroup $Hom(\mathbb{L}_Q^{(N)})$. However, by (8.2.22), it cannot be an algebraic sub-structure of $Hom\left(\mathbb{L}_Q^{(N)}\right)$. ∎

The above theorem shows that, different from the integer-shift group $\mathfrak{B}$, a subgroup of the automorphism group $Aut\left(\mathbb{L}_Q\right)$, our restricted-shift families $\mathfrak{B}^{(N)}$ have no nice algebraic properties as a subset of $Hom\left(\mathbb{L}_Q^{(N)}\right)$, for $N \in \mathbb{N}_{>1}^\infty$. However, every restricted shift $\beta_e^k \in \mathfrak{B}^{(N)}$ acts nicely on $\mathbb{L}_Q^{(N)}$, as a $*$-homomorphism.

8.3 Free Probability on $\mathbb{L}_Q^{(N)}$ Affected by $\mathfrak{B}^{(N)}$

In this section, we consider how the restricted-shift family $\mathfrak{B}^{(N)}$ deform the original free-distributional data on the free product Banach $*$-probability space $\mathbb{L}_Q^{(N)}$ of (8.1.8).

Theorem 8.7. *Let $\beta_e^k \in \mathfrak{B}^{(N)}$ be a restricted shift on $\mathbb{L}_Q^{(N)}$, and let $U_l \in \mathcal{S}^{(N)}$ be a semicircular element of $\mathbb{L}_Q^{(N)}$, for $l \in \{1, ..., N\}$, for $N \in \mathbb{N}_{>1}^\infty$. Then the free distribution of $W_l = \beta_e^k(U_l)$ is either the semicircular law, or the zero free distribution in $\mathbb{L}_Q^{(N)}$.*

Proof. Under hypothesis, for any $N \in \mathbb{N}_{>1}^{\infty}$,

$$W_l = \begin{cases} U_{lek} \in \mathcal{S}^{(N)} & \text{if } 1 \leq lek \leq N \\ O, & \text{otherwise,} \end{cases}$$

in $\mathbb{L}_Q^{(N)}$, by (8.2.1), (8.2.2), (8.2.9) and (8.2.19). So, if $W_l = U_{lek} \in \mathcal{S}^{(N)}$, then it is semicircular, while, if $W_l = O$, then it follows the zero free distribution in $\mathbb{L}_Q^{(N)}$. ∎

The above theorem characterizes how the action of the restricted-shift family $\mathfrak{B}^{(N)}$ deform the semicircular law induced by $\mathcal{S}^{(N)}$; the semicircularity is deformed to be either the semicircular law, or the zero free distribution. So, one can have the following generalized result.

Theorem 8.8. *Let $U_{l_1}, ..., U_{l_s}$ be semicircular elements of $\mathcal{S}^{(N)}$, for*

$$I_s = (l_1, ..., l_s) \in \{1, ..., N\}^n,$$

for $n \in \mathbb{N}$, in $\mathbb{L}_Q^{(N)}$, and let $\beta_e^k \in \mathfrak{B}^{(N)}$ be a restricted shift on $\mathbb{L}_Q^{(N)}$. Define a free (non-reduced, or reduced) word $X[I_s]$ by

$$X[I_s] = \prod_{t=1}^{s} U_{l_t} \in \mathbb{L}_Q^{(N)}. \tag{8.3.1}$$

Then one has either

$$\tau\left(\beta_e^k(X[I_s])\right) = \tau(X[I_s]), \ \text{satisfying (6.2.15)} \tag{8.3.2}$$

or

$$\tau\left(\beta_e^k(X[I_s])\right) = 0.$$

Proof. Let $X[I_s]$ be in the sense of (8.3.1). Then it is a well-defined free random variable of $\mathbb{L}_Q^{(N)}$, as a free (non-reduced, or reduced) word in $\mathcal{S}^{(N)}$, by (8.1.8).
Assume first that there exists at least one entry l_p in the s-tuple I_s such that

$$\beta_e^k\left(U_{l_p}\right) = O,$$

up to (8.2.1), or (8.2.2). Then, by the multiplicativity of $\beta_e^k \in \mathfrak{B}^{(N)}$,

$$\beta_e^k(X[I_s]) = \beta_e^k(U_{t_1}) \cdots \beta_e^k(U_{l_p}) \cdots \beta_e^k(U_{l_s}) = O$$

in $\mathbb{L}_Q^{(N)}$, implying that

$$\tau\left(\beta_e^k(X[I_s])\right) = 0.$$

Meanwhile, if

$$\beta_e^k(U_{l_t}) \neq O \text{ in } \mathbb{L}_Q^{(N)}, \text{ for all } t = 1, ..., s,$$

then

$$\tau \left(\beta_e^k \left(X[I_s] \right) \right) = \tau \left(X[I_s] \right),$$

by (6.3.2), (6.3.3), (7.3.2), (7.3.3), (8.2.1) and (8.2.2), because β_e^k preserves the free "reduced" word of $X[I_s]$ (as an operator) to the same-length free reduced word $\beta_e^k \left(X[I_s] \right)$ with the same free-ness in $\mathbb{L}_Q^{(N)}$. Therefore, the free-distributional data (8.3.2) holds.   ∎

The above theorem generalizes Theorem 8.7 by (8.3.2). But, the proof of Theorem 8.8 illustrates that the free-distributional data (8.3.2) is dictated by the free-probabilistic information of Theorem 8.7, too.

References

1. Ahsanullah, M.: Some inferences on semicircular distribution. J. Stat. Theo. Appl. **15**(3), 207–213 (2016)
2. Bercovici, H., Voiculescu, D.: Superconvergence to the central limit and failure of the cramer theorem for free random variables. Probab. Theo. Related Fields **103**(2), 215–222 (1995)
3. Bozejko, M., Ejsmont, W., Hasebe, T.: Noncommutative probability of type D. Internat. J. Math. **28**(2), 1750010 (2017)
4. Bozheuiko, M., Litvinov, E.V., Rodionova, I.V.: An extended anyon fock space and non-commutative meixner-type orthogonal polynomials in the infinite-dimensional case. Uspekhi Math. Nauk. **70**(5), 75–120 (2015)
5. Cho, I.: Free semicircular families in free product banach ∗-algebras induced by p-adic number fields over primes p. Compl. Anal. Oper. Theo. **11**(3), 507–565 (2017)
6. Cho, I.: Semicircular-like laws and the semicircular law induced by orthogonal projections, Compl. Anal. Oper. Theory **12**(1) (2018). https://doi.org/10.1007/s11785-018-0781-x
7. Cho, I., Dong, J.: Catalan numbers and free distributions of mutually free multi semicircular elements. Adv. Appl. Stat. Sci. (2018, submitted)
8. Cho, I., Jorgensen, P.E.T.: Semicircular elements induced by p-adic number fields. Opuscula Math. **35**(5), 665–703 (2017)
9. Connes, A.: Noncommutative Geometry. Academic Press, San Diego (1994). ISBN 0-12-185860-X
10. Gillespie, T.: Prime number theorems for Rankin-Selberg L-functions over number fields. Sci. China Math. **54**(1), 35–46 (2011)
11. Halmos, P.R.: A Hilbert Space Problem Books. Graduate Texts in Mathematics, vol. 19. Springer, Heidelberg (1982). ISBN 978-0387906850
12. Meng, B., Guo, M.: Operator-valued semicircular distribution and its asymptotically free matrix models. J. Math. Res. Exposit. **28**(4), 759–768 (2008)
13. Nourdin, I., Peccati, G., Speicher, R.: Multi-dimensional semicircular limits on the free Wigner chaos. Progr. Probab. **67**, 211–221 (2013)
14. Pata, V.: The central limit theorem for free additive convolution. J. Funct. Anal. **140**(2), 359–380 (1996)
15. Radulescu, F.: Random matrices, amalgamated free products and subfactors of the C^*-algebra of a free group of nonsingular index. Invent. Math. **115**, 347–389 (1994)

16. Shor, P.: Quantum information theory: results and open problems. In: Geom. Funct. Anal (GAFA), Special Volume: GAFA2000, pp. 816–838 (2000)
17. Speicher, R.: Combinatorial theory of the free product with amalgamation and operator-valued free probability theory. Am. Math. Soc. Mem. **132**(627) (1998)
18. Vladimirov, V.S., Volovich, I.V., Zelenov, E.I.: p-adic analysis and mathematical physics. In: Series Soviet & East European Mathematics, vol. 1. World Scientific (1994). ISBN 978-981-02-0880-6
19. Voiculescu, D., Dykema, K., Nica, A.: Free random variables. CRM Monograph Series, vol. 1 (1992)
20. Yin, Y., Bai, Z., Hu, J.: On the semicircular law of large-dimensional random quaternion matrices. J. Theo. Probab. **29**(3), 1100–1120 (2016)
21. Yin, Y., Hu, J.: On the limit of the spectral distribution of large-dimensional random quaternion covariance matrices. Random Mat. Theo. Appl. **6**(2), 1750004 (2017)

Several Explicit and Recurrent Formulas for Determinants of Tridiagonal Matrices via Generalized Continued Fractions

Feng Qi[1,2] , Wen Wang[3] , Bai-Ni Guo[4(✉)] , and Dongkyu Lim[5(✉)]

[1] School of Mathematical Sciences, Tianjin Polytechnic University,
Tianjin 300387, China
[2] College of Mathematics and Physics, Inner Mongolia University for Nationalities,
Tongliao 028043, Inner Mongolia, China
qifeng618@gmail.com, qifeng618@hotmail.com, qifeng618@qq.com
https://qifeng618.wordpress.com
[3] School of Mathematics and Statistics, Hefei Normal University,
Anhui 230601, China
wwen2014@mail.ustc.edu.cn
[4] School of Mathematics and Informatics, Henan Polytechnic University,
Jiaozuo 454010, Henan, China
bai.ni.guo@gmail.com, bai.ni.guo@hotmail.com
[5] Department of Mathematics Education, Andong National University,
Andong 36729, Republic of Korea
dgrim84@gmail.com, dklim@andong.ac.kr

Abstract. In the paper, by the aid of mathematical induction and some properties of determinants, the authors present several explicit and recurrent formulas of evaluations for determinants of general tridiagonal matrices in terms of finite generalized continued fractions and apply these newly-established formulas to evaluations for determinants of the Sylvester matrix and two Sylvester type matrices.

Keywords: Determinant · Tridiagonal matrix · Induction · Explicit formula · Recurrent formula · Finite generalized continued fraction · Sylvester matrix · Sylvester type matrix

2010 Mathematics Subject Classification: Primary 15A15 · Secondary 11C20 · 15B05 · 15B99 · 65F40

1 Introduction

A finite generalized continued fraction is of the form

$$p_0 + \cfrac{q_1}{p_1 + \cfrac{q_2}{p_2 + \cfrac{q_3}{\ddots \cfrac{}{\;\ddots\; \cfrac{q_{m-1}}{p_{m-2} + \cfrac{q_{m-1}}{p_{m-1} + \frac{q_m}{p_m}}}}}}},$$

This paper was typeset using $\mathcal{AMS}$-LATEX.

© Springer Nature Switzerland AG 2021
Z. Hammouch et al. (Eds.): SM2A 2019, LNNS 168, pp. 233–248, 2021.
https://doi.org/10.1007/978-3-030-62299-2_15

where $p_0, p_1, \ldots, p_m$ and $q_1, q_2, \ldots, q_m$ can be any complex numbers or functions. It can also be written equivalently as

$$p_0 + \underset{\ell=1}{\overset{m}{K}} \frac{q_\ell}{p_\ell} = p_0 + \sum_{\ell=1}^{m} \frac{q_\ell|}{|p_\ell} = p_0 + \frac{q_1}{p_1+} \frac{q_2}{p_2+} \cdots \frac{q_{m-1}}{p_{m-1}+} \frac{q_m}{p_m}.$$

In this paper, we will use the second compact form above. For more information on the theory of continued fractions, please refer to the papers [7, 14] and closely related references therein.

In general, a tridiagonal matrix of order n is defined for $n \in \mathbb{N}$ by

$$D_n = \left(e_{i,j}\right)_{1 \le i,j \le n} - \begin{pmatrix} \alpha_1 & \beta_1 & 0 & 0 & \cdots & 0 & 0 & 0 & 0 \\ \gamma_1 & \alpha_2 & \beta_2 & 0 & \cdots & 0 & 0 & 0 & 0 \\ 0 & \gamma_2 & \alpha_3 & \beta_3 & \cdots & 0 & 0 & 0 & 0 \\ 0 & 0 & \gamma_3 & \alpha_4 & \cdots & 0 & 0 & 0 & 0 \\ \vdots & \vdots & \vdots & \vdots & \ddots & \vdots & \vdots & \vdots & \vdots \\ 0 & 0 & 0 & 0 & \cdots & \alpha_{n-3} & \beta_{n-3} & 0 & 0 \\ 0 & 0 & 0 & 0 & \cdots & \gamma_{n-3} & \alpha_{n-2} & \beta_{n-2} & 0 \\ 0 & 0 & 0 & 0 & \cdots & 0 & \gamma_{n-2} & \alpha_{n-1} & \beta_{n-1} \\ 0 & 0 & 0 & 0 & \cdots & 0 & 0 & \gamma_{n-1} & \alpha_n \end{pmatrix}, \qquad (1.1)$$

where

$$e_{i,j} = \begin{cases} \alpha_i, & 1 \le i = j \le n; \\ \beta_i, & 1 \le i = j - 1 \le n - 1; \\ \gamma_j, & 1 \le j = i - 1 \le n - 1; \\ 0, & \text{otherwise}. \end{cases}$$

In the papers [15, 16, 18], the determinant $|D_n|$ and some special cases were discussed, computed, and applied to several problems in analytic combinatorics and analytic number theory. In the papers [2, 5, 6, 9, 15, 16, 18], there are some computation of the inverse and determinant of the general tridiagonal matrix D_n. For more information about this topic, please refer to the papers [4, 8, 12, 13] and closely related references therein.

Let $n \ge 2$ and

$$P_n = \left(p_{i,j}\right)_{1 \le i,j \le n} = \begin{pmatrix} a_1 & c_1 & 0 & 0 & \cdots & 0 & 0 & 0 & 0 \\ a_2 & b_2 & c_2 & 0 & \cdots & 0 & 0 & 0 & 0 \\ a_3 & 0 & b_3 & c_3 & \cdots & 0 & 0 & 0 & 0 \\ a_4 & 0 & 0 & b_4 & \cdots & 0 & 0 & 0 & 0 \\ \vdots & \vdots & \vdots & \vdots & \ddots & \vdots & \vdots & \vdots & \vdots \\ a_{n-3} & 0 & 0 & 0 & \cdots & b_{n-3} & c_{n-3} & 0 & 0 \\ a_{n-2} & 0 & 0 & 0 & \cdots & 0 & b_{n-2} & c_{n-2} & 0 \\ a_{n-1} & 0 & 0 & 0 & \cdots & 0 & 0 & b_{n-1} & c_{n-1} \\ a_n & 0 & 0 & 0 & \cdots & 0 & 0 & 0 & b_n \end{pmatrix}, \qquad (1.2)$$

where

$$
p_{i,j} = \begin{cases}
a_i, & 1 \leq i \leq n, j = 1; \\
b_i, & 2 \leq i = j \leq n; \\
c_i, & 1 \leq i = j - 1 \leq n - 1; \\
0, & \text{otherwise.}
\end{cases}
$$

In this paper, by the help of mathematical induction and some properties of determinants, we will present several explicit and recurrent formulas for evaluations of two determinants $|P_n|$ and $|D_n|$ and will apply these newly-established formulas to evaluations for determinants of the Sylvester matrix and two Sylvester type matrices.

2   Explicit and Recurrent Formulas for $|P_n|$

Right now we start off to present explicit and recurrent formulas for $|P_n|$.

Theorem 2.1. *Let $n \geq 2$ and $b_k \neq 0$ for $2 \leq k \leq n$. Then the determinant $|P_n|$ can be computed recurrently by*

$$
|P_n| = \lambda_{1,n} \prod_{k=2}^{n} b_k, \tag{2.1}
$$

where

$$
\lambda_{k,n} = a_k - \frac{c_k}{b_{k+1}} \lambda_{k+1,n}, \quad 1 \leq k \leq n - 1 \tag{2.2}
$$

and $\lambda_{n,n} = a_n$.

Proof. When $n = 2$, it is easy to see that

$$
|P_2| = \begin{vmatrix} a_1 & c_1 \\ a_2 & b_2 \end{vmatrix} = a_1 b_2 - a_2 c_1
$$

and

$$
\lambda_{1,2} \prod_{k=2}^{2} b_k = \lambda_{1,2} b_2 = \left(a_1 - \frac{c_1 \lambda_{2,2}}{b_2} \right) b_2 = \left(a_1 - \frac{c_1 a_2}{b_2} \right) b_2 = a_1 b_2 - a_2 c_1 = |P_2|.
$$

This means that the formula (2.1) is valid for $n = 2$.

Assume that the formula (2.1) validates for $n = m - 1$, equivalently speaking,

$$
|P_{m-1}| = \lambda_{1,m-1} \prod_{k=2}^{m-1} b_k.
$$

When $n = m$, expanding the determinant $|P_m|$ according to the first rank and utilizing the assumption for $n = m - 1$ give

$$|P_m| = a_1 \prod_{k=2}^{m} b_k - c_1 \begin{vmatrix} a_2 & c_2 & 0 & \cdots & 0 & 0 & 0 & 0 \\ a_3 & b_3 & c_3 & \cdots & 0 & 0 & 0 & 0 \\ a_4 & 0 & b_4 & \cdots & 0 & 0 & 0 & 0 \\ \vdots & \vdots & \vdots & \ddots & \vdots & \vdots & \vdots & \vdots \\ a_{m-3} & 0 & 0 & \cdots & b_{m-3} & c_{m-3} & 0 & 0 \\ a_{m-2} & 0 & 0 & \cdots & 0 & b_{m-2} & c_{m-2} & 0 \\ a_{m-1} & 0 & 0 & \cdots & 0 & 0 & b_{m-1} & c_{m-1} \\ a_m & 0 & 0 & \cdots & 0 & 0 & 0 & b_m \end{vmatrix}$$

$$= a_1 \prod_{k=2}^{m} b_k - c_1 \lambda_{2,m} \prod_{k=3}^{m} b_k = \left(a_1 - \frac{c_1}{b_2} \lambda_{2,m} \right) \prod_{k=2}^{m} b_k = \lambda_{1,m} \prod_{k=2}^{m} b_k.$$

By mathematical induction, we derive the formula (2.1). The proof of Theorem 2.1 is complete.

Theorem 2.2. *For $n \geq 2$, the determinant $|P_n|$ can be computed explicitly by*

$$|P_n| = a_1 \prod_{k=2}^{n} b_k - \sum_{k=2}^{n} (-1)^k \left(\prod_{\ell=1}^{k-1} c_\ell \prod_{m=k+1}^{n} b_m \right) a_k. \tag{2.3}$$

Proof. From the recurrent relation (2.2), it follows that

$$\lambda_{1,n} = a_1 - \frac{c_1}{b_2} \lambda_{2,n}$$

$$= a_1 - \frac{c_1}{b_2} \left(a_2 - \frac{c_2}{b_3} \lambda_{3,n} \right)$$

$$= a_1 - \frac{c_1}{b_2} \left[a_2 - \frac{c_2}{b_3} \left(a_3 - \frac{c_3}{b_4} \lambda_{4,n} \right) \right]$$

$$= \cdots$$

$$= a_1 - \frac{c_1}{b_2} \left[a_2 - \frac{c_2}{b_3} \left(a_3 - \frac{c_3}{b_4} \left[a_4 - \cdots - \frac{c_{\ell-1}}{b_\ell} \left(a_\ell - \frac{c_\ell}{b_{\ell+1}} \lambda_{\ell+1,n} \right) \right] \right) \right]$$

$$= \cdots$$

$$= a_1 - \frac{c_1}{b_2} \left[a_2 - \frac{c_2}{b_3} \left(a_3 - \frac{c_3}{b_4} \left[a_4 - \cdots - \frac{c_{\ell-1}}{b_\ell} \left(a_\ell - \cdots \right. \right. \right.$$

$$\left. \left. \left. - \frac{c_{n-3}}{b_{n-2}} \left[a_{n-2} - \frac{c_{n-2}}{b_{n-1}} \left(a_{n-1} - \frac{c_{n-1}}{b_n} \lambda_{n,n} \right) \right] \right) \right] \right)$$

$$= a_1 - \frac{c_1}{b_2} \left[a_2 - \frac{c_2}{b_3} \left(a_3 - \frac{c_3}{b_4} \left[a_4 - \cdots - \frac{c_{\ell-1}}{b_\ell} \left(a_\ell - \cdots \right. \right. \right.$$

$$\left. \left. \left. - \frac{c_{n-3}}{b_{n-2}} \left[a_{n-2} - \frac{c_{n-2}}{b_{n-1}} \left(a_{n-1} - \frac{c_{n-1}}{b_n} a_n \right) \right] \right) \right] \right)$$

$$= a_1 - \frac{c_1}{b_2}\left(a_2 - \frac{c_2}{b_3}\left[a_3 - \frac{c_3}{b_4}\left(a_4 - \cdots - \frac{c_{\ell-1}}{b_\ell}\left[a_\ell - \cdots\right.\right.\right.\right.$$
$$\left.\left.\left.\left. - \frac{c_{n-3}}{b_{n-2}}\left(a_{n-2} - \frac{c_{n-2}}{b_{n-1}}a_{n-1} + \frac{c_{n-2}c_{n-1}}{b_{n-1}b_n}a_n\right)\right]\right)\right]\right)$$

$$= a_1 - \frac{c_1}{b_2}\left(a_2 - \frac{c_2}{b_3}\left[a_3 - \frac{c_3}{b_4}\left(a_4 - \cdots - \frac{c_{\ell-1}}{b_\ell}\left[a_\ell - \cdots\right.\right.\right.\right.$$
$$\left.\left.\left.\left. - \left(\frac{c_{n-3}}{b_{n-2}}a_{n-2} - \frac{c_{n-3}c_{n-2}}{b_{n-2}b_{n-1}}a_{n-1} + \frac{c_{n-3}c_{n-2}c_{n-1}}{b_{n-2}b_{n-1}b_n}a_n\right)\right]\right)\right]\right)$$

$$= \cdots$$

$$= a_1 - \sum_{k=2}^{n}(-1)^k\left(\prod_{\ell=2}^{k}\frac{c_{\ell-1}}{b_\ell}\right)a_k$$

for $n \geq 2$. Substituting this result into (2.1) and simplifying lead to (2.3). The proof of Theorem 2.2 is complete.

Remark 2.1. Applying $a_k = k$, $b_k = k$, and $c_k = k$ to the explicit formula (2.3) in Theorem 2.2 reveals

$$\begin{vmatrix} 1 & 1 & 0 & 0 & \cdots & 0 & 0 & 0 & 0 \\ 2 & 2 & 2 & 0 & \cdots & 0 & 0 & 0 & 0 \\ 3 & 0 & 3 & 3 & \cdots & 0 & 0 & 0 & 0 \\ 4 & 0 & 0 & 4 & \cdots & 0 & 0 & 0 & 0 \\ \vdots & \vdots & \vdots & \vdots & \ddots & \vdots & \vdots & \vdots & \vdots \\ n-3 & 0 & 0 & 0 & \cdots & n-3 & n-3 & 0 & 0 \\ n-2 & 0 & 0 & 0 & \cdots & 0 & n-2 & n-2 & 0 \\ n-1 & 0 & 0 & 0 & \cdots & 0 & 0 & n-1 & n-1 \\ n & 0 & 0 & 0 & \cdots & 0 & 0 & 0 & n \end{vmatrix} = \frac{1-(-1)^n}{2}n!.$$

3 Explicit and Recurrent Formulas for $|D_n|$

Now we are in a position to present explicit and recurrent formulas for $|D_n|$.

Theorem 3.1. *For $n \in \mathbb{N}$, the determinant $|D_n|$ can be explicitly and recurrently computed by*

$$|D_n| = \alpha_1\alpha_2 + (\alpha_1 - \beta_1\gamma_1)\prod_{m=3}^{n}\left[\alpha_m + \overset{m-2}{\underset{\ell=1}{K}}\frac{(-\beta_{m-\ell}\gamma_{m-\ell})}{\alpha_{m-\ell}}\right] \tag{3.1}$$

$$- \sum_{k=3}^{n}\left[\prod_{\ell=1}^{k-1}(\beta_\ell\gamma_\ell)\right]\frac{\prod_{m=k+1}^{n}\left[\alpha_m + K_{\ell=1}^{m-2}\frac{(-\beta_{m-\ell}\gamma_{m-\ell})}{\alpha_{m-\ell}}\right]}{\prod_{m=2}^{k-1}\left[\alpha_m + K_{\ell=1}^{m-2}\frac{(-\beta_{m-\ell}\gamma_{m-\ell})}{\alpha_{m-\ell}}\right]}$$

and

$$|D_n| = \eta_{1,n}\left(\alpha_2 + \prod_{k=3}^{n}\left[\alpha_k + \overset{k-2}{\underset{\ell=1}{K}}\frac{(-\beta_{k-\ell}\gamma_{k-\ell})}{\alpha_{k-\ell}}\right]\right), \tag{3.2}$$

where $K_{\ell=q}^p$ for $p < q$ is understood to be zero,

$$\eta_{1,n} = -1 - \frac{\beta_1}{\alpha_2}\eta_{2,n}, \quad \eta_{2,n} = \gamma_1 - \frac{\beta_2}{\alpha_3 - \frac{\beta_2\gamma_2}{\alpha_2}}\eta_{3,n}, \quad \eta_{3,n} = -\frac{\gamma_1\gamma_2}{\alpha_2} - \frac{\beta_3}{\alpha_4 - \frac{\beta_3\gamma_3}{\alpha_3 - \frac{\beta_2\gamma_2}{\alpha_2}}}\eta_{4,n},$$

$$\eta_{k,n} = (-1)^k \frac{\prod_{\ell=1}^{k-1}\gamma_\ell}{\alpha_2 + \prod_{\ell=3}^{k-1}[\alpha_\ell + K_{m=1}^{\ell-2}\frac{(-\beta_{\ell-m}\gamma_{\ell-\ell})}{\alpha_{\ell-m}}]} - \frac{\beta_k}{\alpha_{k+1} + K_{\ell=1}^{k-1}\frac{(-\beta_{k-\ell+1}\gamma_{k-\ell+1})}{\alpha_{k-\ell+1}}}\eta_{k+1,n},$$

for $4 \le k \le n-1$, and

$$\eta_{n,n} = (-1)^n \frac{\prod_{\ell=1}^{n-1}\gamma_\ell}{\alpha_2 + \prod_{k=3}^{n-1}[\alpha_k + K_{\ell=1}^{k-2}\frac{(-\beta_{k-\ell}\gamma_{k-\ell})}{\alpha_{k-\ell}}]}.$$

Proof. The determinant $|D_n|$ of the tridiagonal matrix D_n in (1.1) can be rewritten as

$$|D_n| = \begin{vmatrix} \alpha_1 & \beta_1 & 0 & 0 & \cdots & 0 & 0 & 0 & 0 \\ \gamma_1 & \alpha_2 & \beta_2 & 0 & \cdots & 0 & 0 & 0 & 0 \\ -\frac{\gamma_1\gamma_2}{\alpha_2} & 0 & \alpha_3 - \frac{\beta_2\gamma_2}{\alpha_2} & \beta_3 & \cdots & 0 & 0 & 0 & 0 \\ 0 & 0 & \gamma_3 & \alpha_4 & \cdots & 0 & 0 & 0 & 0 \\ \vdots & \vdots & \vdots & \vdots & \ddots & \vdots & \vdots & \vdots & \vdots \\ 0 & 0 & 0 & 0 & \cdots & \alpha_{n-3} & \beta_{n-3} & 0 & 0 \\ 0 & 0 & 0 & 0 & \cdots & \gamma_{n-3} & \alpha_{n-2} & \beta_{n-2} & 0 \\ 0 & 0 & 0 & 0 & \cdots & 0 & \gamma_{n-2} & \alpha_{n-1} & \beta_{n-1} \\ 0 & 0 & 0 & 0 & \cdots & 0 & 0 & \gamma_{n-1} & \alpha_n \end{vmatrix}$$

$$= \begin{vmatrix} \alpha_1 & \beta_1 & 0 & 0 & \cdots & 0 & 0 & 0 & 0 \\ \gamma_1 & \alpha_2 & \beta_2 & 0 & \cdots & 0 & 0 & 0 & 0 \\ -\frac{\gamma_1\gamma_2}{\alpha_2} & 0 & \alpha_3 - \frac{\beta_2\gamma_2}{\alpha_2} & \beta_3 & \cdots & 0 & 0 & 0 & 0 \\ \frac{\gamma_1\gamma_2\gamma_3}{\alpha_2\alpha_3 - \beta_2\gamma_2} & 0 & 0 & \alpha_4 - \frac{\alpha_2\beta_3\gamma_3}{\alpha_2\alpha_3 - \beta_2\gamma_2} & \cdots & 0 & 0 & 0 & 0 \\ \vdots & \vdots & \vdots & \vdots & \ddots & \vdots & \vdots & \vdots & \vdots \\ 0 & 0 & 0 & 0 & \cdots & \alpha_{n-3} & \beta_{n-3} & 0 & 0 \\ 0 & 0 & 0 & 0 & \cdots & \gamma_{n-3} & \alpha_{n-2} & \beta_{n-2} & 0 \\ 0 & 0 & 0 & 0 & \cdots & 0 & \gamma_{n-2} & \alpha_{n-1} & \beta_{n-1} \\ 0 & 0 & 0 & 0 & \cdots & 0 & 0 & \gamma_{n-1} & \alpha_n \end{vmatrix}$$

$$= \cdots$$

$$= \begin{vmatrix} a_1 & \beta_1 & 0 & 0 & \cdots & 0 & 0 & 0 & 0 \\ a_2 & b_2 & \beta_2 & 0 & \cdots & 0 & 0 & 0 & 0 \\ a_3 & 0 & b_3 & \beta_3 & \cdots & 0 & 0 & 0 & 0 \\ a_4 & 0 & 0 & b_4 & \cdots & 0 & 0 & 0 & 0 \\ \vdots & \vdots & \vdots & \vdots & \ddots & \vdots & \vdots & \vdots & \vdots \\ a_{n-3} & 0 & 0 & 0 & \cdots & b_{n-3} & \beta_{n-3} & 0 & 0 \\ a_{n-2} & 0 & 0 & 0 & \cdots & 0 & b_{n-2} & \beta_{n-2} & 0 \\ a_{n-1} & 0 & 0 & 0 & \cdots & 0 & 0 & b_{n-1} & \beta_{n-1} \\ a_n & 0 & 0 & 0 & \cdots & 0 & 0 & 0 & b_n \end{vmatrix},$$

where

$$b_2 = \alpha_2, \quad b_3 = \alpha_3 - \frac{\beta_2\gamma_2}{b_2}, \quad b_4 = \alpha_4 - \frac{\beta_3\gamma_3}{b_3}, \quad \ldots, \quad b_{n-3} = \alpha_{n-3} - \frac{\beta_{n-4}\gamma_{n-4}}{b_{n-4}},$$

$$b_{n-2} = \alpha_{n-2} - \frac{\beta_{n-3}\gamma_{n-3}}{b_{n-3}}, \quad b_{n-1} = \alpha_{n-1} - \frac{\beta_{n-2}\gamma_{n-2}}{b_{n-2}}, \quad b_n = \alpha_n - \frac{\beta_{n-1}\gamma_{n-1}}{b_{n-1}}$$

and

$$a_1 = \alpha_1, \quad a_2 = \gamma_1, \quad a_3 = -\frac{\gamma_2}{b_2}a_2, \quad a_4 = -\frac{\gamma_3}{b_3}a_3, \quad \ldots, \quad a_{n-3} = -\frac{\gamma_{n-4}}{b_{n-4}}a_{n-4},$$

$$a_{n-2} = -\frac{\gamma_{n-3}}{b_{n-3}}a_{n-3}, \quad a_{n-1} = -\frac{\gamma_{n-2}}{b_{n-2}}a_{n-2}, \quad a_n = -\frac{\gamma_{n-1}}{b_{n-1}}a_{n-1}.$$

The sequences b_k and a_k for $k \geq 3$ can be formulated by finite generalized continued fractions

$$b_k = \alpha_k - \cfrac{\beta_{k-1}\gamma_{k-1}}{\alpha_{k-1} - \cfrac{\beta_{k-2}\gamma_{k-2}}{\alpha_{k-2} - \cfrac{\beta_{k-3}\gamma_{k-3}}{\alpha_{k-3} - \cfrac{\beta_{k-4}\gamma_{k-4}}{\ddots \quad \cfrac{\beta_3\gamma_3}{\alpha_4 - \cfrac{\beta_2\gamma_2}{\alpha_3 - \frac{\beta_2\gamma_2}{\alpha_2}}}}}} = \alpha_k + \mathop{K}_{\ell=1}^{k-2} \frac{(-\beta_{k-\ell}\gamma_{k-\ell})}{\alpha_{k-\ell}}$$

and

$$a_k = (-1)^k \frac{\prod_{\ell=1}^{k-1}\gamma_\ell}{\prod_{\ell=2}^{k-1}b_\ell}.$$

Using (2.3) results in

$$|D_n| = \left(b_2 + \prod_{k=3}^n B_k\right)a_1 - \left(\beta_1 \prod_{m=3}^n B_m\right)a_2 - \sum_{k=3}^n (-1)^k \left(\prod_{\ell=1}^{k-1}\beta_\ell \prod_{m=k+1}^n B_m\right)a_k$$

$$= \alpha_1\left(\alpha_2 + \prod_{k=3}^n\left[\alpha_k + \mathop{K}_{\ell=1}^{k-2}\frac{(-\beta_{k-\ell}\gamma_{k-\ell})}{\alpha_{k-\ell}}\right]\right) - \beta_1\gamma_1\prod_{m=3}^n\left[\alpha_m + \mathop{K}_{\ell=1}^{m-2}\frac{(-\beta_{m-\ell}\gamma_{m-\ell})}{\alpha_{m-\ell}}\right]$$

$$- \sum_{k=3}^n \prod_{\ell=1}^{k-1}\beta_\ell \prod_{m=k+1}^n\left[\alpha_m + \mathop{K}_{\ell=1}^{m-2}\frac{(-\beta_{m-\ell}\gamma_{m-\ell})}{\alpha_{m-\ell}}\right] \frac{\prod_{\ell=1}^{k-1}\gamma_\ell}{\prod_{\ell=2}^{k-1}\left[\alpha_\ell + \mathop{K}_{i=1}^{\ell-2}\frac{(-\beta_{\ell-i}\gamma_{\ell-i})}{\alpha_{\ell-i}}\right]}$$

which can be rearranged as (3.1).

Making use of (2.1) and (2.2) yields (3.2). The proof of Theorem 3.1 is complete.

4   Discussions

In this section, we discuss our main results and related ones by several remarks.

Remark 4.1. In [3, p. 1018], it was stated that J. J. Sylvester found in 1854 that

$$|M_n(s)| = \begin{vmatrix} s & 1 & 0 & 0 \cdots 0 & 0 & 0 & 0 \\ n & s & 2 & 0 \cdots 0 & 0 & 0 & 0 \\ 0 & n-1 & s & 3 \cdots 0 & 0 & 0 & 0 \\ 0 & 0 & n-2 & s \cdots 0 & 0 & 0 & 0 \\ \vdots & \vdots & \vdots & \vdots \ddots \vdots & \vdots & \vdots & \vdots \\ 0 & 0 & 0 & 0 \cdots s & n-2 & 0 & 0 \\ 0 & 0 & 0 & 0 \cdots 3 & s & n-1 & 0 \\ 0 & 0 & 0 & 0 \cdots 0 & 2 & s & n \\ 0 & 0 & 0 & 0 \cdots 0 & 0 & 1 & s \end{vmatrix} = \prod_{k=0}^{n}(s+n-2k).$$

An application of (3.1) to $|M_n(s)|$ yields

$$|M_n(s)| = s^2 + (s-n)\prod_{m=3}^{n}\left[s + \underset{\ell=1}{\overset{m-2}{K}}\frac{-(m-\ell)(n-m+\ell+1)}{s}\right]$$

$$-\sum_{k=3}^{n}\left[\prod_{\ell=1}^{k-1}\ell(n-\ell+1)\right]\frac{\prod_{m=k+1}^{n}\left[s + K_{\ell=1}^{m-2}\frac{-(m-\ell)(n-m+\ell+1)}{s}\right]}{\prod_{m=2}^{k-1}\left[s + K_{\ell=1}^{m-2}\frac{-(m-\ell)(n-m+\ell+1)}{s}\right]}$$

$$= s^2 + (s-n)\prod_{m=3}^{n}\left[s + \underset{\ell=1}{\overset{m-2}{K}}\frac{-(m-\ell)(n-m+\ell+1)}{s}\right]$$

$$-n!\sum_{k=3}^{n}\frac{(k-1)!}{(n-k+1)!}\frac{\prod_{m=k+1}^{n}\left[s + K_{\ell=1}^{m-2}\frac{-(m-\ell)(n-m+\ell+1)}{s}\right]}{\prod_{m=2}^{k-1}\left[s + K_{\ell=1}^{m-2}\frac{-(m-\ell)(n-m+\ell+1)}{s}\right]}$$

$$\triangleq s^2 + (s-n)\prod_{m=3}^{n}S(s;m,n) - n!\sum_{k=3}^{n}\frac{(k-1)!}{(n-k+1)!}\frac{\prod_{m=k+1}^{n}S(s;m,n)}{\prod_{m=2}^{k-1}S(s;m,n)}.$$

Now we try to explicitly compute

$$S(s;m,n) = s + \underset{\ell=1}{\overset{m-2}{K}}\frac{-(m-\ell)(n-m+\ell+1)}{s}.$$

When $m = 3$, it is easy to obtain that

$$S(s;3,n) = \frac{s^2 - 2(n-1)}{s} \triangleq \frac{\beta_1}{\alpha_1}.$$

When $m = 4$, employing the above result for $S(s;3,n)$, we can acquire

$$S(s;4,n) = \frac{\beta_1 s - 3(n-2)\alpha_1}{\beta_1} = \frac{s(s^2 - 5n + 8)}{s^2 - 2n + 2} \triangleq \frac{\beta_2}{\alpha_2}.$$

If assuming $S(s;k+1,n) = \frac{\beta_{k-1}}{\alpha_{k-1}}$, then, by mathematical induction, we have

$$S(s;k+2,m) = \frac{\beta_k}{\alpha_k} = \frac{\beta_{k-1}s - \alpha_{k-1}(k+1)(n-k)}{\beta_{k-1}}.$$

Note that $\alpha_{k-1} = \beta_{k-2}$. Then

$$\beta_k - \beta_{k-1}s + \beta_{k-2}(k+1)(n-k) = 0.$$

Further replacing k by $k+2$ results in

$$\beta_{k+2} - \beta_{k+1}s + (k+3)(n-k-2)\beta_k = 0.$$

By the approach utilized in [15, Theorem 3.1], the characteristic equation is

$$t^2 - st + (k+3)(n-k-2) = 0$$

which has solutions

$$t = \frac{s \pm \sqrt{s^2 - 4(k+3)(n-k-2)}}{2}.$$

Consequently, it follows that

$$\beta_k = A\left(\frac{s + \sqrt{s^2 - 4(k+3)(n-k-2)}}{2}\right)^{k-1} + B\left(\frac{s - \sqrt{s^2 - 4(k+3)(n-k-2)}}{2}\right)^{k-1},$$

where

$$A = -\frac{2s^3 - 2(5n-8) - (s^2 - 2n + 2)(s + \sqrt{s^2 - 20n + 80})}{2\sqrt{s^2 - 20n + 80}}$$

and

$$B = \frac{2s^3 - 2(5n-8) - (s^2 - 2n + 2)(s - \sqrt{s^2 - 20n + 80})}{2\sqrt{s^2 - 20n + 80}}.$$

In a word, we provide an alternative expression for the Sylvester determinant $|M_n(s)|$.

Remark 4.2. In [3], by virtue of left eigenvector method, the determinants

$$|M_n(s,t)| = \begin{vmatrix} s & 1 & 0 & 0 & \cdots & 0 & 0 & 0 \\ n & s+t & 2 & 0 & \cdots & 0 & 0 & 0 \\ 0 & n-1 & s+2t & 3 & \cdots & 0 & 0 & 0 \\ 0 & 0 & n-2 & s+3t & \cdots & 0 & 0 & 0 \\ \vdots & \vdots & \vdots & \vdots & \ddots & \vdots & \vdots & \vdots \\ 0 & 0 & 0 & 0 & \cdots & s+(n-2)t & n-1 & 0 \\ 0 & 0 & 0 & 0 & \cdots & 2 & s+(n-1)t & n \\ 0 & 0 & 0 & 0 & \cdots & 0 & 1 & s+nt \end{vmatrix}$$

$$= \prod_{k=0}^{n}\left(s + \frac{nt}{2} + \frac{n-2k}{2}\sqrt{t^4 + 4}\right),$$

and

$$|M_n(s,t;x,y)| = \begin{vmatrix} s & x & 0 & 0 & \cdots & 0 & 0 & 0 \\ nv & s+t & 2x & 0 & \cdots & 0 & 0 & 0 \\ 0 & (n-1)y & s+2t & 3x & \cdots & 0 & 0 & 0 \\ 0 & 0 & (n-2)y & s+3t & \cdots & 0 & 0 & 0 \\ \vdots & \vdots & \vdots & \vdots & \ddots & \vdots & \vdots & \vdots \\ 0 & 0 & 0 & 0 & \cdots & s+(n-2)t & (n-1)x & 0 \\ 0 & 0 & 0 & 0 & \cdots & 2y & s+(n-1)t & nx \\ 0 & 0 & 0 & 0 & \cdots & 0 & y & s+nt \end{vmatrix}$$

$$= \prod_{k=0}^{n}\left(s + \frac{nt}{2} + \frac{n-2k}{2}\sqrt{t^4+4xy}\right)$$

of tridiagonal matrices similar to the Sylvester matrix were collected and calculated. These evaluations can be computed alternatively by Theorem 3.1.

Remark 4.3. The condition $b_k \neq 0$ for $2 \leq k \leq n$ in Theorem 2.1 is removed off in Theorem 2.2. Therefore, the explicit formula (2.3) is better than the recurrent formulas (2.1) and (2.2).

Remark 4.4. The explicit formula (2.3) can be simply reformulated as

$$|P_n| = \sum_{k=1}^{n}(-1)^{k+1}\left(\prod_{\ell=1}^{k-1}c_\ell \prod_{m=k+1}^{n}b_m\right)a_k,$$

where the empty product is understood to be 1 as usual.

Remark 4.5. Let

$$U_n = (u_{i,j})_{1\leq i,j\leq n} = \begin{pmatrix} \alpha_1 & \gamma_1 & 0 & 0 & \cdots & 0 & 0 & 0 & \tau_1 \\ \alpha_2 & \beta_2 & \gamma_2 & 0 & \cdots & 0 & 0 & 0 & \tau_2 \\ \alpha_3 & 0 & \beta_3 & \gamma_3 & \cdots & 0 & 0 & 0 & \tau_3 \\ \alpha_4 & 0 & 0 & \beta_4 & \cdots & 0 & 0 & 0 & \tau_4 \\ \vdots & \vdots & \vdots & \vdots & \ddots & \vdots & \vdots & \vdots & \vdots \\ \alpha_{n-3} & 0 & 0 & 0 & \cdots & \beta_{n-3} & \gamma_{n-3} & 0 & \tau_{n-3} \\ \alpha_{n-2} & 0 & 0 & 0 & \cdots & 0 & \beta_{n-2} & \gamma_{n-2} & \tau_{n-2} \\ \alpha_{n-1} & 0 & 0 & 0 & \cdots & 0 & 0 & \beta_{n-1} & \gamma_{n-1} \\ \alpha_n & 0 & 0 & 0 & \cdots & 0 & 0 & 0 & \beta_n \end{pmatrix},$$

where

$$u_{i,j} = \begin{cases} \alpha_i, & 1 \leq i \leq n, j = 1; \\ \beta_i, & 2 \leq i = j \leq n; \\ \gamma_i, & 1 \leq i = j-1 \leq n-1; \\ \tau_i, & 1 \leq i \leq n-2, j = n; \\ 0, & \text{otherwise.} \end{cases}$$

The determinant $|U_n|$ can be rewritten as

$$|U_n| = \begin{vmatrix} \alpha_1 & \gamma_1 & 0 & 0 & \cdots & 0 & 0 & 0 & \tau_1 \\ \alpha_2 & \beta_2 & \gamma_2 & 0 & \cdots & 0 & 0 & 0 & \tau_2 \\ \alpha_3 & 0 & \beta_3 & \gamma_3 & \cdots & 0 & 0 & 0 & \tau_3 \\ \alpha_4 & 0 & 0 & \beta_4 & \cdots & 0 & 0 & 0 & \tau_4 \\ \vdots & \vdots & \vdots & \vdots & \ddots & \vdots & \vdots & \vdots & \vdots \\ \alpha_{n-3} & 0 & 0 & 0 & \cdots & \beta_{n-3} & \gamma_{n-3} & 0 & \tau_{n-3} \\ \alpha_{n-2} & 0 & 0 & 0 & \cdots & 0 & \beta_{n-2} & \gamma_{n-2} & \tau_{n-2} \\ \alpha_{n-1} & 0 & 0 & 0 & \cdots & 0 & 0 & \beta_{n-1} & \gamma_{n-1} \\ \alpha_n & 0 & 0 & 0 & \cdots & 0 & 0 & 0 & \beta_n \end{vmatrix}$$

$$= \begin{vmatrix} \alpha_1 - \frac{\alpha_n \tau_1}{\beta_n} & \gamma_1 & 0 & 0 & \cdots & 0 & 0 & 0 & 0 \\ \alpha_2 - \frac{\alpha_n \tau_2}{\beta_n} & \beta_2 & \gamma_2 & 0 & \cdots & 0 & 0 & 0 & 0 \\ \alpha_3 - \frac{\alpha_n \tau_3}{\beta_n} & 0 & \beta_3 & \gamma_3 & \cdots & 0 & 0 & 0 & 0 \\ \alpha_4 - \frac{\alpha_n \tau_4}{\beta_n} & 0 & 0 & \beta_4 & \cdots & 0 & 0 & 0 & 0 \\ \vdots & & \vdots & \vdots & \ddots & \vdots & \vdots & \vdots & \vdots \\ \alpha_{n-3} - \frac{\alpha_n \tau_{n-3}}{\beta_n} & 0 & 0 & 0 & \cdots & \beta_{n-3} & \gamma_{n-3} & 0 & 0 \\ \alpha_{n-2} - \frac{\alpha_n \tau_{n-2}}{\beta_n} & 0 & 0 & 0 & \cdots & 0 & \beta_{n-2} & \gamma_{n-2} & 0 \\ \alpha_{n-1} - \frac{\alpha_n \gamma_{n-1}}{\beta_n} & 0 & 0 & 0 & \cdots & 0 & 0 & \beta_{n-1} & 0 \\ \alpha_n & 0 & 0 & 0 & \cdots & 0 & 0 & 0 & \beta_n \end{vmatrix}$$

$$= \beta_n \begin{vmatrix} \alpha_1 - \frac{\alpha_n \tau_1}{\beta_n} & \gamma_1 & 0 & 0 & \cdots & 0 & 0 & 0 \\ \alpha_2 - \frac{\alpha_n \tau_2}{\beta_n} & \beta_2 & \gamma_2 & 0 & \cdots & 0 & 0 & 0 \\ \alpha_3 - \frac{\alpha_n \tau_3}{\beta_n} & 0 & \beta_3 & \gamma_3 & \cdots & 0 & 0 & 0 \\ \alpha_4 - \frac{\alpha_n \tau_4}{\beta_n} & 0 & 0 & \beta_4 & \cdots & 0 & 0 & 0 \\ \vdots & & \vdots & \vdots & \ddots & \vdots & \vdots & \vdots \\ \alpha_{n-3} - \frac{\alpha_n \tau_{n-3}}{\beta_n} & 0 & 0 & 0 & \cdots & \beta_{n-3} & \gamma_{n-3} & 0 \\ \alpha_{n-2} - \frac{\alpha_n \tau_{n-2}}{\beta_n} & 0 & 0 & 0 & \cdots & 0 & \beta_{n-2} & \gamma_{n-2} \\ \alpha_{n-1} - \frac{\alpha_n \gamma_{n-1}}{\beta_n} & 0 & 0 & 0 & \cdots & 0 & 0 & \beta_{n-1} \end{vmatrix}.$$

An application of Theorem 2.1 and 2.2 and Remark 4.4 straightforwardly yields

$$|U_n| = \sum_{k=1}^{n-2} (-1)^{k+1} (\alpha_k \beta_n - \tau_k \alpha_n) \prod_{\ell=1}^{k-1} \gamma_\ell \prod_{m=k+1}^{n-1} \beta_m + (-1)^n (\alpha_{n-1} \beta_n - \tau_{n-1} \alpha_n) \prod_{\ell=1}^{n-2} \gamma_\ell$$

and

$$|U_n| = \Lambda_{1,n-1} \prod_{k=2}^{n} \beta_k,$$

where

$$\Lambda_{k,n-1} = \alpha_k - \frac{\alpha_n}{\beta_n} \tau_k - \frac{\gamma_k}{\beta_{k+1}} \Lambda_{k+1,n-1}, \quad 1 \leq k \leq n-2$$

and $\Lambda_{n-1,n-1} = \alpha_{n-1} - \frac{\alpha_n}{\beta_n} \gamma_{n-1}.$

Remark 4.6. The determinant $|P_n|$ of P_n in (1.2) can be rearranged as

$$
|P_n| =
\begin{vmatrix}
a_1 & c_1 & 0 & 0 & \cdots & 0 & 0 & 0 & 0 \\
a_2 & b_2 & c_2 & 0 & \cdots & 0 & 0 & 0 & 0 \\
a_3 & 0 & b_3 & c_3 & \cdots & 0 & 0 & 0 & 0 \\
a_4 & 0 & 0 & b_4 & \cdots & 0 & 0 & 0 & 0 \\
\vdots & \vdots & \vdots & \vdots & \ddots & \vdots & \vdots & \vdots & \vdots \\
a_{n-3} & 0 & 0 & 0 & \cdots & b_{n-3} & c_{n-3} & 0 & 0 \\
a_{n-2} & 0 & 0 & 0 & \cdots & 0 & b_{n-2} & c_{n-2} & 0 \\
a_{n-1} & 0 & 0 & 0 & \cdots & 0 & 0 & b_{n-1} & c_{n-1} \\
0 & 0 & 0 & 0 & \cdots & 0 & 0 & -\frac{a_n b_{n-1}}{a_{n-1}} & b_n - \frac{a_n c_{n-1}}{a_{n-1}}
\end{vmatrix}
$$

$$
= \cdots
$$

$$
=
\begin{vmatrix}
a_1 & c_1 & 0 & \cdots & 0 & 0 & 0 \\
0 & b_2 - \frac{a_2 c_1}{a_1} & c_2 & \cdots & 0 & 0 & 0 \\
0 & -\frac{a_3 b_2}{a_2} & b_3 - \frac{a_3 c_2}{a_2} & \cdots & 0 & 0 & 0 \\
0 & 0 & -\frac{a_4 b_3}{a_3} & \cdots & 0 & 0 & 0 \\
\vdots & \vdots & \vdots & \ddots & \vdots & \vdots & \vdots \\
0 & 0 & 0 & \cdots & c_{n-3} & 0 & 0 \\
0 & 0 & 0 & \cdots & b_{n-2} - \frac{a_{n-2} c_{n-3}}{a_{n-3}} & c_{n-2} & 0 \\
0 & 0 & 0 & \cdots & -\frac{a_{n-1} b_{n-2}}{a_{n-2}} & b_{n-1} - \frac{a_{n-1} c_{n-2}}{a_{n-2}} & c_{n-1} \\
0 & 0 & 0 & \cdots & 0 & -\frac{a_n b_{n-1}}{a_{n-1}} & b_n - \frac{a_n c_{n-1}}{a_{n-1}}
\end{vmatrix}
$$

$$
= a_1
\begin{vmatrix}
b_2 - \frac{a_2 c_1}{a_1} & c_2 & \cdots & 0 & 0 & 0 \\
-\frac{a_3 b_2}{a_2} & b_3 - \frac{a_3 c_2}{a_2} & \cdots & 0 & 0 & 0 \\
0 & -\frac{a_4 b_3}{a_3} & \cdots & 0 & 0 & 0 \\
\vdots & \vdots & \ddots & \vdots & \vdots & \vdots \\
0 & 0 & \cdots & c_{n-3} & 0 & 0 \\
0 & 0 & \cdots & b_{n-2} - \frac{a_{n-2} c_{n-3}}{a_{n-3}} & c_{n-2} & 0 \\
0 & 0 & \cdots & -\frac{a_{n-1} b_{n-2}}{a_{n-2}} & b_{n-1} - \frac{a_{n-1} c_{n-2}}{a_{n-2}} & c_{n-1} \\
0 & 0 & \cdots & 0 & -\frac{a_n b_{n-1}}{a_{n-1}} & b_n - \frac{a_n c_{n-1}}{a_{n-1}}
\end{vmatrix}
$$

$$
\triangleq a_1 |Q_{n-1}|.
$$

Therefore, by virtue of Theorems 2.1 and 2.2, we derive that the determinant $|Q_{n-1}|$ satisfies

$$
|Q_{n-1}| = \frac{|P_n|}{a_1} = \frac{\lambda_{1,n}}{a_1} \prod_{k=2}^{n} b_k \tag{4.1}
$$

and

$$
|Q_{n-1}| = \frac{|P_n|}{a_1} = \prod_{k=2}^{n} b_k - \frac{1}{a_1} \sum_{k=2}^{n} (-1)^k \left(\prod_{\ell=1}^{k-1} c_\ell \prod_{m=k+1}^{n} b_m \right) a_k. \tag{4.2}
$$

Further letting

$$
\begin{cases}
\alpha_k = b_{k+1} - \dfrac{a_{k+1}c_k}{a_k}, & 1 \le k \le n-1 \\[2mm]
\beta_k = c_{k+1}, & 1 \le k \le n-2 \\[2mm]
\gamma_k = -\dfrac{a_{k+2}b_{k+1}}{a_{k+1}}, & 1 \le k \le n-2
\end{cases}
\tag{4.3}
$$

in Eqs. (4.1) and (4.2) reveals

$$
|D_{n-1}| = \frac{\lambda_{1,n}}{a_1} \prod_{k=2}^{n} b_k
\tag{4.4}
$$

and

$$
|D_{n-1}| = \prod_{k=2}^{n} b_k - \frac{1}{a_1} \sum_{k=2}^{n} (-1)^k \left(\prod_{\ell=1}^{k-1} c_\ell \prod_{m=k+1}^{n} b_m \right) a_k.
\tag{4.5}
$$

From the second equality in (4.3), it is not difficult to see that $c_k = \beta_{k-1}$ for $2 \le k \le n-1$. If we can derive another relations from (4.3) to express a_k for $1 \le k \le n$ and b_k for $2 \le k \le n$ in terms of α_k for $1 \le k \le n-1$, β_k for $1 \le k \le n-2$, and γ_k for $1 \le k \le n-2$, then, by substituting these relations into (4.4) and (4.5), an alternative and explicit expression for evaluation of $|D_n|$ would be concluded. This is an open problem and we leave it to the interested readers.

Remark 4.7. In [1, Lemma 1.1] and [11, Lemma 2.1], it was acquired that

$$
\begin{vmatrix}
\tau_1 & \tau_2 & \tau_3 & \tau_4 & \cdots & \tau_{n-2} & \tau_{n-1} & \tau_n \\
\alpha & \beta & 0 & 0 & \cdots & 0 & 0 & 0 \\
\gamma & \alpha & \beta & 0 & \cdots & 0 & 0 & 0 \\
0 & \gamma & \alpha & \beta & \cdots & 0 & 0 & 0 \\
\vdots & \vdots & \vdots & \vdots & \ddots & \vdots & \vdots & \vdots \\
0 & 0 & 0 & 0 & \cdots & \beta & 0 & 0 \\
0 & 0 & 0 & 0 & \cdots & \alpha & \beta & 0 \\
0 & 0 & 0 & 0 & \cdots & \gamma & \alpha & \beta
\end{vmatrix}
= \sum_{k=1}^{n} (-1)^{k-1} \tau_k b^{n-k} (\beta\gamma)^{(k-1)/2} U_{k-1}\left(\frac{\alpha}{2\sqrt{\beta\gamma}} \right),
\tag{4.6}
$$

where $U_k(s)$ is the kth Chebyshev polynomials of the second kind, which can be generated [19, 20, 22] by

$$
\frac{1}{1 - 2st + t^2} = \sum_{k=0}^{\infty} U_k(s) t^k, \quad |s| < 1, \quad |t| < 1.
$$

Taking $\tau_1 = \tau_2 = \cdots = \tau_{n-1} = 0$ and $\tau_n = 1$ and reformulating, the formula (4.6) becomes

$$
\begin{vmatrix}
\alpha & \beta & 0 & \cdots & 0 & 0 \\
\gamma & \alpha & \beta & \cdots & 0 & 0 \\
0 & \gamma & \alpha & \cdots & 0 & 0 \\
\vdots & \vdots & \vdots & \ddots & \vdots & \vdots \\
0 & 0 & 0 & \cdots & \alpha & \beta \\
0 & 0 & 0 & \cdots & \gamma & \alpha
\end{vmatrix}_{n \times n}
= (\beta\gamma)^{n/2} U_n\left(\frac{\alpha}{2\sqrt{\beta\gamma}} \right).
\tag{4.7}
$$

This is different from

$$
\begin{vmatrix}
\alpha & \beta & 0 & \cdots & 0 & 0 \\
\gamma & \alpha & \beta & \cdots & 0 & 0 \\
0 & \gamma & \alpha & \cdots & 0 & 0 \\
\vdots & \vdots & \vdots & \ddots & \vdots & \vdots \\
0 & 0 & 0 & \cdots & \alpha & \beta \\
0 & 0 & 0 & \cdots & \gamma & \alpha
\end{vmatrix}_{n \times n}
=
\begin{cases}
\dfrac{\left(\alpha + \sqrt{\alpha^2 - 4\beta\gamma}\,\right)^{n+1} - \left(\alpha - \sqrt{\alpha^2 - 4\beta\gamma}\,\right)^{n+1}}{2^{n+1}\sqrt{\alpha^2 - 4\beta\gamma}}, & \alpha^2 \neq 4\beta\gamma \\[4mm]
(n+1)\left(\dfrac{\alpha}{2}\right)^n, & \alpha^2 = 4\beta\gamma
\end{cases}
$$

(4.8)

and

$$
\begin{vmatrix}
\alpha & \beta & 0 & \cdots & 0 & 0 \\
\gamma & \alpha & \beta & \cdots & 0 & 0 \\
0 & \gamma & \alpha & \cdots & 0 & 0 \\
\vdots & \vdots & \vdots & \ddots & \vdots & \vdots \\
0 & 0 & 0 & \cdots & \alpha & \beta \\
0 & 0 & 0 & \cdots & \gamma & \alpha
\end{vmatrix}_{n \times n}
= \prod_{j=1}^{n} \left(\beta + 2\alpha\sqrt{\dfrac{\gamma}{\alpha}} \, \cos \dfrac{j\pi}{n+1} \right)
$$

(4.9)

which are established and collected in [15, pp. 130] and [18, Theoem 4].

Comparing (4.7) with (4.8) and (4.9), taking $\beta = \gamma = 1$ and $\alpha = 2s$, and simplifying yield

$$
U_n(s) = \prod_{j=1}^{n} \left(1 + 2\sqrt{2s} \, \cos \dfrac{j\pi}{n+1} \right)
$$

$$
= \begin{cases}
\dfrac{\left(s + \sqrt{s^2 - 1}\,\right)^{n+1} - \left(s - \sqrt{s^2 - 1}\,\right)^{n+1}}{2^{n+1}\sqrt{s^2 - 1}}, & s^2 \neq 1 \\[4mm]
(n+1)s^n, & s^2 = 1
\end{cases}
$$

which are alternative explicit formulas for the Chebyshev polynomials of the second kind $U_n(s)$.

Remark 4.8. On 21 September 2019, we were reminded of the paper [10] in which an alternative explicit formula for elements of the inverse of a tridiagonal matrix and an efficient and fast computing method to obtain elements of the inverse of a tridiagonal matrix by backward continued fractions were investigated.

Remark 4.9. Theorem 2.2 in this paper has been applied in the proof of [17, Theorem 3.3].

Remark 4.10. This paper is a revised version of the preprint [21].

Acknowledgements. The authors appreciate anonymous referees for their careful corrections to and valuable comments on the original version of this paper.

Funding. The fourth author was supported by the National Research Foundation of Korea under Grant NRF-2018R1D1A1B07041846, South Korea.

Conflict of Interest. No potential conflict of interest was reported by the authors.

References

1. Bozkurt, D., Da Fonseca, C.M., Yılmaz, F.: The determinants of circulant and skew-circulant matrices with tribonacci numbers. Math. Sci. Appl. E-Notes **2**(2), 67–75 (2014)
2. Chen, Z.R.: Inversion of general tridiagonal matrices. J. Numer. Methods Comput. Appl. **8**(3), 158–164 (1987). (Chinese)
3. Chu, W.: Fibonacci polynomials and Sylvester determinant of tridiagonal matrix. Appl. Math. Comput. **216**, 1018–1023 (2010). https://doi.org/10.1016/j.amc.2010.01.089
4. Coelho, D.F.G., Dimitrov, V.S., Rakai, L.: Efficient computation of tridiagonal matrices largest eigenvalue. J. Comput. Appl. Math. **330**, 268–275 (2018). https://doi.org/10.1016/j.cam.2017.08.008
5. El-Mikkawy, M., Karawia, A.: Inversion of general tridiagonal matrices. Appl. Math. Lett. **19**(8), 712–720 (2006). https://doi.org/10.1016/j.aml.2005.11.012
6. El-Shehawey, M.A., El-Shreef, Gh.A., Al-Henawy, A.Sh.: Analytical inversion of general periodic tridiagonal matrices. J. Math. Anal. Appl. **345**(1), 123–134 (2008). https://doi.org/10.1016/j.jmaa.2008.04.002
7. Euler, L.: An essay on continued fractions. Math. Systems Theory **18**(4), 295–328 (1985). https://doi.org/10.1007/BF01699475. Translated from the Latin by B. F. Wyman and M. F. Wyman
8. Fischer, C.F., Usmani, R.A.: Properties of some tridiagonal matrices and their application to boundary value problems. SIAM J. Numer. Anal. **6**(1), 127–142 (1969). https://doi.org/10.1137/0706014
9. Huang, Y., McColl, W.F.: Analytical inversion of general tridiagonal matrices. J. Phys. A **30**(22), 7919–7933 (1997). https://doi.org/10.1088/0305-4470/30/22/026
10. Kılıç, E.: Explicit formula for the inverse of a tridiagonal matrix by backward continued fractions. Appl. Math. Comput. **197**(1), 345–357 (2008). https://doi.org/10.1016/j.amc.2007.07.046
11. Kırklar, E., Yılmaz, F.: A general formula for determinants and inverses of r-circulant matrices with third order recurrences. Math. Sci. Appl. E-Notes **7**(1), 1–8 (2019)
12. Kouachi, S.: Eigenvalues and eigenvectors of some tridiagonal matrices with non-constant diagonal entries. Appl. Math. (Warsaw) **35**(1), 107–120 (2008). https://doi.org/10.4064/am35-1-7
13. Kouachi, S.: Eigenvalues and eigenvectors of tridiagonal matrices. Electron. J. Linear Algebra **15**, 115–133 (2006). https://doi.org/10.13001/1081-3810.1223. Article 8
14. Panti, G.: Slow continued fractions, transducers, and the Serret theorem. J. Number Theory **185**, 121–143 (2018). https://doi.org/10.1016/j.jnt.2017.08.034
15. Qi, F., Čerňanová, V., Semenov, Y.S.: Some tridiagonal determinants related to central Delannoy numbers, the Chebyshev polynomials, and the Fibonacci polynomials. Politehn. Univ. Bucharest Sci. Bull. Ser. A Appl. Math. Phys. **81**(1), 123–136 (2019)
16. Qi, F., Čerňanová, V., Shi, X.-T., Guo, B.-N.: Some properties of central Delannoy numbers. J. Comput. Appl. Math. **328**, 101–115 (2018). https://doi.org/10.1016/j.cam.2017.07.013
17. Qi, F., Huang, C.-J.: Computing sums in terms of beta, polygamma, and Gauss hypergeometric functions. Rev. R. Acad. Cienc. Exactas Fís. Nat. Ser. A Mat. RACSAM **114**, 9 p. (2020). Article no. 191. https://doi.org/10.1007/s13398-020-00927-y

18. Qi, F., Liu, A.-Q.: Alternative proofs of some formulas for two tridiagonal deter-minants. Acta Univ. Sapientiae Math. **10**(2), 287–297 (2018). https://doi.org/10.2478/ausm-2018-0022

19. Qi, F., Niu, D.-W., Lim, D.: Notes on explicit and inversion formulas for the Chebyshev polynomials of the first two kinds. Miskolc Math. Notes **20**(2), 1129–1137 (2019). https://doi.org/10.18514/MMN.2019.2976

20. Qi, F., Niu, D.-W., Lim, D., Guo, B.-N.: Some properties and an application of multivariate exponential polynomials. Math. Methods Appl. Sci. **43**(6), 2967–2983 (2020). https://doi.org/10.1002/mma.6095

21. Qi, F., Wang, W., Lim, D., Guo, B.-N.: Some formulas for determinants of tridi-agonal matrices in terms of finite generalized continued fractions. HAL preprint (2019). https://hal.archives-ouvertes.fr/hal-02372394

22. Qi, F., Zou, Q., Guo, B.-N.: The inverse of a triangular matrix and several iden-tities of the Catalan numbers. Appl. Anal. Discrete Math. **13**(2), 518–541 (2019). https://doi.org/10.2298/AADM190118018Q

Author Index

© Springer Nature Switzerland AG 2021
Z. Hammouch et al. (Eds.): SM2A 2019, LNNS 168, p. 249, 2021.
https://doi.org/10.1007/978-3-030-62299-2

Printed by Printforce, the Netherlands